GAME THEORETICAL APPLICATIONS TO ECONOMICS AND OPERATIONS RESEARCH

THEORY AND DECISION LIBRARY

General Editors: W. Leinfellner (*Vienna*) and G. Eberlein (*Munich*)

Series A: Philosophy and Methodology of the Social Sciences

Series B: Mathematical and Statistical Methods

Series C: Game Theory, Mathematical Programming and Operations Research

Series D: System Theory, Knowledge Engineering and Problem Solving

SERIES C: GAME THEORY, MATHEMATICAL PROGRAMMING AND OPERATIONS RESEARCH

VOLUME 18

Scope: Particular attention is paid in this series to game theory and operations research, their formal aspects and their applications to economic, political and social sciences as well as to socio-biology. It will encourage high standards in the application of game-theoretical methods to individual and social decision making.

The titles published in this series are listed at the end of this volume.

GAME THEORETICAL APPLICATIONS TO ECONOMICS AND OPERATIONS RESEARCH

edited by

T. PARTHASARATHY
Indian Statistical Institute

B. DUTTA
Indian Statistical Institute

J. A. M. POTTERS
Catholic University Nijmegen

T. E. S. RAGHAVAN
University of Illinois

D. RAY
Boston University

and

A. SEN
Indian Statistical Institute

KLUWER ACADEMIC PUBLISHERS
BOSTON / DORDRECHT / LONDON

A C.I.P. Catalogue record for this book is available from the Library of Congress.

ISBN 0-7923-4712-9

Published by Kluwer Academic Publishers,
P.O. Box 17, 3300 AA Dordrecht, The Netherlands.

Sold and distributed in the U.S.A. and Canada
by Kluwer Academic Publishers,
101 Philip Drive, Norwell, MA 02061, U.S.A.

In all other countries, sold and distributed
by Kluwer Academic Publishers Group,
P.O. Box 322, 3300 AH Dordrecht, The Netherlands.

Printed on acid-free paper

Printed in the Netherlands

TABLE OF CONTENTS

CHAPTER III. NONCOOPERATIVE GAMES

CHAPTER IV. LINEAR COMPLEMENTARITY PROBLEMS AND GAME THEORY

CHAPTER V. ECONOMIC AND OR APPLICATIONS

PREFACE

This volume contains papers that were presented at the International Conference on Game Theory and Economic Applications held at the Indian Institute of Science, Bangalore during January 2-6, 1996. The Conference was sponsored jointly by the Indian Institute of Science, the Indian Statistical Institute and the Jawaharlal Nehru Centre for Advanced Research.

About one hundered participants from all over the world attended the Conference where papers were presented on wide variety of topics: Decision Theory, Cooperative and non cooperative game theory and Economic and Operations Research Applications. Participants were invited to contribute their papers for publication in the conference proceedings, and submission were refereed according to the usual standard of high quality journals in these fields.

We thank all the participants of the Conference, the contributors to this volume and the referees of the submitted papers. We are extremely grateful to Kluwer Academic Publishers for their unstinted cooperation at all stages of the production of this volume.

We gratefully acknowledge the following persons for their help at various stages of the conference : V.S.Borkar, M.K.Ghosh, B.G.Raghavendra, Guruswami Babu, B.K.Pal, T.S.Arthanari, M.Usha, G.Ravindran, Dilip Mukherjee, Stef Tijs and other secretarial staff from Indian Institute of Sciences, Indian Statistical Institute, and Jawharlal Nehru Centre.

We gratefully acknowledge the generous financial support provided for the Conference by the Indian Institute of Science, the Indian Statistical Institute, Jawaharlal Centre for Advanced Research respectively as well as the travel support provided to several participants, by the Indo-US cooperative Science Program, National Science Foundation, Washington D.C. and by International Centre for Theoretical Physics, Trieste.

We are extremely grateful to Drs: S.R.Mohan and S.K. Neogy and Mr. Amit.K.Biswas and Mr.B.Ganesan who organised the entire collection of accepted papers in LaTeX format. It is no exaggeration to say that this volume would not have seen the light of the day without their help.

T.Parthasarathy
J.A.M.Potters
D.Ray

B.Dutta
T.E.S.Raghavan
A.Sen

INTRODUCTION

The papers in the volume are classified in five different chapters. The first four chapters are devoted respectively to the theory of two-person games, linear complmentarity problem and game theory, cooperative and noncooperative games. The fifth chapter contains diverse applications of these various theories. Taken together, they exhibit the rich versatility of these theories and lively interaction between the, mathematical theory of games and significant economic and operatons researchproblems.

1 Two-person games

Helmes and Srinivasan consider the problem of estimating an unknown paramneter vector θ through a vector y which can be observed. More precisely, the question they address is to find a linear combination of the data y which minimises the minimum risk of all such procedures. A solution to this problem is offered through fractional programming. They also present an efficient method to solve the fractional programming problem in some special cases.

Bapat and Stef Tijs consider a matrix game in which the pay-off matrix is the vertex-edge incidence matrix of either a directed or undirected graph. For the directed incidence matrix game, they derive results on the value and the structure of optimal strategies when the graph has no directed cycle. The problem of determining strategies for the undirected incidence matrix games is shown to be related to the theory of 2-matchings.

Parthasarathy, Ravindran and Sabatini study injectivity of cubic linear mappings which is related to the (real) Jacobian Conjecture. They derive results using results from completely mixed games due to Kalplansky.

Srijit Mishra and Krishna Kunar consider the problem of obtaining a pure strategy equilibrium in matrix games with random pay-offs. In that context they generalise the notion of separation of diagonals due to von Neumann and Morgestern and give a set of necessary and sufficient conditions for the game to have mixed strategy equilibrium.

2 Cooperative Games

Peter Sudholter gives a survey on modified nucleolus of a game, its definition, interpretation and a list of elementary properties. In the later half of his paper, he discusses the notion of modified kernel as well as modified bargaining set of a game.

Theo Driessen and Yuki Funaki discuss the relationship between the prenucleolus and a new value, called ENPAC-value. The authors give several alternative sufficent conditons for the equality of the ENPAC-value to the prenucleolus.

Elena Yanovskaya considers three solutions, namely ϵ Core, (pre)nucleolus, and (pre)kernel for cooperative games for nontansferable utilities. She studies these solution concept with the help of excess function. It is shown that both the prenucleolus and the prekernel do not posses the reduced game property for all excess functions satisfying Kalai's condition. Axiomatic characterisation of a core and the collection of ϵ cores are given for some excess functions.

Yuki Funaki and Theo Driessen focus their attention on a uniform treatement of a special type of one point solution for coperaative games, called the egalitarian non-individual contribution (ENIC) value. The main goal of the authors is to provide an aximatic characterisation of the ENIC-Value in general to construct four particular ENIC-values.

3 Noncooperative Games

Bergantinos and Potters attaches for each normalized NTU-game (N,v), a relevant strategy game Γ_u. Then they show that there is a nice coreespondence between core allocations of (N,v) and Nash equilibria Γ_u. Further a relation is described between the pay-off function of the strategic game and remainder map considered by Driessen and Tijs.

Ali Khan, Kali Rath and Yeneng Sun present an example of a nonatomic game without pure Nash Equilibria. They also present a theorem on the existence of pure strategy Nash Equilibria in nonatomic games in which the set of players is modelled on nonatomic Loeb measure space.

Abraham Neyman and Sylvain Sorin in their paper on equilibria in repeated games, show that every two person game incomplete information in which the information to both players is identical and deterministic has an equilibrium.

Vemeulen, Potters and Jansen introduce a new kind of perturbations for normal form games and they investigate the stability of these perturbations. The CQ sets obtained in this manner satisfy the Kohlberg-Mertens program except invariance. In order to overcome this problem, the authors modify their solution concept in such a way that all properties formulated by Kohlberg-Mertens are satisfied.

4 Linear Complementarity Problem and Game Theory

Amit and Murthy introduce a 'chain rule' and show that no Q_0-matrix can satisfy this rule. They use this rule to answer a certain conjecture due to Stone for 5x5 matrices. Thus chain rule is quite handy in many situations to decide whether a given matrix is Q_0 or not.

Mohan, Neogy and Parthasarathy consider n-person stochastic games, with finite state and action spaces, in which player n controls the law of motion and where each plater wants to minimise his limiting average expected costs. For such games the authors show that stationary equilibrium strategies can be computed by applying Lemke's algorithm to solve a related linear complementarity problem. This result is quite useful from the point of view of algorithms.

Roman Snajder and Seetharama Gowda investigate the Lipschitz continuity of the solution map in the settings of horizontal, vertical and mixed linear complementarity problems. In each of the settings, they show that the solution map is (globally) Lipschitzian if and only if the solution map is single valued. These generalise a similar result of Murthy, Parthasarathy and Sabatini proved in the LCP setting.

5 Economic & OR Applications

Guillermo Owen considers the folowing: a bookie (pari-mutuel system), faced by several bettors with different subjective probabilities, has the problem of choosing pay-off odds so as to avoid the risk of loss. It is shown under some conditions equilibrium set of pay-off odds exists. Some examples are worked out in detail.

Hubert Chin describes a heuristic approach for finding the nucleolus of assignment games using genetic algorithms. It is not clear how the algorithm proposed here compares with the other known algorithms.

Raghavan gives a nice survey on algorithms to compute nucleolus for structured cooperative games. This is based on the current algorithms available to calculate nucleolus effectively for (i) General games (Studied by Potters Reijnerse and Ansing) (ii) Assignment ganes (Raghavan and Solymosi) (iii) Tree games (Maschler, Owen, Granot and Zhao) (iv) Interval games (stidued by Driessen, Solymosi and Aarts)

Bettina Klaus considers the problem of reallocating the total endowment of an infinitely divisible good among agents with single-peaked preferences and study several properties of reallocation rules such as individual rationality, endowment monotonicity, envy freeness and bilateral consistency. Main result is the proof that individual rationality and endowment monotonicity imply Pareto Optimality. The result is then used to give two characterisations of the uniform reallocation rule.

Ahmet Alkan considers a model of sealed bid auctions with resale. The policy question whether the seller would fare better under the multiprice rule (where winners pay their actual bids) or the uniprice rule (where winners pay the highest losing bid) has been discussed since the 60's and seen a recent revival. While theory has mostly recommended the uniprice rule, the results of the present author recommend the multiprice rule.

Alan Richards and Nirvikar Sing analyses the impact of a two-level game for water allocations. Nash bargaining theory is used to derive several propositions on the consequences of different bargaining rules for water allocations. The effect on international negotiations of the ability to commit to having domestic negotiations is also examined. The authors cite several live examples of two-level games over water.

Meenakshi Rajeev considers the role of money as a medium of exchange in a competitive set up. Her set-up is derived from the frame-work of Kiyotaki and Wright. She examines how monetized trading post set-up manifest itself through (the agent's) behaviour.

COMPUTING LINEAR MINI-MAX ESTIMATORS

K.Helmes[1] and C. Srinivasan

Abstract: *Consider a vector of data $y = \theta + \varepsilon$, $y \in \mathbb{R}^n$, where $\varepsilon = (\varepsilon_i)_{1 \le i \le n}$ is an error vector which satisfies the "usual" conditions, and $\theta \in \Theta \subset \mathbb{R}^n$. The problem is to find/compute a vector m for which $\langle m, y\rangle$ minimizes the maximal risk among all linear estimators, i.e. m is a solution of $\min_{m \in \mathbb{R}^n} \max_{\theta \in \Theta} E\left[\left(\langle \ell, \theta\rangle - \langle m, \theta\rangle\right)^2\right]$, $\ell \in \mathbb{R}^n$ given. A solution of this mini-max problem is determined by the solution of the fractional optimization problem $\max_{\theta \in \Theta} \frac{\langle \ell, \theta\rangle^2}{1 + \| \theta \|^2}$. We present an efficient method to solve the fractional programming problem for the special case when Θ is a symmetric, bounded set described by linear inequalities, i.e. $\Theta = \{\theta \mid A\theta \le b\}$. We also report on studies where the new method has been compared with direct approaches to the non-linear optimization problem.*

1 Introduction

The following estimation problem has been extensively studied in the literature, cf., for instance, [2] – [16], [18] – [20], and the references cited therein. Consider a vector of data

$$y = \theta + \varepsilon, \tag{1.1}$$

where $y \in \mathbb{R}^n$, θ is an unknown vector, but $\theta \in \Theta$, Θ a given subset of $\mathbb{R}^n$, and $\varepsilon = (\varepsilon_i)_{1 \le i \le n}$ is a sequence of uncorrelated random variables with mean value zero and variance equal to one. The problem is to find a (linear) estimator for the number $\langle \ell, \theta\rangle = \sum_{i=1}^{n} \ell_i \theta_i$, where $\ell \in \mathbb{R}^n$ is given; more precisely, the question is to specify a linear expression of the data y, $\langle m^*, y\rangle = \sum_{i=1}^{n} m_i^* y_i$, which minimizes the maximal risk of all such procedures, i.e. m^* is a solution of the mini-max problem

$$\min_{m \in \mathbb{R}^n} \max_{\theta \in \Theta} E\left[\left(\langle \ell, \theta\rangle - \langle m, y\rangle\right)^2\right] = \min_{m \in \mathbb{R}^n} \max_{\theta \in \Theta} \left(\langle \ell - m, \theta\rangle^2 + \|m\|^2\right). \tag{1.2}$$

The fundamental result concerning this particular problem is the following mini-max theorem. Here we shall only give a special version of a more general result; for generalizations, modifications and a proof of the (general) theorem see [11], [7], [7] and [16].

[1] supported in part by NSF Grant DMS 9404 990

Theorem 1.1 *Let Θ be a convex compact set which is symmetric about the origin. Then*

$$\min_{m\in\mathbb{R}^n}\max_{\theta\in\Theta} E\left[\left(\langle \ell,\theta\rangle - \langle m,y\rangle\right)^2\right] = \min_{m\in\mathbb{R}^n}\max_{\theta\in\Theta}\left(\langle \ell - m,\theta\rangle^2 + \|m\|^2\right)$$
$$= \max_{\theta\in\Theta}\frac{\langle \ell,\theta\rangle^2}{1+\|\theta\|^2}\,. \qquad (1.3)$$

The up-shot of this result is the fact that the mini-max problem has been reduced to a fairly simple fractional programming problem. While there is an extensive literature on theoretical analyses of the linear mini-max estimation problem the aspect of numerically computing the optimal estimator has been somewhat "neglected", cf., however, [20]. The algorithm proposed in [20] is based on a direct approach to the mini-max problem, viz. computing the left-hand-side of Eq. (1.3). Here we consider algorithms to evaluate the right-hand-side of Eq. (1.3).

A first idea is to simply use any non-linear optimizer to solve the fractional programming problem. This approach has the benefit of being applicable no matter what kind of functionals describe the set Θ. A second idea is to solve a sequence of related maximization problems (in general, each problem is a *degenerate* quadratic optimization problem). To this end, consider the functional, $\lambda \in \mathbb{R}$,

$$\Phi(\lambda) = \max_{\theta\in\Theta}\left\{\langle \ell,\theta\rangle^2 - \lambda\left(1+\|\theta\|^2\right)\right\}. \qquad (1.4)$$

If can be easily shown that if λ^* is a root of Φ, $\Phi(\lambda^*) = 0$, then λ^* is the value of the fractional programming problem described above. For the special case when Θ is described by linear constraints one can do better. For this case we propose an algorithm based on a change of variables. A series of numerical tests have shown that this algorithm is usually "better"than any of the other ones.

The paper is organized as follows. In Section 2 we describe the change of variables for the special case where $\ell = e_n = (0, 0, \ldots, 0, 1)$, and describe a numerically efficient procedure how the case of a general vector ℓ can be reduced to the special case $\ell = e_n$. In the final section of the manuscript we provide some test results.

2 The change of variable approach

¿From now on we shall assume the set Θ to be defined as the set all solution vectors for a linear system of inequalities, i.e.

$$\Theta = \left\{\theta \,\middle|\, A\theta \le b\right\}, \qquad (2.1)$$

where A and b are such that Θ is a compact, symmetric and, trivially, convex set. Assuming Θ to be symmetric implies that for every row $A_{[i]}$ of A the vector $-A_{[i]}$ is also a row of A, and that the corresponding co-ordinates of b are the same.
While assuming the prior knowledge about Θ to be captured by a set of linear inequalities is certainly a loss of generality, it can be justified for mainly three reasons:

(1) the appearance of linear constraints in many applications, e.g. for integral estimation, cf. [17], (2) the fact that non-linear constraints could be approximated using linear constraints and, by the way, are approximated this way at each iteration by any non-linear programming solver, and (3) the algorithm described below.

To find the linear mini-max estimator for the special case under consideration we need to solve the following optimization problem (cf. Introduction):

$$\max_{A\theta \leq b} \left\{ \frac{\langle \ell, \theta \rangle^2}{1 + ||\theta||^2} \right\}. \tag{2.2}$$

Below we shall use the following notation: If not otherwise stated any vector is written as a coloumn matrix. If A is a $p \times q$-matrix then $A^{[i]} \in \mathbb{R}^p$, $1 \leq i \leq q$, denotes the i-th coloumn of A, and A can be written as:

$$A = \left[A^{[i]}, \ldots, A^{[q]} \right].$$

Theorem 2.1 *Let $\Theta = \{A\theta \leq b\}$ be a compact, symmetric (and convex) set, and let $\ell \neq 0$; without loss of generality assume $\ell_n < 0$. Put $a = \ell - ||\ell|| e_n$, $q = \frac{a}{||a||}$ and $H = id - 2qq^T$. Define $\tilde{A} = AH^T$. Then*

$$\max_{A\theta \leq b} \left\{ \frac{\langle \ell, \theta \rangle^2}{1 + ||\theta||^2} \right\} = \left[1 + \min_{\mathcal{A}u \geq \beta} \left\{ ||u||^2 \right\} \right]^{-1}, \tag{2.3}$$

where $\mathcal{A} = \left[-\tilde{A}^{[1]}, \ldots, -\tilde{A}^{[n-1]}, +b \right]$ *and* $\beta = +\tilde{A}^{[n]}$.

Proof. By construction $H\ell = e_n$, and H is a symmetric and orthogonal matrix. When checking $H\ell = e_n$ note that $2\langle a, \ell \rangle = ||a||^2$, and that the assumption $\ell_n < 0$ holds. Define

$$\nu := H\theta. \tag{2.4}$$

Then

$$\max_{A\theta \leq b} \left\{ \frac{\langle \ell, \theta \rangle^2}{1 + ||\theta||^2} \right\} = \max_{\tilde{A}\nu \leq b} \left\{ \frac{\langle e_n, \nu \rangle^2}{1 + ||\nu||^2} \right\} = \max_{\tilde{A}\nu \leq b} \left\{ \frac{\nu_n^2}{1 + ||\nu||^2} \right\}. \tag{2.5}$$

To exclude a trivial case, we may assume the maximum to be different from zero. Let us write any feasible ν as $\nu = (x, \xi)$, where $x \in \mathbb{R}^{n-1}$, $\xi \in \mathbb{R}$. Taking the reciprocal of $\nu_n^2 / \left(1 + ||\nu||^2 \right)$, we obtain the minimization problem:

$$\min_{\tilde{A}\nu \leq b} \left\{ \frac{1 + ||\nu||^2}{\nu_n^2} \right\} = \min_{\substack{\tilde{A}\nu \leq b \\ \nu = (x, \xi)}} \left\{ \frac{1}{\xi^2} + \sum_{i=1}^{n-1} \frac{\nu_i^2}{\xi^2} + 1 \right\}. \tag{2.6}$$

Now change variables as follows. Let

$$\zeta := 1/\xi, \ ,\xi \neq 0, \quad \text{and} \quad z = \frac{x}{\xi} \in I\!R^{n-1}. \tag{2.7}$$

Furthermore, for $\nu = (x, \xi)$, $\xi > 0$ (without loss of generality), multiply the vector inequality $\tilde{A}\nu \leq b$ by $1/\xi$ to obtain:

$$\frac{\nu_1}{\xi}\tilde{A}^{[1]} + \ldots + \frac{\nu_{n-1}}{\xi}\tilde{A}^{[n-1]} + \tilde{A}^{[n]} \ \leq \ \frac{1}{\xi}b.$$

Rearranging this linear combination, and using the variables $u = (z, \zeta)$ we see the equivalence of the maximization problem and the minimum distance problem, $u \neq 0$,

$$\min_{\mathcal{A}u \geq \beta} \left\{ \|u\|^2 \right\}, \tag{2.8}$$

where $\mathcal{A}$ und β are described in Theorem 2.1.

Thus we may use any quadratic optimization solver for linear constraints to find the linear mini-max estimator; for instance, Wolfe's modified simplex algorithm would do; for more up-to-date methods cf. [1] and [13]. Given a solution of the minimization problem, i.e.

$$w^* = \|u^*\|^2 = \min_{\mathcal{A}u \geq \beta} \left\{ \|u\|^2 \right\},$$

the best linear mini-max estimator for the case of a quadratic loss function, and its mini-max risk is given as follows:

Theorem 2.2 *Let $u^* = (z^*, \zeta^*)$ be a solution of the minimum distance problem* (2.8), *and let w^* denote the minimal distance. Put*

$$\xi^* = \frac{1}{\zeta^*}, \ x^* = \frac{z^*}{\zeta^*}, \ \nu^* = (x^*, \xi^*) \ \textit{and} \ \theta^* = H^T \nu^*. \tag{2.9}$$

The linear minimax estimator is defined by the linear form $\langle m^*, y \rangle$ where

$$m^* = \frac{\langle \ell, \theta^* \rangle}{1 + \|\theta^*\|^2} \theta^*, \tag{2.10}$$

and the mini-max risk equals $\frac{1}{1+w^*}$.

We are now in the position to describe the "change-of-variables algorithm" for computing the linear mini-max estimator for the case of linear constraints and a quadratic loss function.

Algorithm. Given the mini-max problem described in the theorem stated in the Introduction, see Eq. (1.3).

Step 1: Define the Householder matrix H as given in Theorem 2.1. Put $\tilde{A} = AH^T$, and consider the transformed fractional programming problem described by Eq. (2.5).

Step 2: Solve the minimum distance problem (2.8); see Theorem 2.1 for the notation used.

Step 3: Use the transformation rules stated in Theorem 2.2 to go back from the solution of the minimum distance problem to a solution of the original problem.

3 Some test results[1]

Here we provide some of the numerical results which we used to compare the three different ideas (algorithms) for computing linear mini-max estimators; all of them are using exististing optimization packages, f.i. SAS, NAG, IMSL and GINO. Here we shall only report on two (unrelated) test cases, viz.

case I: The vector $\ell = e_n$, and the linear system $A\theta \le b$ is given by, $2m$ is the number of inequalities,

$$a_{ij} = \begin{cases} (-1)^{i+j}(i+j-1) \,, & 1 \le i \le 2m-1 \quad \text{and } i \text{ is odd} \\ -a_{ij} \,, & 2 \le i \le 2m \qquad \text{and } i \text{ is even} \end{cases}$$

$$b_i = 10p \,, \qquad i = 2p-1 \quad \text{and} \quad i = 2p \,, \quad 1 \le p \le m.$$

case II: The set of inequalities $A\theta \le b$ are defined as linear approximations to an ellipsoid, and $\ell = (1, 1, \ldots, 1)$. For the case of an ellipsoid there is an analytical solution for the linear minimax estimation problem. So we considered case II, among other reasons, to be able to check our numerical computations.

The maximization and minimization problems (2.2) and (2.8) respectively, were solved using different size linear systems of case I. The results in Table 1 (test runs using GINO) show that the minimization problem almost always yielded an optimal solution in fewer iterations than the maximization problem. The two cases, 50 × 50 and 100 × 50, which have the "max" outperforming the "min" are due to the fact that the solution obtained to the "max problem" was clearly not optimal. The same qualitative behaviour has been observed when we employed SAS-NLP and the NAG-library. Table 2 provides similar results for case I with a slightly modified objective function, i.e. we assumed the variance of the noise to be $\frac{2}{3}$ instead of 1. For case II we have analyzed a large number of test-runs using SAS-NLP. We have considered systems with 10^5 inequalities approximating ellipsoids in $I\!R^4$ and higher. The observations we made were the same as in case I.

[1]We thank Tom Powers who helped us with some test runs.

Table 1: Comparison of Max and Min problems for Case I

Typ	Size	Iterations
Max	20×15	22
Min	20×15	4
Max	20×50	21
Min	20×50	7
Max	50×50	4
Min	50×50	7
Max	100×50	6
Min	100×50	9
Max	50×100	18
Min	50×100	7
Max	10×10	11
Min	10×10	7

Table 2: Comparison of Max and Min problems for Case I with noise -variance $\frac{2}{3}$

Typ	Size	Iterations
Max	20×15	20
Min	20×15	8
Max	20×50	25
Min	20×50	6
Max	50×50	24
Min	50×50	6
Max	100×50	46
Min	100×50	7
Max	50×100	39
Min	50×100	6
Max	10×10	13
Min	10×10	7
Max	100×70	53
Min	100×70	6
Min	100×80	6

Finally, table 3 shows that, depending on the package used, e.g. GINO compared with NAG, the number of iterations for the minimization problem would even be smaller. The qualitative behaviour of number of iterations for the max problem and the min problem is, however, the same irrespectively of the package used.

Table 3: Number of iterations for min-problems

Size	Gino	Nag
Min 20 × 15	8	2
Min 20 × 50	6	3
Min 50 × 50	6	3
Min 100 × 50	7	3
Min 50 × 100	6	3
Min 10 × 10	7	2
Min 100 × 70	6	3
Min 100 × 80	6	3

Acknowledgements: We thank an anonymous referee for some useful comments.

References

[1] Bertsekas D.P. (1995). Nonlinear Programming. Athena Scientific, Belmont.

[2] Drygas, H. (1991). On an extension of the Girko inequality in linear minimax estimation. Proceeding of the Probastat '91 Conference, Bratislava/CSFR.

[3] Drygas, H. (1993). Spectral methods in linear minimax estimation. Kasseler Mathematische Schriften 4/93. To be published in the Proceedings of the Oldenburg Minimax Workshop (3./4.8.1992).

[4] Drygas, H. and Lauter (1993). On the representation of the linear min max estimator in the convex linear model. Kasseler Mathematische Schriften No. 7/93.

[5] Gaffke, N. and Heiligers (1989). Bayes, admissible, and minimax linear estimators in linear models with restricted parameter space. *Statistics* 20, 487–508.

[6] Gaffke, N. and Heiligers (1991). Note on a Paper by P. Alson. *Statistics* 22, 3–8.

[7] Helmes, K. and Srinivasan (1992). Linear minimax estimation. Springer Verlay *Lecture Notes in Economics and Mathematics,* 389, 9–23.

[8] Helmes, K. and Christopeit (1996). Linear estimation with ellipsoidal constraints. To appear in *Acta Applicandae Mathematicae*, vol. 43, No. 1, 3–15.

[9] Hoffmann, K. (1979). Characterization of minimax linear estimators in linear regression. *Math. Operationsforsch. u. Statistik, Ser. Statist.*, 20, 19–26.

[10] Ibragimov, A.D. and Hasminski (1987). Estimation of linear functionals in Gaussian noise. *Theory Prob. Appl.,* vol. 32, No. 1, 30–39.

[11] Ibragimov, A.D. and Hasminski (1984). On non-parametric estimation of the value of a linear function in Gaussian white noise. *Theory of Probability and Its Applications,* Vol. XXIX, No. 1, 18–32.

[12] Lauter, H. (1975). A minimax linear estimator for linear parameters under restrictions in form of inequalities. *Math. Operationsforsch. u. Statist.*, 6, Heft 5, 689–695.

[13] Nash, S.G. and Sofer (1996). Linear and Nonlinear Programming. McGraw-Hill, New York.

[14] Pilz, J. (1986). Minimax linear regression estimation with symmetric parameter restrictions. *J. Statist. Plann. Inference* 13, 297–318.

[15] Pilz, J. (1991). Bayesian estimation and experimental design in linear regression models. 2nd edition, Wiley, Chichester, New York.

[16] Pinelis, I.F. (1988). On minimax risk. *Theory Prob. Appl.*, vol. 33, 104–109.

[17] Ritter, K. (1995). Asymptotic optimality of regular sequence design. To appear in *Annals of Statistics.*

[18] Stahlecker, P. and Drygas (1992). Representation theorems in linear minimax estimation. Report No. V-85-92, University of Oldenburg.

[19] Stahlecker, P., Janner and Schmidt (1991). Linear-affine Minimax-Schatzer unter Ungleichungsrestriktionen. *Allg. Statist. Archiv,* 75, 245–264.

[20] Stahlecker, P. and Trenkler (1991). Linear and ellipsoidal restrictions in linear regression. *Statistics* 22, 163–176.

Kurt Helmes
Institute for Operations Research
Humbolt
University of Berlin
Berlin, Germany

C.Srinivasan
Dept. of Statistics
University of Kentucky
Kentucky, USA

INCIDENCE MATRIX GAMES

R. B. Bapat[1] and Stef Tijs

Abstract: *We consider the two-person zero-sum game in which the strategy sets for Players I and II consist of the vertices and the edges of a directed graph respectively. If Player I chooses vertex v and Player II chooses edge e, then the payoff is zero if v and e are not incident and otherwise it is* 1 *or* −1 *according as e originates or terminates at v. We obtain an explicit expression for the value of this game and describe the structure of optimal strategies. A similar problem is considered for undirected graphs and it is shown to be related to the theory of 2-matchings in graphs.*

1 Introduction

A *two-person zero-sum game*, also known as a *matrix game*, consists of two players, each with finitely many pure strategies. We denote the players as Player I and Player II and let their strategy sets be $\{1, \ldots, m\}$ and $\{1, \ldots, n\}$ respectively. If Player I chooses strategy i and Player II chooses strategy j then the payoff to Player I from Player II is a_{ij}. The $m \times n$ matrix $A = [a_{ij}]$ is called the *payoff matrix*. A mixed strategy consists of a probability distribution over the set of pure strategies. If $x = (x_1, \ldots, x_m)$ and $y = (y_1, \ldots, y_n)$ are mixed strategies for Player I and II respectively, then $E_A(x, y) = \sum_{i=1}^{m} \sum_{j=1}^{n} a_{ij} x_i y_j$ is the expected payoff to Player I.

We will assume that the reader is familiar with the fundamental aspects of matrix games, see, for example, [6, 7].

In particular, the well-known Minimax Theorem due to von Neumann asserts the following: Let A be an $m \times n$ payoff matrix. Then there exists a real number called the *value* of A, denoted $val(A)$, and mixed strategies $x^0 = (x_1^0, \ldots, x_m^0)$ and $y^0 = (y_1^0, \ldots, y_n^0)$ for Players I and II respectively such that

$$\min_j \sum_{i=1}^{m} a_{ij} x_i^0 = val(A) = \max_i \sum_{j=1}^{n} a_{ij} y_j^0.$$

The strategies x^0, y^0 are said to be *optimal* for Players I,II respectively.

We will also make use of the basic concepts from Graph Theory without defining them explicitly, see, for example, [1,4].

The main purpose of this paper is to solve (i.e., to determine the value and optimal strategies of) two very natural games related to graphs. In the first game the strategy sets consist of the vertices and the edges of a directed graph. If Player I chooses vertex v and Player II chooses edge e, then the payoff is zero if v and e are not incident and otherwise it is 1 or -1 according as e originates or terminates at v. Observe that if the graph is the directed cycle on three vertices, then this game reduces to the well known "Stone-Paper-Scissors" game [6]. For a different generalization of the Stone-Paper-Scissors game see [3].

[1]The present research was carried out while this author was visiting Tilburg University.

In Section 2 we obtain a simple, explicit expression for the value of this game. We also determine the structure of optimal strategies.

In Section 3 we consider the same game but with an undirected graph. The payoff is thus zero if v and e are not incident and is 1 otherwise.

The value of such games is related to some fundamental graph theoretic notions such as the matching number and the vertex covering number. The problem of determining the optimal strategies is shown to be intimately related to the theory of 2-matchings [5], which is a well-studied area of Graph Theory.

2 The directed incidence matrix game

We now introduce some terminology and notation. Let $G = (V, E)$ be a directed graph with at least one edge. We assume, unless stated otherwise, that $V = \{v_1, \dots, v_m\}$ and $E = \{e_1, \dots, e_n\}$. The (vertex-edge) incidence matrix of G is the $m \times n$ matrix $A = [a_{ij}]$ defined as follows: $a_{ij} = 0$ if v_i and e_j are not incident; otherwise, $a_{ij} = 1$ or -1 according as e_i originates or terminates at v_i.

The following result will be used in the sequel. The proof is easy and hence is omitted.

Lemma 1 *Let the matrix A be the direct sum of the matrices $A_1, A_2, \dots, A_k$, where $val(A_i) > 0$, for $i = 1, 2, \dots, k$, i.e.,*

$$A = \begin{bmatrix} A_1 & 0 & \cdots & 0 \\ 0 & A_2 & \cdots & 0 \\ \vdots & \vdots & \ddots & \vdots \\ 0 & 0 & \cdots & A_k \end{bmatrix}.$$

Then

$$val(A) = \left\{ \sum_{i=1}^{k} \frac{1}{val(A_i)} \right\}^{-1}.$$

A vertex $v \in V$ is called a *source* if it has indegree zero and a *sink* if it has outdegree zero. A *star* is a connected graph in which all vertices, except one which is called the *centre,* have degree 1.

Lemma 2 *Let $G = (V, E)$ be a directed graph with the $m \times n$ incidence matrix A. Then $0 \le val(A) \le 1$. Furthermore, $val(A) = 0$ if G has a directed cycle and $val(A) = 1$ if G is a star with the centre being a source.*

Proof: As usual, let $V = \{v_1, \dots, v_m\}$ and $E = \{e_1, \dots, e_n\}$. Let $\tilde{e_j}, j = 1, \dots, n$ denote the pure strategy for Player II which chooses edge e_j with probability 1. The strategy $x = (\frac{1}{m}, \dots, \frac{1}{m})^T$ for Player I satisfies $E_A(x, \tilde{e_j}) = 0, j = 1, \dots, n$. Thus $val(A) \ge 0$. Since $a_{ij} \le 1$ for all i, j, it is clear that $val(A) \le 1$.

Now suppose G has a directed cycle C with k vertices. Consider the strategy $y = (y_1, \dots, y_n)^T$ for Player II given by

$$y_i = \begin{cases} \frac{1}{k}, & \text{if } e_i \text{ is an edge of } C \\ 0, & \text{otherwise.} \end{cases}$$

Then $E_A(\tilde{v_i}, y) = 0, i = 1, \dots, m$, where $\tilde{v_i}$ is the pure strategy for Player I which chooses vertex v_i with probability 1, and thus $val(A) \le 0$. It follows that $val(A) = 0$ in this case.

Now suppose G is a star with v as the centre and suppose v has indegree zero. We assume, without loss of generality, that $v = v_1$. Then A has a saddle point at the position $(1,1)$ and therefore $val(A) = a_{11} = 1$. ■

We say that G is *acyclic* if it has no directed cycles. Assume now that G is acyclic and for each $v \in V$, let $P(v)$ denote a directed path, originating at v and having maximum possible length. Clearly such a path must terminate at a sink. Note that there may be more than one such path but we choose and fix one. Let $\rho(v)$ denote the length of $P(v)$. If v is a sink then we set $\rho(v) = 0$. For each arc $e \in E$, let $\eta(e)$ denote the number of vertices v such that e is an arc of the path $P(v)$.

Lemma 3 $\sum_{v \in V} \rho(v) = \sum_{e \in E} \eta(e)$.

Proof: Let B be the $m \times n$ matrix defined as follows. The rows of B are indexed by V and the columns of B by E. If $v \in V, e \in E$, then the (v,e)-entry of B is 1 if $e \in P(v)$ and 0 otherwise. Then observe that $\{\rho(v) : v \in V\}$ are the row sums of B and $\{\eta(e) : e \in E\}$ are the column sums of B. Since the sum of the row sums must equal that of the column sums, the result is proved. ■

The following is the main result of this section.

Theorem 4 *Let $G = (V, E)$ be an acyclic directed graph with the $m \times n$ incidence matrix A. Then*

$$val(A) = \frac{1}{\sum_{v \in V} \rho(v)}$$

is the value of the matrix game A with the optimal strategies $\{\frac{1}{\sum_{v \in V} \rho(v)}\rho(v) : v \in V\}$ for Player I and $\{\frac{1}{\sum_{v \in V} \rho(v)}\eta(e) : e \in E\}$ for Player II.

Proof: As usual, let $V = \{v_1, \ldots, v_m\}$ and $E = \{e_1, \ldots, e_n\}$. We assume that the graph is connected, since otherwise, we may prove the result for each connected component and then apply Lemma 1.

Fix $j \in \{1, \ldots, n\}$ and suppose that the arc e_j is from v_ℓ to v_k. We have

$$\frac{1}{\sum_{v \in V} \rho(v)} \sum_{i=1}^{m} \rho(v_i) a_{ij} = \frac{1}{\sum_{v \in V} \rho(v)} \{\rho(v_\ell) - \rho(v_k)\}. \tag{1}$$

Note that $\rho(v_\ell) \geq \rho(v_k) + 1$ and therefore it follows from (1) that

$$\frac{1}{\sum_{v \in V} \rho(v)} \sum_{i=1}^{m} \rho(v_i) a_{ij} \geq \frac{1}{\sum_{v \in V} \rho(v)}. \tag{2}$$

Now fix $i \in \{1, \ldots, m\}$ and let

$$U = \{j : e_j \text{ originates at } v_i\}, W = \{j : e_j \text{ terminates at } v_i\}.$$

We have

$$\frac{1}{\sum_{v \in V} \rho(v)} \sum_{j=1}^{n} a_{ij} \eta(e_j) = \frac{1}{\sum_{v \in V} \rho(v)} \left\{ \sum_{j \in U} \eta(e_j) - \sum_{j \in W} \eta(e_j) \right\}. \tag{3}$$

If $U = \emptyset$, i.e., if v_i is a sink, then the right hand side of (3) is clearly nonpositive. Now suppose that $U \neq \emptyset$. Observe that for any vertex $v \neq v_i$, the path $P(v)$ either contains exactly one edge from U and one edge from W or has no intersection with either U or W.

Thus for any $v \neq v_i$, the path $P(v)$ either makes a contribution of 1 both to $\sum_{j\in U} \eta(e_j)$ and $\sum_{j\in W} \eta(e_j)$ or does not contribute to either of these terms. Also the path $P(v_i)$ makes a contribution of 1 to $\sum_{j\in U} \eta(e_j)$ but none to $\sum_{j\in W} \eta(e_j)$. Thus if v_i is not a sink, then

$$\sum_{j\in U} \eta(e_j) - \sum_{j\in W} \eta(e_j) = 1.$$

Therefore we conclude that for $i \in \{1, \ldots, m\}$,

$$\frac{1}{\sum_{v\in V} \rho(v)} \sum_{j=1}^{n} a_{ij}\eta(e_j) \leq \frac{1}{\sum_{v\in V} \rho(v)}. \tag{4}$$

The result is proved in view of (2) and (4). ∎

As a corollary of Theorem 4 we now obtain a converse of (the second part of) Lemma 2. The *if* parts in the next result were proved in Lemma 2, while the *only if* parts follow from Theorem 4.

Lemma 5 *Let $G = (V, E)$ be a directed graph with the $m \times n$ incidence matrix A. Then $val(A) = 0$ if and only if G has a directed cycle and $val(A) = 1$ if and only if G is a star with the centre being a source.*

The next result gives bounds on the value of the incidence matrix of a directed, acyclic, connected graph.

Theorem 6 *Let $G = (V, E)$ be a connected, acyclic, directed graph with the $m \times n$ incidence matrix A. Then*

$$\frac{1}{\binom{m}{2}} \leq val(A) \leq 1.$$

Proof: It was shown in Lemma 2 that $val(A) \leq 1$. The lower bound will be proved by induction on the number of vertices. As before, for each $v \in V$, let $P(v)$ denote a directed path, originating at v and having maximum possible length, and let $\rho(v)$ denote the length of $P(v)$. In view of Theorem 4, we must show that

$$\sum_{v\in V} \rho(v) \leq \frac{m(m-1)}{2}. \tag{5}$$

Clearly, (5) holds for $m = 2, n = 1$. We assume that (5) holds for graphs with fewer than $m(\geq 3)$ vertices and proceed by induction.

Let v_0 be a source and let e_0 be an arc originating at v_0 which is contained in $P(v_0)$. For each vertex v of G other than v_0, $P(v)$ is the path of maximum possible length in $G \setminus \{v_o\}$, the graph obtained from G by removing v_0. Also, we clearly have $\rho(v_0) \leq m - 1$. By the induction assumption,

$$\sum_{v\in V\setminus v_0} \rho(v) \leq \binom{m-1}{2}.$$

Thus

$$\sum_{v\in V} \rho(v) \leq m - 1 + \frac{(m-1)(m-2)}{2} = \frac{m(m-1)}{2}$$

and the proof is complete. ∎

We remark that the lower bound in Theorem 6 is attained precisely when G is a directed path on m vertices.

We now consider the structure of optimal strategies.

Theorem 7 *Let $G = (V, E)$ be a directed, acyclic graph. Then the optimal strategy for Player I is unique.*

Proof: As before we assume that $V = \{v_1, \ldots, v_m\}$ and $E = \{e_1, \ldots, e_n\}$, and let A be the incidence matrix of G. Suppose $\{\phi(v), v \in V\}$ is optimal for Player I. Let $v_i \in V$ be a sink. We claim that $\phi(v_i) = 0$. To see this, let $\{y_1, \ldots, y_n\}$ be optimal for Player II. If $\phi(v_i) > 0$, then we must have

$$\sum_{j=1}^{n} a_{ij} y_j = val(A). \tag{6}$$

Since v_i is a sink, $a_{ij} \leq 0, j = 1, \ldots, n$, whereas, by Theorem 4, $val(A) > 0$. This contradicts (6) and the claim is proved.

Now let $v \in V$ be a vertex which is not a sink and let $v = u_0, u_1, \ldots, u_{k-1}, u_k = w$ be a path of maximum length originating at v. Since ϕ is optimal,

$$\phi(u_i) - \phi(u_{i+1}) \geq val(A), i = 0, 1, \ldots, k-1.$$

Thus

$$\sum_{i=0}^{k-1} \{\phi(u_i) - \phi(u_{i+1})\} \geq k val(A),$$

and hence

$$\phi(v) - \phi(w) \geq \rho(v) val(A).$$

Since w must necessarily be a sink, $\phi(w) = 0$ by our earlier claim and hence

$$\phi(v) \geq k val(A) = \rho(v) val(A). \tag{7}$$

Thus

$$1 = \sum_{v \in V} \phi(v) \geq val(A) \sum_{v \in V} \rho(v) = 1,$$

where the last equality follows by Theorem 4. Thus equality must hold in (7) and the result is proved. ■

The assumption that G be acyclic is necessary in Theorem 7 as can be seen from the next example.

Let $G = (V, E)$ where $V = \{v_1, v_2, v_3, v_4\}$ and $E = \{v_1 v_2, v_2 v_3, v_3 v_1, v_2 v_4\}$. Then it can be verified that the strategies $(\frac{1}{3}, \frac{1}{3}, \frac{1}{3}, 0)$ and $(\frac{1}{4}, \frac{1}{4}, \frac{1}{4}, \frac{1}{4})$ are both optimal for Player I.

Let us concentrate now on the dimension of the optimal strategy space of Player II, for an $m \times n$ incidence matrix. A pure strategy is called *essential* if it is used with positive probability in some optimal strategy, otherwise it is *inessential*. Let d_1, d_2 be the dimensions of the optimal strategy spaces $\mathcal{O}_1$ and $\mathcal{O}_2$ of Players I, II respectively, and let f_1, f_2 be the number of essential strategies for Players I,II respectively. Then according to a well-known result of Bohnenblust, Karlin and Shapley [2], $f_1 - d_1 = f_2 - d_2$.

Now let $G = (V, E)$ be a directed, acyclic graph with the $m \times n$ incidence matrix A. By Theorem 7, the optimal strategy for Player I is unique. Also any vertex which is not a sink is essential for Player I. Let s be the number of sinks in G and let t be the number of inessential strategies (i.e., arcs) for Player II. Then we conclude by the result of Bohnenblust, Karlin and Shapley that the dimension of the optimal strategy space of Player II is

$$n - m + s - t.$$

Example 8 *Let $G = (V, E)$ be the directed graph with vertices v_1, v_2, v_3, v_4 and edges $e_1 = (v_1, v_2), e_2 = (v_2, v_3), e_3 = (v_2, v_4)$, and $e_4 = (v_1, v_3)$. Let A be the corresponding incidence matrix, so*

$$A = \begin{bmatrix} 1 & 0 & 0 & 1 \\ -1 & 1 & 1 & 0 \\ 0 & -1 & 0 & -1 \\ 0 & 0 & -1 & 0 \end{bmatrix}.$$

Then $val(A) = \frac{1}{3}$ and $\mathcal{O}_1 = \{(\frac{2}{3}, \frac{1}{3}, 0, 0)\}$. Thus $m = n = 4, s = 2, t = 1$, and $d_1 = 0$. Therefore $\mathcal{O}_2$ must be one-dimensional. In fact, it can be verified that $\mathcal{O}_2$ is precisely the line segment joining $\{(\frac{1}{3}, 0, \frac{2}{3}, 0)\}$ and $\{(\frac{1}{3}, \frac{2}{3}, 0, 0\})$.

3 The undirected incidence matrix game

In this section we consider undirected graphs. Let $G = (V, E)$ be a graph with m vertices and n edges. The incidence matrix A of G is an $m \times n$ matrix with a_{ij} being zero if the i-th vertex is not incident with the j-th edge and one otherwise. Our main interest in this section is in obtaining the value and the optimal strategies for the matrix game with the payoff matrix A.

We recall some graph-theoretic terminology and notation, see, for example, Lovász and Plummer [5]. A set of edges constitute a *matching* if no two edges in the set are incident with a common vertex. The maximum cardinality of a matching in G is the *matching number* of G, denoted by $\nu(G)$. A set of vertices of G is a *vertex cover* if they are collectively incident with all the edges in G. The minimum cardinality of a vertex cover is the *vertex covering number* of G denoted by $\tau(G)$.

We first prove a simple preliminary result.

Lemma 9 *Let $G = (V, E)$ be a graph with the $m \times n$ incidence matrix A. Then*

$$\frac{1}{\tau(G)} \leq val(A) \leq \frac{1}{\nu(G)}. \tag{8}$$

Proof: Let $\tau(G) = k, \nu(G) = \ell$, and suppose that vertices numbered $i_1, \ldots, i_k$ form a cover and that the edges numbered $j_1, \ldots, j_\ell$ form a matching. If Player I chooses vertices $i_1, \ldots, i_k$ uniformly with probability $\frac{1}{k}$, then against any pure strategy of Player II he is guaranteed a payoff of at least $\frac{1}{k}$. Similarly, if Player II chooses edges $j_1, \ldots, j_\ell$ uniformly with probability $\frac{1}{\ell}$, then against any pure strategy of Player I he loses at most $\frac{1}{\ell}$. These two observations together give (8). ■

Recall that a graph is said to have a *perfect matching* if it has a matching in which the edges are collectively incident with all the vertices. A graph is *Hamiltonian* if it has a cycle, called a *Hamilton cycle*, containing every vertex exactly once.

In the next result we identify some classes of graphs for which the value of the corresponding game is easily determined.

Theorem 10 *Let $G = (V, E)$ be a graph with the $m \times n$ incidence matrix A.*

(i) If G is bipartite then $val(A) = \frac{1}{\nu(G)}$.

(ii) If G is the path on m vertices then $val(A) = \frac{m}{2}$ if m is even and $\frac{m-1}{2}$ if m is odd.

(iii If G has a perfect matching then $val(A) = \frac{2}{m}$.

(iv) If G is Hamiltonian then $val(A) = \frac{2}{m}$.

(v) If G is the complete graph on m vertices, then $val(A) = \frac{2}{m}$.

Proof: If G is bipartite, then by the well-known König Theorem (see, for example, [5]), $\nu(G) = \tau(G)$ and the result follows by Lemma 9. Therefore (i) is proved. Since a path is bipartite, (ii) follows in view of the observation that the matching number of a path on m vertices is $\frac{m}{2}$ if m is even and $\frac{m-1}{2}$ if m is odd. Similarly, if G has a perfect matching, then $\nu(G) = \tau(G) = \frac{m}{2}$ and (iii) is proved.

To prove (iv), first suppose that G is a cycle on m vertices. Then $n = m$ and the strategies for Players I, II which choose all pure strategies with equal probability, namely $\frac{1}{m}$, are easily seen to be optimal. Thus $val(A) = \frac{2}{m}$.

Now suppose G is Hamiltonian. Then the value of G is at least equal to the value of the game corresponding to a Hamilton cycle in G and thus, $val(A) \geq \frac{2}{m}$ by the preceding observation. Furthermore, if Player II chooses only the edges in the Hamilton cycle with equal probability then against any pure strategy of Player I he loses at most $\frac{2}{m}$. Therefore (iv) is proved.

Finally, (v) follows since a complete graph is clearly Hamiltonian. ■

We now turn to arbitrary graphs, not necessarily covered by Theorem 10. In this case the problem of determining the value and optimal strategies is closely related to the theory of 2-matchings and a theorem of Tutte [8]. We now indicate how one can determine the structure of optimal strategies using Tutte's theorem.

We first need some definitions, see [5], p.214. Let $G = (V, E)$ be a graph. A *2-matching* of G is an assignment of weights $0, 1$ or 2 to the edges of G such that the sum of the weights of edges incident with any vertex is at most 2. The sum of the weights in a 2-matching is called the *size* of the 2-matching. The maximum size of a 2-matching in G is denoted by $\nu_2(G)$.

A *2-cover* of G is an assignment of weights $0, 1$ or 2 to the vertices of G such that the sum of the weights of the two endpoints of any edge is at least 2. The sum of the weights in a 2-cover is called the *size* of the 2-cover. The minimum size of a 2-cover in G is denoted by $\tau_2(G)$.

We now state a result due to Tutte [8], see [5].

Theorem 11 *If G is any graph, then $\nu_2(G) = \tau_2(G)$.*

Theorem 11 directly leads to the following result about the structure of optimal strategies in the game with the matrix being the incidence matrix of a graph. The proof is omitted.

Theorem 12 *Let $G = (V, E)$ be a graph with the $m \times n$ incidence matrix A. Then there exists a positive integer k such that $val(A) = \frac{2}{k}$. Furthermore, there exist partitions $V = V_1 \cup V_2 \cup V_3, E = E_1 \cup E_2 \cup E_3$ such that for Player I, choosing vertices in V_1 uniformly with probability $\frac{1}{k}$ and vertices in V_2 uniformly with probability $\frac{2}{k}$ is optimal. Similarly for Player II, choosing edges in E_1 uniformly with probability $\frac{1}{k}$ and edges in E_2 uniformly with probability $\frac{2}{k}$ is optimal.*

Example 13 *Consider the graph $G = (V, E)$ with the vertex set $V = \{v_1, v_2, \ldots, v_7\}$ and edges $E = \{v_1v_2, v_2v_3, v_1v_3, v_3v_4, v_4v_5, v_4v_6, v_5v_6, v_5v_7\}$. The incidence matrix of G is given by*

$$A = \begin{bmatrix} 1 & 0 & 1 & 0 & 0 & 0 & 0 & 0 \\ 1 & 1 & 0 & 0 & 0 & 0 & 0 & 0 \\ 0 & 1 & 1 & 1 & 0 & 0 & 0 & 0 \\ 0 & 0 & 0 & 1 & 1 & 0 & 1 & 0 \\ 0 & 0 & 0 & 0 & 1 & 1 & 0 & 1 \\ 0 & 0 & 0 & 0 & 0 & 1 & 1 & 0 \\ 0 & 0 & 0 & 0 & 0 & 0 & 0 & 1 \end{bmatrix}.$$

The value of the game is $\frac{2}{7}$. An optimal strategy for Player I is to choose each of the vertices v_1, v_2, v_3 with probability $\frac{1}{7}$ and vertices v_4, v_5 each with probability $\frac{2}{7}$. An optimal strategy for Player II is to choose edges v_1v_2, v_2v_3, v_1v_3 each with probability $\frac{1}{7}$ and edges v_4v_6, v_5v_7 each with probability $\frac{2}{7}$.

References

1. J. A. Bondy and U. S. R. Murty (1976), *Graph Theory with Applications*, The Macmillan Press Ltd., London.

2. H.F. Bohnenblust, S. Karlin and L.S. Shapley (1950), Solutions of discrete two-person games, in *Contributions to The Theory of Games, Volume 1,* H. W. Kuhn and A. W. Tucker ed, Princeton University Press, Princeton, pp. 51-72.

3. David C. Fisher and Jennifer Ryan (1955), Tournament games and positive tournaments, *Journal of Graph Theory,* 19(2), 217-236.

4. L. Lovász (1979), Combinatorial Problems and Exercises, North-Holland, Amsterdam.

5. L. Lovász and M. D. Plummer(1986), Matching Theory, Elsevier Science Publ. Co., New York.

6. Guillermo Owen (1982), Game Theory, Academic Press, New York.

7. T. Parthasarathy and T.E.S. Raghavan (1971), Some Topics in Two Person Games, Amer. Elsevier. Publ. Co., New York.

8. W. T. Tutte (1953), The 1-factors of oriented graphs, *Proc. Amer. Math. Soc.*, 4, 922-931.

R. B. Bapat
Indian Statistical Institute
New Delhi-110016
India

Stef Tijs
CentER and Econometrics Department
Tilburg University, P. O. Box 90153
5000, Tilburg University, The Netherlands

COMPLETELY MIXED GAMES AND REAL JACOBIAN CONJECTURE[1]

T. Parthasarathy G. Ravindran and M. Sabatini [2]

Abstract: *In this paper, we consider Cubic Linear Mapping $F : R^n \to R^n$, where $F_i = X_i + (AX)_i^3$, $X \in R^n, i = 1, 2, \cdots n$ A is an $n \times n$ matrix and study the injectivity of F when A is a P_0 matrix or when A is a Z matrix. In the sequel, we also derive many interesting results related to cubic linear mappings .In our proofs, we make use of concepts from game theory.*

1 Introduction

The problem of proving global injectivity of mappings in R^n have received considerable attention since Hadamard proved homeomorphism for proper maps having a nonvanishing Jacobian. In particular, polynomial mappings have been of great interest because of its applicability to Dynamical Systems, Mathematical Economics, Differential Equations etc. In this paper we restrict our attention to real cubic linear polynomial maps and establish univalence for such maps. Our motivation for considering such maps comes from a paper due to Druzkowski [4].

A map $F : R^n \to R^n, F(X) = (F_1(X), \cdots, F_n(X))$ is called a polynomial map if for each $i = 1, 2, \cdots, n,$ F_i is a real polynomial in n variables $X = (X_1, X_2, \cdots, X_n); X \in R^n$.

The question is : How do we recognize whether such a polynomial map F is invertible and the inverse to be a polynomial ?

Keller conjectured that if the Jacobian of the above mapping is a non-zero constant, then F is injective and the inverse is a polynomial. It is not known, whether the above question has an affirmative answer in general for any n. The original conjecture of Keller was in the complex case and is still open even when $n = 2$! A related conjecture is the following.

Real Jacobian Conjecture (RJC): Let F be a polynomial map from $R^n \to R^n$. Suppose $|J(x)| \neq 0$ for every $x \in R^n$, then F is one-one.

Note that the RJC is not the real case of Keller's Conjecture. However, recently, Pinchuk [13] has given a counter example to RJC. In view of this example, we can hope to consider only special cases of RJC by placing further conditions on the mapping F and prove injectivity.

Several Authors have contributed to the Jacobian Conjecture (both real and Complex) having a non-zero constant Jacobian and RJC by placing further conditions on the Jacobian matrix. They include Bass et al,Campbell, Druzkowski, van den Essen and Parthasarathy. van den Essen and Parthasarathy [5] asserted global univalence for polynomial mappings in C^n having a Jacobian whose leading principal minors are non-vanishing constants. Campbell [2,3] proved univalence for such maps in the real case. Druzkowsky-Bass etal [4,1] reduced Jacobian conjecture to proving univalence for Cubic Linear Mappings. Others who have

[1]Dedicated to Professor Gary Meisters on the occasion of his sixtyfifth birthday

[2]This work was done while this author was visiting ISI, Delhi and he would like to thank Indian Statistical Institute, for the hospitality.

contributed in global univalence of general mappings by placing conditions on the Jacobian matrix include Gale-Nikaido [6], Meisters and Olech [8], Parthasarathy [10] and Ravindran [14]. For a good review on this subject see [10].

Our motivation for considering cubic linear mappings comes from a reduction theorem [1] which we state below. We need the following definition.

Definition 1.1 (Cubic Linear Mapping) CLM : A mapping $F = (F_1, F_2, \cdots, F_n) : R^n \to R^n$ is said to be a cubic linear mapping, if $F_i = X_i + (AX)_i^3 = X_i + (a_{i1}X_1 + a_{i2}X_2 + \cdots + a_{in}X_n)^3, i = 1, 2 \cdots n$, where $A = [a_{ij}]$ is a n × n matrix and $X = (X_1, X_2, \cdots, X_n) \in R^n$.

We will refer to Cubic Linear Mappings in R^n as CLM.

Theorem 1.1 (Reduction Theorem) [1]

In order to verify the Jacobian Conjecture (for every $n \geq 2$) it is sufficient to check Jacobian Conjecture (for every $n \geq 2$) only for polynomial mappings $F = (F_1, F_2, \cdots, F_n)$ of the cubic linear form: $F_i(X) = X_i + (AX)_i^3, \quad i = 1, 2, \cdots, n$ and A is a n × n nilpotent matrix.

In view of the above theorem, we consider only Cubic linear mappings in general, (i.e.) not necessarily having constant Jacobian but only nonsingular Jacobian and prove injectivity when A satisfies certain conditions. In the next section, We state and prove our main results and devote the last section for important remarks and open problems. In proving our results we make use of concepts from Game Theory. For results relating to Game theory please refer [11,12].

2 P_0 Matrices and Cubic Linear Mappings

In this section we prove univalence for cubic linear mappings when A is a P_0 matrix and derive several results based on this. We also give alternate proofs of some results due to Druzkowski [4]. In all our results, we shall assume F to be a cubic linear polynomial and J to be its Jacobian. We call a matrix A , a P (P_0) -matrix if all its principal minors are positive (non-negative). A is called a Z- matrix if all its off-diagonal entries are non-positive. We begin this section by proving some interesting results relating to cubic linear mappings.

Lemma 2.1 : If two odd cubic functions take the same value at two points then their derivatives vanish at the same point.

Proof:Let $\Phi_i(t) = k_i + b_i t + d_i t^3$. and $\Phi(t) = (\Phi_1(t), \ \Phi_2(t))$. $\Phi(t_1) = \Phi(t_2), \ t_1 \neq t_2$ implies $-b_i/d_i = t_1^2 + t_1.t_2 + t_2^2$

$\Phi_i'(t^*) = b_i + 3d_i t^2 = 0$ implies $(t^*)^2 = -b_i/3d_i = 1/3(t_1^2 + t_1 t_2 + t_2^2$ which is independent of i.

Lemma 2.2 : Let F be CLM and if the Jacobian is nonsingular for all $x \in R^n$ then the determinant of the Jacobian is positive for all $x \in R^n$.

Proof: The Jacobian matrix can be easily seen to be

$$J = \begin{bmatrix} 1 + 3(Ax)_1^2 a_{11} & 3(Ax)_1^2 a_{12} & \cdots & 3(Ax)_1^2 a_{1n} \\ 3(Ax)_2^2 a_{21} & 1 + 3(Ax)_2^2 a_{22} & \cdots & 3(Ax)_2^2 a_{2n} \\ \vdots & \vdots & \vdots & \vdots \\ 3(Ax)_n^2 a_{n1} & 3(Ax)_n^2 a_{n2} & \cdots & 1 + 3(Ax)_n^2 a_{nn} \end{bmatrix}$$

Note that for $x_1 = x_2 = \cdots = x_n = 0$, $|J| = 1$ and if $|J| < 0$ for some $x \in R^n$ then $|J|$ will be zero for some x since the determinant is a continuous function which leads to a contradiction. □

Remark 2.1: The above lemma shows that it is enough to prove univalence for cubic linear mappings having a Jacobian whose determinant is positive.

Proposition 2.3: Let F be a CLM with nonvanishing Jacobian. F is injective on lines passing thro' the origin.

Proof:Take a line through origin O. Consider $F(tx) = tx + t^3(Ax)^3$.

Let $F(t_1x) = F(t_2x)$, then by lemma 2.1 ,$dF_j/dt(t^*) = 0$ for all j for some t^* implies J_F is singular which is a contradiction. □

The above proposition has been observed by Oda for algebraically closed fields.

Now we prove our main theorem and derive its consequences.

Proposition 2.4 : Let F be a CLM. If A is a P_0 matrix then F is one-one .

Proof : Since A is a P_0-matrix, by the sign reversal property we know $\forall x \in R^n$, $\exists$ an i such that $x_i \neq 0$ and $(Ax)_i x_i \geq 0$.

Let $x \neq y$, $x, y \in R^n$ be any two points.

$$\begin{aligned} F_j(x) - F_j(y) &= (x_j - y_j) + (Ax)_j^3 - (Ay)_j^3 \\ &= (x - y)_j + (Ax - Ay)_j((Ax)_j^2 + (Ax)_j(Ay)_j + (Ay)_j^2 \\ &= (x - y)_j + (A(x - y))_j \cdot [((Ax)_j + \frac{(Ay)_j}{2})^2 + \frac{3}{4} \cdot (Ay)_j^2] \end{aligned}$$

Since A is P_0, there exists an i such that $x_i - y_i \neq 0$ and $(x - y)_i \cdot [A(x - y)_i] \geq 0$ for some i and $x_i - y_i \neq 0$.

Multiplying the i-th equation by $x_i - y_i$ we see that

$$(x_i - y_i)^2 + (x_i - y_i)[(A(x - y))_i \cdot (\geq 0))]$$

is positive. This means that $F_i(x) \neq F_i(y)$ for some i, for any $x \neq y$.

Remark 2.2: It is easy to check that if A is a P_0-matrix then the Jacobian is a P-matrix and hence univalence prevails by Gale-Nikaido's [6] result. But the above proof is given directly using the sign reversal property.

Now we extend the exposition of Proposition 2.3 to lines not passing through the origin, that is, we try to find out conditions under which injectivity prevails on lines not passing through the origin.

If $x \neq y \in R^n$, set $z = y - x$. The line thro x,y is x+tz. and$F(x + tz) = x + tz + (Ax + tAz)^3$, where $F_j(x + tz) = x_j + tz_j + ((Ax)_j + t(Az)_j)^3$. We define the scalars k_j, b_j, c_j, and d_j to be the coeeficients of t^0, t^1, t^2 and t^3 respectively in the above expansion.

Proposition 2.5 : Assume F to be a CLM with a nonvanishing Jacobian.

a)If for some j, $z_j \neq 0, z_j(Az)_j \geq 0$, then F is injective on $x + tz$

b) If $x + tz$ intersects the null space of A, then F is injective on $x + tz$.

Proof:

a) Let us compute $d/dtF_j()$ and impose the restriction that its discriminant is $\Delta_j \leq 0$. This means that $z_j(a_jz) \geq 0$.This is similar to the P_0 property discussed in Proposition 2.4.

b)Assume that Δ_j changes sign for every j. Since the Jacobian is nonvanishing, there is at least one nonconstant component. For any j there exists a change of parameter $t \rightarrow t - c_j/3d_j$ that transforms the original cubic function F_j into an odd cubic function. We look for conditions underwhich $c_j/3d_j$ is independent of j (for the nonconstant components), so that we can apply lemma 2.1 in order to have injectivity on such a line. Note that $c_j = 3(Ax)_j(Az)_j^2$ and $d_j = (Az)_j^3$. Hence the condition is, there exists $h \in R$ such that $A(x + hz) = 0$. that is , the line $(x + tz)$ intersects the null space of A. Hence F is injective on every line intersecting the null space of A.

Our next proposition shows further that the nonsingularity of the Jacobian restricts the mapping to be a P-function if we assume A to be Z-matrix. In proposition 2.8 , we go on to show the same property holds even if we just assume A to be nonsingular.

Proposition 2.6 : Let F be CLM. If $A \in Z$, then $|J| \neq 0$ iff A is P_0.

Proof : If A is P_0, then J is a P-matrix and hence $|J| \neq 0$. To prove the converse, since $A \in Z$, the Jacobian is also a Z- matrix and since $|J| \neq 0, v(J) \neq 0$ (refer [11]). For $x = 0$, $v(J) > 0$ and hence $v(J)$ is positive $\forall\ x \in R^n$ (otherwise, if $v(J) < 0$ for some $x \in R^n$, since the value function is continuous, this will imply $v(J)$ is zero for some x, which means J will be singular). This means J is a completely mixed game for all $x \in R^n$ and is a P-matrix for all $x \in R^n$ [11]. Hence if any principal minor of A is negative, $v(A)$ cannot be nonnegative ($v(A) \geq 0$ will imply A is P_0). Therefore $v(A)$ is negative, that is, $\exists x \geq 0, \sum x_i = 1$ such that $Ax < 0$ and since $J = I + 3[diag(Ax)^2].A$ where $diag(Ax)^2$ is a diagonal matrix whose i-th diagonal element is $(Ax)_i^2$ and $v(A) < 0$, for some $y = \lambda x$ (λ , a suitably large scalar) $v(J(y)) < 0$ which is a contradiction to the fact that J is P matrix $\forall x \in R^n$. This completes the proof of Proposition 2.6. □

Remark 2.3: In proving the above result we have used the fact that if $v(A) < 0$ for any matrix then $v(I + DA) < 0$ for all sufficiently large positive diagonal matrices.

Corollary 2.7 Let F be CLM. if $A \in Z$, $|J| \neq 0$, then F is one-one.

Now we have the following interesting result.

Proposition 2.8 : Let F be CLM. If $|J| \neq 0$ and if $|A| \neq 0$, then A is P_0 .

Proof: We already know from lemma 2.2, that $|J| > 0\ \forall\ x \in R^n$. Now we will show that all the principal minors of A are non-negative. On the contrary, suppose $|A_{\alpha\alpha}| < 0$ where $A_{\alpha\alpha}$ is the leading principal minor gotten by deleting the last $n - k$ rows and columns from A. Since A is nonsingular, $\exists$ an $x \in R^n$ such that

$$Ax = \begin{pmatrix} \mathbf{1_k} \\ \mathbf{0_{n-k}} \end{pmatrix} \qquad *$$

where $\mathbf{1_k}$ is a $k \times 1$ vector of ones and $\mathbf{0_{n-k}}$ is a $(n - k) \times 1$ vector of zeros .

Now $|J(x)| = |I + 3[diag(Ax)^2] \cdot A|$ where diag $(Ax)^2$ is a diagonal matrix whose i-th diagonal element is $(Ax)_i^2$ and x is a solution for $*$.

Let λ be a sufficiently large positive scalar.Now $|J(\lambda x)| = |I + \Lambda\ A|$ where Λ is a diagonal matrix with $(\lambda)_{ii} = \lambda$ for i=1, 2...k and zero for i=k+1, ..., n. Since $|A_{\alpha\alpha}| < 0, \alpha = 1, 2, \cdots, k$, for a suitable $\lambda \cdot x$, we can show by expansion of $|I + \Lambda\ A|$ that $|J(\lambda x)| < 0$,.This will be a contradiction to $|J(x)| > 0\ \forall\ x \in R^n$. Hence A is P_0 matrix. □

Using Propositions 2.4 and 2.8 we can obtain the following result due to Druzkowski.

Corollary 2.9: Let F be CLM. if $|J| \neq 0$ and $|A| \neq 0$ then F is one-one.

The following result is due to Druzkowski, we give a direct proof for the same.

Proposition 2.10: Let F be CLM and suppose $|J_F| \neq 0$, and Rank$(A) = 1$, then F is one-one.

Proof:Let $a_1, a_2, \cdots, a_n$ denote the rows of A.Assume w.l.o.g. $a_1 \neq 0$, then Rank(A)=1 implies $a_i = t_i a_1$ and $t_1 = 1$ and it is enough to prove our result for $t_i \neq 0\ \forall\ i$.

Note that $F_i(x) = x_i + t_i^3(a_1.x)^3$. and

$F_i(x) - F_i(y) = (x_i - y_i) + t_i^3[(a_1.x)^3 - (a_1.y)^3]$, i=1, 2..., n.

and w.l.o.g.$x_i \neq y_i$ for all i.

Now $F(x) = F(y)$ implies $x_i - y_i = t_i^3(x_1 - y_1)$ i=1, 2, ..n.

and $F_1(x) - F_1(y) = x_1 - y_1 + (a_1(x - y))[(a_1x)^2 + (a_1x)(a_1y) + (a_1y)^2]$

$$= x_1 - y_1 + (a_{11}(x_1 - y_1) + a_{12}t_2^3(x_1 - y_1) + \cdots + a_{1n}t_n^3(x_1 - y_1)) \cdot [\geq 0]$$

. Let $a_{11} + a_{12}t_2^3 + ... + a_{1n}t_n^3 =, say, \kappa$. It can be easily seen that $|J|$=3 $\kappa(a_1x)^2 + 1$ .Since $|J(x)| = 3\kappa \cdot (a_1x)^2 + 1 > 0$ for all $x \Rightarrow \kappa > 0$.

Therefore $(F_1(x) - F_1(y))(x_1 - y_1) > 0$. which is a contradiction to $F(x) = F(y) \Rightarrow F$ is one-one. □

3 Remarks and Open Problems

In this section we make some important remarks and suggest open problems. To this end we establish the following.

Proposition 3.1: Let F be CLM. If $A < 0$, then Jacobian vanishes for some $x \in R^n$. 7

Proof : It is easy to check that

$$J = I + 3diag(Ax)^2 \cdot A$$

where $diag(Ax)^2$ is diagonal matrix as defined. (in lemma 2.2)

Hence J is a Z-matrix and for $x = 0, J = I$ which mean $v(J)$ is positive at x=0. Since $A < 0$, for a suitable vector x^*, sufficiently large, $3(diagAx)^2 A$ will be a large negative matrix and hence value of J will be less than zero. Since the value function is continuous with respect to the entries of the matrix, value of J is zero for some y. By the property of Z-matrix it follows J will be singular at y [J is a Z-matrix with value zero and hence completely mixed at y. Therefore J is singular.] □

Proposition 3.2: Let $G_i = x_i + (SASx)_i^3$ and $F_i = x_i + (Ax)_i^3$. Then $|J_F| \neq 0 \Leftrightarrow |J_G| \neq 0$ where $F = (F_1, F_2, \cdots, F_n.)$, $G = (G_1, G_2, \cdots, G_n.)$ and S is any signature matrix (i.e.) diagonal matrix with ± 1 as its entries.

Proof: Follows from the definition of Cubic Linear mappings. (i.e.)

$$G_i = (S_\circ F_\circ S(x))_i = x_i + (SASx)_i^3$$

It is trivial to see that $J_{S_\circ F_\circ S} = |S||J_F||S| = |J_F|$ and the result follows. □

Corollary 3.3: Let F be CLM. If A is an N-matrix or $almostP$ then the Jacobian vanishes. □

An N matrix is a matrix whose principal minors are negative. An almost P-matrix is the inverse of an N-matrix.

Remark 3.1: Note that the proof of corollaries also follows from Theorem 2.8. $|J| \neq 0$ is a crucial assumption even if $|A| \neq 0$ as suggested by the following simple example. Let $F = X - X^3$, Here the Jacobian vanishes. Also note that $F(0) = F(1)$ and F is not one-one. But A is a single element -1.

Remark 3.2: In view of Proposition 3.1, Theorem 1, in Yu's [17] paper is vaccuous since $J(F)$ will vanish whenever $F = X - N$ where N is a polynomial of degree greater than or equal to two and having non-negative co-efficients. Theorem 14 [17] also becomes vaccuous as $I - J(N)$ will also be singular by our Proposition 3.1 above.

Remark 3.3: One may ask whether injectivity prevails if we assume $|J| \neq 0, J \geq 0$ (all partial derivatives are non-negative). In general, the answer is negative as given by the following example [14].

Let $F = [f_1(u,v), f_2(u,v), f_3(u,v,w)]$
where

$$\begin{aligned}
f_1(u,v) &= e^{2u} - v^2 + 3 \\
f_2(u,v) &= 4ve^{2u} - v^3 \\
f_3(u,v,w) &= (10 + e^{2u})(e^v + e^{-v})(e^{100w} - e^{-100w})
\end{aligned}$$

Consider $G = A_\circ^{-1} F_\circ A(x)$
where

$$A^{-1} = \begin{bmatrix} 1 & 1 & 1 \\ -1 & 0 & 1 \\ 0 & -1 & 1 \end{bmatrix}$$

We can check the partial derivatives of G to be non-negative and $|J_G| \neq 0$ but $F(0,2,0) = F(0,-2,0) = (0,0,0) \Rightarrow G$is not one-one $\square$

This leads us the following **Conjecture**: Is the above result true for polynomial mappings. In particular, assume F to be Cubic linear mapping having nonvanishing Jacobian, Is F one-one if $A \geq 0$.

Acknowledgement: We would like to thank Professors: L.A.Campbell, L.M. Druzkowskii and A.van den Essen for several useful suggestions.

References

1. Bass, H. Connell, E.H. and Wright, D.L.(1982) "The Jacobian conjecture : Reduction of Degree and formal expansion of the inverse", Bull. AMS. 7, 287-330 .

2. Campbell, L.A.(1993) "Decomposing Samuelson Maps", Lin. Alg and its applns 187, 227-238.

3. Campbell, L.A.(1994) "Rational Samuelson maps are univalent", J of Pure and Applied Algebra 92, 227-240 .

4. Druzkowski, L.M. (1983) "An Effective approach to Keller's Jacobian conjecture", Math. Ann. 264, 303-313.

5. van den Essen. A and Parthasarathy. T.(1992) "Polynomial maps and a conjecture of Samuelson", Lin. Alg. and its applns. 177, 191-195 .

6. Gale, D and Nikaido, H.(1965) "The Jacobian Matrix and global univalence of mappings", Math. Ann. 159, 81-93 .

7. Kaplansky, I.(1945) "A contribution to von Neumann's theory of games", Annals of Math 46, 474-479 .

8. Meisters, G.H. and Olech, C. (1990) "A Jacobian condition for injectivity of differentiable maps", Annals. Math. Polon LI 249-254 .

9. Olech, C., Parthasarathy, T. and Ravindran, G.(1991) "Almost N-matrices and linear complementarity problem", Lin. Alg and its applns. 145, 107-125.

10. Parthasarathy, T.(1983) "On Global Univalence Theorems", Lecture Notes in Mathematics No. 977, Springer-Verlag. Berlin.

11. Parthasarathy, T. and Ravindran, G.(1986) "The Jacobian matrix, Global Univalence and completely mixed games", Math. O.R. 11, 663-671.

12. Parthasarathy, T. and Ravindran, G. (1990)"N-matrices", Lin. Alg and its applns. 139, 89-102.

13. Pinchuk, S. (1994) "A Counter example to the real Jacobian Conjecture", Mathematische Zeitshrift, 217, 1-4.

14. Ravindran, G. (1986) "Global Univalence and completely mixed games". Ph.D. thesis, Indian Statistical Institute, New Delhi.

15. Sabatini, M. (1993) "An extension to Hadamard global inverse function Theorem in the plane", Nonlinear Analysis, J.M.A. 20, 1069-1077.

16. Samuelson, P.A. (1953) "Prices of factors and goods in general equilibrium" Rev. Econ. Studies 21, 1-20.

17. Yu , J.T. (1995) "On the Jacobian Conjecture: Reduction of coefficients" , J of Algebra 171, 515-523.

T. Parthasarathy
Indian Statistical Institute
7 SJS Sansanwal Marg
New Delhi 110016.
India

G. Ravindran
Indian Statistical Institute
8th Mile, Mysore Road
RV College PO, Bangalore 560059
India

M.Sabatini
Department of Mathematics
University of Trento
Povo, Italy

PROBABILITY OF OBTAINING A PURE STRATEGY EQUILIBRIUM IN MATRIX GAMES WITH RANDOM PAYOFFS

Srijit Mishra and T. Krishna Kumar

Abstract: *If the payoffs in an mXn zero-sum matrix game are drawn randomly from a finite set of numbers, N, then the probability of obtaining a pure strategy equilibrium, p, will be a weighted sum of the probabilities of obtaining a pure strategy equilibrium, p_s, with s distinct payoffs, the weights, q_s, being the probabilities of obtaining s distinct payoffs from N. However, as $N \to \infty$ the probability $q_{mn} \to 1$. In this limiting case $p = p_{mn}$. Although p_{mn} has been derived by Goldman (1957) and Papavassilopoulos (1995), our method is more general. We show that $p_{mn} = \Sigma_t p^t_{mn}$ where p^t_{mn} denotes the probability of obtaining a pure strategy equilibrium for the tth $(t = 1, ..., s(= mn))$ ordinal payoff, the ordinality being the rank when the payoffs are put in an ascending order.*

Further, we introduce the notion of **separation of arrays**, $S(r^k, c^l)$, *which is a necessary and sufficient condition for the equilibrium of an mXn zero-sum matrix game to be associated with a mixed strategy solution. This generalizes the notion of* **sepatation of diagonals** *for 2X2 zero-sum matrix games derived by Von Neumann and Morgenstern (1953).*

It can be easily verified that as m or n increases p_{mn} decreases. Then given the importance of strong equilibrium, which is always a pure strategy equilibrium, a possible behaviourial interpretation is that players may prefer to play games with less number of strategies.

1. INTRODUCTION

This is a follow up of Mishra and Kumar (1994). We analyze the solution of a stochastic game in which payoffs are drawn randomly from a finite set of numbers. We derive the probability of obtaining an equilibrium with pure strategies. We also derive the necessary and sufficient conditions for obtaining an equilibrium with mixed strategies.

If the payoffs are drawn randomly[1] from a finite set of numbers, N, then one can derive the probability of obtaining a pure strategy equilibrium. We observe that the probability of obtaining a pure strategy equilibrium, p, in an mXn matrix game depends upon the number of distinct payoffs $s(s = 1, ..., mn)$. This probability is a weighted sum of the probabilities of obtaining a pure strategy equilibrium given that there are s distinct payoffs, the weights being the probabilities of having s distinct payoffs. However, as N tends to infinity the probability of all payoffs being distinct $(s = mn)$ tend to unity.

Under the assumption that N is infinite Goldman (1957), and more recently, Papavassilopoulos (1995) derived the general formula regarding the probability of obtaining a pure strategy equilibrium. However, the process of going to the limit through the exact proba-

[1] The characterization of random games given in Dresher (1970) holds.

bility for a finite N is a distinguishing feature of our paper.[2] Further, our method is more general because it can be used to arrive at the probability of obtaining a pure strategy equilibrium associated with the tth $(t = 1, ..., s)$ ordinal payoff, the ordinality being determined according to the rank after all the s distinct payoffs are put in an ascending order.

Having derived the probability for obtaining a equilibrium for zero-sum games Papavassilopoulos (1995) goes on to derive the probability of obtaining a pure strategy equilibrium for non zero-sum games for two or more players when the payoffs are drawn randomly. Instead of going into the non-zero sum game territory we till the ground further in the zero-sum game field. We derive the necessary and sufficient conditions for zero-sum mXn matrix games to have a mixed strategy equilibrium. This is done through the generalization of the notion of separation of diagonals given by Neumann and Morgenstern (1953: p.173) for 2X2 games.[3]

2. PURE STRATEGY EQUILIBRIA IN mXn MATRIX GAMES

In a mXn matrix game denoted by $A = a_{ij}$ $(i = 1, ..., m; j = 1, ..., n)$ where a_{ij} are the payoffs, the probability of obtaining a pure strategy equilibrium when the payoffs are drawn randomly depends upon whether all the payoffs are distinct (the game is strictly ordinal) or the payoffs are tied (the game is weakly ordinal). Strict and weak ordinality are discussed by Powers (1990). The probability of obtaining a pure strategy equilibrium, p, depends on the probability of obtaining a pure strategy equilibrium in a game with s distinct payoffs, p_s, and the probability of obtaining a game with s distinct payoffs, q_s, when the payoffs are drawn randomly from a set of N finite numbers. More precisely,

$$p = \Sigma_s p_s q_s; \; s = 1, ..., mn \tag{1}$$

where
$p_s = \Sigma_t p_s^t = (\Sigma_t w_s^t / h_s)$; p_s^t is the probability of obtaining a pure strategy equilibrium for the tth $(t = 1, .., s)$ ordinal payoff, h_s is the number of games possible with a set of s distinct payoffs and w_s^t is the number of games from these h_s games where the tth ordinal payoff has a pure strategy equilibrium; and

$$q_s = (u_s / N^{mn}); \; u_s = (h_s . NC_s) \text{ and } \Sigma u_s = N^{mn} \tag{1a}$$

It is difficult to give formulae in the general case for w_s^t and h_s, and hence, for p_s and q_s. Therefore, it is difficult to describe the precise nature of $\Sigma p_s q_s$.[4] In the present exercise we limit ourselves to strictly ordinal games, $s = mn$, and derive p_{mn} and q_{mn}.

In a strictly ordinal mXn matrix game, A, no two a_{ij}'s are equal. Now, if all the payoffs are put in an ascending order and ranked ordinally then there will be $t (t = 1, ..., mn)$ distinct ordinal payoffs. Further, it may be mentioned that the probability of obtaining

[2] The similarities and differences between our results and that of Papavassilopoulos are highlighted in Mishra and Kumar (1997).

[3] The authors derived the probabilities independently and presented the results at the Annual Conference of the Indian Econometric Society in May 1994. An anonymous referee of a Journal brought their attention to the earlier result proved by Goldman, and R.B. Bapat has drawn their attention to the recent paper by Papavassilopoulos.

[4] For $2X2$ games, the precise nature of p_s and q_s; $s = 1, ..., 4$ was discussed in detail in Mishra and Kumar (1994). It is the challenge posed by the second author to the first to prove this result for $2X2$ games that lead us in proving that and extending it to the mXn matrix games.

a pure strategy equilibrium, as derived in this exercise, is a summation of probabilities of obtaining the pure strategy equilibrium for the tth ordinal payoff, p^t_{mn}

$$p_{mn} = \Sigma_t p^t_{mn} \qquad (2)$$

As the lowest $(m-1)$ ordinal payoffs cannot be the maximum of a column and the largest $(n-1)$ ordinal payoffs cannot be the minimum of a row it follows that $p^t_{mn} = 0$ for $t < m$ and for $t > (mn - (n-1)$. But, for $t = m, ..., (mn - (n-1))$

$$p^t_{mn} = \{mn(m-1)!(n-1)![(m-1).(n-1)]![(t-1)C_{(m-1)}(nm-t)C_{(n-1)}]\}/(mn)! \qquad (3)$$

where

mn corresponds to the event that the pure strategy equilibrium payoff can be in any of the mn cells;

$(m-1)!$ denotes the number of possible ordering of the (1) payoffs larger than the pure strategy equilibrium payoff;

$(n-1)!$ denotes the number of possible ordering of the $(n-1)$ distinct payoffs smaller than the pure strategy equilibrium payoff;

$[(m-1).(n-1)]!$ denotes the number of possible ordering of the $[(m-1).(n-1)]$ payoffs excluding the row and column containing the pure strategy equilibrium payoff;

$(t-1)C_{(m-1)}$ denotes the possible sets of $(m-1)$ payoffs smaller than the pure strategy equilibrium payoff when the pure strategy equilibrium payoff is the t-th ordinal payoff;

$(nm-t)C_{(n-1)}$ denotes the possible sets of $(n-1)$ payoffs larger than the pure strategy equilibrium payoff when the pure strategy equilibrium is the t-th ordinal payoff;

$(mn)!$ denotes the possible number of ways of ordering the mn distinct payoffs in the mn cells.

From equation (2) we can arrive at the probability of obtaining a pure strategy equilibrium for strictly ordinal games. However, if the payoffs are drawn from a finite set of numbers, N, then it is necessary to find the probability of obtaining a strictly ordinal game. Following equation (1a)

$$q_{mn} = (mn)!.NC_{mn}/N^{mn} \qquad (4)$$

where

$(mn)!$ denotes the total number of games possible from mn distinct payoffs (see h_s in equation (1));

NC_{mn} denotes the number of mn distinct payoffs possible from a set of N numbers;

N^{mn} denotes the total number of mXn games possible from a set of N numbers.

Thus, for finite N, the probability of obtaining a pure strategy equilibrium for strictly ordinal games will be

$$\begin{aligned} p_{mn}q_{mn} &= \Sigma_t p^t_{mn} q_{mn} \\ &= m!n![(m-1).(n-1)]!\Sigma_t[(t-1)C_{(m-1)}(nm-t)C_{(n-1)}].NC_{mn}/N^{mn} \end{aligned} \tag{5}$$

Now it is interesting to look into the probability of obtaining a pure strategy equilibrium as $N \to \infty$. This we show in Theorem 1 after proving Lemma 1.

Lemma 1: In an mXn matrix game as $N \to \infty$ $q_{mn} \to 1$.

Proof: As from equation (1a) $q_s = ((h_s.NC_s)/(\Sigma h_s.NC_s))$, dividing throughout by the largest factor $(h_{mn}.NC_{mn})$ and letting $N \to \infty$ one can see that $q_{mn} \to 1$.

Theorem 1: In an mXn matrix game as $N \to \infty$ $p \to p_{mn}$.

Proof: From the definition of q_s (equation (1a)) it can be noted that in q_s the denominator is $O(N^{mn})$ but, the numerator is $o(N^{mn})$ for $s < mn$.[5] However, for $s = mn$ the numerator is $O(N^{mn})$. This shows that $p \to p_{mn}q_{mn}$. Using Lemma 1, $p \to p_{mn}$.

Following Theorem 1 it can be said that as $N \to \infty$ the games that matter are strictly ordinal games. This shows the importance of strictly ordinal games over weakly ordinal games for two person zero sum games when the payoffs are drawn from an infinite set of numbers.[6]

Such an assumption of infinite N is implicit in the probability for obtaining a pure strategy equilibrium for mXn matrix games derived by Goldman (1957) and Papavassilopoulos (1995). And hence, their method is actually a simpler way of arriving at p_{mn}

$$p_{mn} = m!n!/(m+n-1)! \tag{6}$$

where $[m!n! = (mn.(m-1)!.(n-1)!)]$ is the same as in equation (3); $(m+n-1)!$ denote the number of possible ordering of the (m+n-1) payoffs in the cells of the row and column containing the pure strategy equilibrium payoff.

4. CONDITIONS OF OBTAINING AN EQUILIBRIUM SOLUTION WITH MIXED STRATEGIES

Separation of diagonals[7] is a necessary and sufficient condition for the equilibrium solution of a 2X2 matrix game to be associated with a mixed strategy [Neumann and Morgenstern (1953: p.173)]. They also state that the results can not be generalized for matrix games larger than $2X2$ [Neumann and Morgenstern (1953: p.179)]. Following Mishra (1994) and Mishra and Kumar (1994) we explain the phenomenon in a manner where *separation of diagonals* will be a special case and the results for a mixed strategy can be generalized and

[5]A term is said to be $O(N)$ if as $N \to \infty$ the ratio of that term and N tends to a non-zero constant. A term is said to be o(N) if as $N \to \infty$ the ratio of that term and N tends to zero.

[6]Powers (1990) showed the importance of strictly ordinal games over weakly ordinal games for n-person non-constant sum games when the number of strategies of two or more players goes to infinity.

[7]If all the elements of one diagonal are greater (dominant diagonal) than all the elements of the other diagonal (dominated diagonal) then one can say that there is *separation of diagonals*.

extended to the mXn case. Before introducing this generalization it is necessary to introduce the concepts of a row array of payoffs, r, and a column array of payoffs, c.

A row array of payoffs, r, is defined as

$$r = (c_1, ..., c_j, ..., c_n) \tag{7}$$

where c_j can be any element of the jth column, and this array r is such that it has as its elements one and only one payoff from each column, and one from every column.

Likewise a column array of payoffs, c, is defined as

$$c = (r_1, ..., r_i, ..., r_m) \tag{8}$$

where r_i can be any element of the ith row, and this array c is such that it has as its elements one and only one payoff from each row, and one from every row. If there are m rows and n columns the total number of such row and column arrays are m^n and n^m respectively. Let us denote them as $r^k(k = 1, ..., m^n)$ and $c^l(l = 1, ..., n^m)$.

It follows that in an mXn matrix game there will be a row array consisting the maximum element of all the columns. Let us denote this row array as r^{max}. Similarly there will be a column array consisting of the minimum element of all the rows. Let us denote this column array as c^{min}.

We also introduce the notion of separation of arrays using r^k and c^l. A pair of arrays (r^k, c^l) will be separated if and only if all the elements of r^k are strictly greater than all the elements of c^l, that is, $r_j^k > c_i^l$ for all (i, j). We denote a pair of separated arrays as $S(r^k, c^l)$. Using the above notions we give Theorem 2 and its converse Theorem 3.

Theorem 2: If the equilibrium in an mXn matrix game is associated with a mixed strategy then there exists an $S(r^k, c^l)$.

Proof: If the equilibrium of the game is associated with a mixed strategy then it follows that $min\{column\ maxima\} > max\{row\ minima\}$. This implies that $min\{r^{max}\} > max\{c^{min}\}$, and hence, $S(r^{max}, c^{min})$.

Theorem 3: If there exiss $S(r^k, c^l)$ then the equilibrium of the game is associated with a mixed strategy.

Proof: For any r^k, the values of each element will be less than or equal to the corresponding element in r^{max}. Similarly, for any c^l, the values of each element will be less than or equal to the corresponding element in c^{min}. Now, if $S(r^k, c^l)$ then it implies that $S(r^{max}, c^{min})$. Under $S(r^{max}, c^{min}), min\{r^{max}\} > max\{c_{min}\}$. This implies that $min\{column\ maxima\} > max\{row\ minima\}$, which means that the equilibrium of the game is associated with a mixed strategy.

5. A BEHAVIOURISTIC INTERPRETATION OF OUR RESULTS

Having derived the conditions for a mixed strategy we give some interpretations of a mixed strategy. One interpretation could be that a player would assign probabilities to a number of strategies to keep the other player guessing and in so doing would increase her security level

[Luce and Raiffa (1957:p.75)]. Alternatively, the probabilities assigned could be an n-tuple common belief.[8] According to Harsanyi (1973) the strategies are deterministic but the payoffs are random. Harsanyi shows that under such an assumption pure strategy equilibrium will be a strong equilibrium (which is stable in some sense). Though a larger number of strategies would confound the other player it would be in the interest of both the players to reduce the number of strategies, as by reducing the strategies they increases the probability of obtaining a pure strategy equilibrium. Hence, what they would be looking for is a game with less number of strategies. Yet another interpretation is that the probabilities of the strategies assigned could in fact be the distribution of strategies of the interacting populations [see Rubinstein (1991)].

We suggest some behaviouristic interpretation of our results given the decision mode of the player.

First, let us visualize a situation where a player has to decide the number of strategies he would use for a given number of strategies of the other player. Under the assumption that the payoffs will be drawn randomly a motive for having a strong equilibrium should induce the player to select fewer strategies or in fact a single strategy.

Second, let us consider the situation when the number of strategies for both the players are given. The payoffs and the probability distribution are common knowledge. In such a scenario the motive for strong equilibrium would make the players eliminate from consideration all those strategies (theirs and their opponents) which are associated with a low probability of giving rise to an equilibrium (whether with pure or mixed strategies).

The above two interpretations emphasize the importance of fewer strategies. It is in this regard that we give two examples. One is the emergence of specific and clear provision as against vague and general ones in the written tenancy contracts of South India (Reddy (1996: 134, 184). The second is the preference of *limited contract-enforcing regime* over *contract-enforcing regime* because the latter can leave lot of ambiguities, making legal enforcement mechanism inefficient (see Basu (1992: 347-348)).

Acknowledgements The authors are grateful to R B Bapat, Tilman Borgers, Kyeong Duk Kim, G P Papavassilopoulos, T Parthasarathy, T E S Raghavan and four anonymous referees for their comments on earlier drafts and to the participants of the 30th Annual conference of the Indian Econometric Society and the Second International Conference on Game Theory and Economic Applications who offered useful comments. The authors are thankful to Srideba Nanda for the spontaneity with which he sent some material. Srijit also acknowledges G N Rao for his encouragement. However, the usual disclaimer applies and the authors only blame each other for any residual errors.

References

1. Babu, P Guruswamy (1994), "Common Belief", A paper presented at the 30th Annual Conference of the Indian Econometric Society, University of Mysore, Mysore, May 1-3, 1994.

2. Basu, Kaushik (1992), "Markets, Laws and Governments", Bimal Jalan (Editor) *The Indian Economy: Problems and Prospects*, Penguin, New Delhi, pp.338-355.

[8] Although Rubinstein (1991) uses the word common knowledge we think that the term common belief is more appropriate. For a discussion on common belief see Babu (1994).

3. Dresher, Melvin (1970), "Probability of a Pure Equilibrium Point in Person Games", *Journal of Combinatorial Theory*, Vol.8, pp.134-145.

4. Goldman, A J (1957), "The Probability of a Saddlepoint", *American Mathematical Monthly*, Vol.64, pp.729-730.

5. Harsanyi, John C (1973), "Games with Randomly Disturbed Payoffs: A New Rationale for Mixed Strategy", *International Journal of Game Theory*, Vol.2, pp.1-23.

6. Luce Duncan and Raiffa, H (1957), *Games and Decisions*, John Wiley and Sons.

7. Mishra, Srijit (1994), "A Note on "A Property of Matrix Games with Random Payoffs: A Curiosity Explored"," Centre for Development Studies, November 17, 1994, Mimeo.

8. Mishra, Srijit and Kumar, T Krishna (1994), "A Property of Matrix Games with Random Pay offs: A Curiosity Explored," A paper presented at the 30th Annual Conference of the Indian Econometric Society, University of Mysore, Mysore, May 1-3, 1994.

9. Mishra, Srijit and Kumar, T Krishna (1997), "On the Probability of Existence of Pure Equilibria in Matrix Games", *Journal of Optimization Theory and Applications*, forthcoming.

10. Neumann, John Von and Morgenstern, Oskar (1953), *Theory of Games and Economic Behavior*, Third edition, Princeton University Press, Princeton.

11. Papavassilopoulos, V P (1995), "On the Probability of Existence of Pure Equilibria in Matrix Games", *Journal of Optimization Theory and Applications*, Vol.87, pp.419-439.

12. Powers, I Y (1990), "Limiting Distributions of the Number of Pure Strategy Nash Equilibria in N Person Games", *International Journal of Game Theory*, Vol.19, pp.277-286.

13. Reddy, M Atchi (1996), *Lands and Tenants in South India: A study of Nelore District 1850-1990*, Oxford University Press, Delhi.

14. Rubinstein, Ariel (1991), "Comments on the Interpretation of Game Theory", *Econometrica*, Vol.59, pp.909-924.

Srijit Mishra
Centre for Development Studies
Thiruvananthapuram 695011
India

T. Krishna Kumar
Indian Statistical Institute
Bangalore 560059
India

NONLINEAR SELF DUAL SOLUTIONS FOR TU-GAMES

Peter Sudhölter[1]

Abstract: *For cooperative transferable utility games solution concepts are presented which resemble the core-like solution concepts prenucleolus and prekernel. These modified solutions take into account both, the 'power', i.e. the worth, and the 'blocking power' of a coalition, i.e. the amount which the coalition cannot be prevented from by the complement coalition, in a totally symmetric way. As a direct consequence of the corresponding definitions they are self dual, i.e. the solutions of the game and its dual coincide. Sudhölter's recent results on the modified nucleolus are surveyed. Moreover, an axiomatization of the modified kernel is presented.*

0 Introduction

In a series of papers (Sudhölter (1993,1994,1996a,b)) a new solution concept, the modified nucleolus, for cooperative side payment games with a finite set of players is discussed.

The expression 'modified nucleolus' refers to the strong relationship of this solution to the (pre)nucleolus introduced by Schmeidler (1966).

An imputation belongs to the nucleolus of a game, if it successively minimizes the maximal excesses, i.e. the differences of the worths of coalitions and the aggregated weight of these coalitions with respect to (w.r.t.) the imputation, and the number of coalitions attaining them. For the precise definition Section 2 is referred to. By regarding the excesses as a measure of dissatisfaction the nucleolus obtains an intuitive meaning as pointed out by Maschler, Peleg, and Shapley (1979).

The solution discussed in the recent papers constitutes an attempt to treat all coalitions equally as far as this is possible. Therefore it is natural to regard the differences of excesses as a measure of dissatisfaction leading to the following intuitive definition. A preimputation belongs to the modified nucleolus $\Psi(v)$ of a game v, if it successively minimizes the maximal differences of excesses and the number of coalition pairs attaining them. The modified nucleolus takes into account both the 'power', i.e. the worth, and the 'blocking power' of a coalition, i.e. the amount which the coalition cannot be prevented from by the complement coalition. If the power of a coalition is measured by its worth (as usual), then the blocking power of a coalition should be measured by its worth w.r.t. the dual game. Alike the prenucleolus, which only depends on the worths of the coalitions, the modified nucleolus is a singleton.

To give an example look at the glove game with three players, one of them (player **1**) possessing a unique right hand glove whereas the other players (**2** and **3**) possess one single left hand glove each. The worth of a coalition is the number of pairs of gloves of the coalition (i.e. one or zero). If a coalition has positive worth, then **1** is a member of the coalition, i.e. player **1** is a veto player possessing, in some sense, all of the power. Indeed the (pre)nucleolus assigns one to player **1** and zero to the other players. On the other hand both players **2** and **3**

[1] I am grateful to an anonymous referee for insightful remarks and comments.

together can prevent player **1** from any positive amount by forming a 'syndicate'. Therefore they together have the same blocking power as player **1** has. The modified nucleolus takes care of this fact and assigns 1/2 to the first and 1/4 to each of the other players.

A further motivation to consider the new solution concept is its behaviour on the remarkable class of weighted majority games. For the subclasses of weighted majority constant-sum games on the one hand and for homogeneous games on the other hand the nucleolus (see Peleg (1968)) and the minimimal integer representation (see Ostmann (1987) and Rosenmüller (1987)) respectively can be regarded as canonical representation. Fortunately, the modified nucleolus coincides with the prenucleolus on constant-sum games and, up to normalization, with the weights of the minimal integer representation on homogeneous games. Additionally, it induces a representation for an arbitrary weighted majority game. Therefore the modified nucleolus can be regarded as a canonical representation in the general weighted majority case. For the details Sudhölter (1996b) is referred to.

In general a solution concept which assigns the same preimputations to both, the game and its dual is called **self dual**. Analogously to the prenucleolus the prekernel possesses a self dual modification (see Sudhölter 1993).

This paper is organized as follows:

Section 1 recalls some well-known definitions and necessary notations. In Section 2 the definition and some properties of the modified nucleolus are recalled. The dual game v^* of a game v assigns to each coalition the real number which can be given to it if the worth of the grand coalition is shared and the complement coalition obtains its worth. By looking at complements it turns out that the modified solution concepts of v and v^* coincide (the solutions satisfy **self duality**), this also being a characteristic of the Shapley value. In what follows results of Sudhölter (1996a) are surveyed. The modified nucleolus, e.g., can be viewed as the restriction of the prenucleolus of the dual cover (a certain replication) of the game. The dual cover of a game arises from a game v with player set N by taking the union of two disjoint copies of N to be the new player set and assigning to a coalition S the maximum of the sums of the worths of the intersections of S with the first copy w.r.t. v and the second copy w.r.t. v^* or, conversely, the first copy w.r.t. v^* and the second w.r.t. v. Hence both, the game and its dual, are totally symmetric ingredients of the dual cover. This 'restriction' result enables us to reformulate many properties of the prenucleolus for the modified nucleolus, e.g., the modified nucleolus can be computed by each of the well-known algorithms for the calculation of the prenucleolus (see, e.g., Kopelowitz (1967) or Sankaran (1992)) applied to the dual cover. The coincidence of the pre- and modified nucleolus on constant-sum games is a further interesting property. At the end of this section the behavior of the modified nucleolus on weighted majority games is discussed.

In Sudhölter (1996a) two axiomatizations of the modified nucleolus are presented which are comparable to Sobolev's (1975) characterization of the prenucleolus. In Section 3 one axiomatization of the modified nucleolus is recalled.

In Section 4 two self dual modifications of the prekernel are introduced. The proper modified kernel contains the modified nucleolus and is a subset of the modified kernel. The application to glove games shows that the new solutions concepts do no necessarily coincide.

In the last section an axiomatization of the modified kernel is presented which is similar to Peleg's (1986) axiomatization of the prekernel.

1 Notation and Definitions

A cooperative game with transferable utility — **a game** — is a pair $G = (N, v)$, where N is a finite nonvoid set and

$$v : 2^N \rightarrow I\!R, \; v(\emptyset) = 0$$

is a mapping. Here $2^N = \{S \subseteq N\}$ is the set of **coalitions** of G.

If $G = (N, v)$ is a game, then N is the **grand** coalition or the set of **players** and v is called **characteristic** (or **coalitional**) **function** of G. Since the nature of G is determined by the characteristic function, v is called game as well.

If $G = (N, v)$ is a game, then the **dual** game (N, v^*) of G is defined by

$$v^*(S) = v(N) - v(N \setminus S)$$

for all coalitions S. The set of **feasible payoff vectors** of G is denoted by

$$X^*(N, v) = X^*(v) = \{x \in I\!R^N \mid x(N) \leq v(N)\},$$

whereas

$$X(N, v) = X(v) = \{x \in I\!R^N \mid x(N) = v(N)\}$$

is the set of **preimputations** of G (also called set of **Pareto optimal** feasible payoffs of G). Here

$$x(S) = \Sigma_{i \in S} x_i \quad (x(\emptyset) = 0)$$

for each $x \in I\!R^N$ and $S \subseteq N$. Additionally, let x_S denote the restriction of x to S, i.e.

$$x_S = (x_i)_{i \in S} \in I\!R^S,$$

whereas $A_S = \{x_S \mid x \in A\}$ for $A \subseteq I\!R^N$. For disjoint coalitions $S, T \subseteq N$ and $x \in I\!R^N$ let $(x_S, x_T) = x_{S \cup T}$.

A solution concept σ **on** a set Γ of games is a mapping that associates with every game $(N, v) \in \Gamma$ a set $\sigma(N, v) = \sigma(v) \subseteq X^*(v)$.

If $\bar{\Gamma}$ is a subset of Γ, then the canonical restriction of a solution concept σ on Γ is a solution concept on $\bar{\Gamma}$. We say that σ is a solution concept on $\bar{\Gamma}$, too. If Γ is not specified, then σ is a solution concept on every set of games.

Some convenient and well-known properties of a solution concept σ on a set Γ of games are as follows.

(1) σ is **anonymous** (satisfies **AN**), if for each $(N, v) \in \Gamma$ and each bijective mapping $\tau : N \rightarrow N'$ with $(N', \tau v) \in \Gamma$

$$\sigma(N', \tau v) = \tau(\sigma(N, v))$$

holds (where $(\tau v)(T) = v(\tau^{-1}(T))$, $\tau_j(x) = x_{\tau^{-1} j}$ $(x \in I\!R^N, j \in N', T \subseteq N')$).

In this case v and τv are **equivalent** games.

(2) σ satisfies the **equal treatment property (ETP)**, if for every $x \in \sigma(N, v)$ $(v \in \Gamma)$ interchangeable players $i, j \in N$ are treated equally, i.e. $x_i = x_j$. Here i and j are **interchangeable**, if $v(S \cup \{i\}) = v(S \cup \{j\})$ for $S \subseteq N \setminus \{i, j\}$.

(3) σ **respects desirability** if for every $(N, v) \in \Gamma$ every $x \in \sigma(N, v)$ satisfies $x_i \geq x_j$ for a player i who is at least as desirable as player j. Here i is **at least as desirable** as j if $v(S \cup \{i\}) \geq v(S \cup \{j\})$ for $S \subseteq N \setminus \{i, j\}$.

(4) σ satisfies the **null player property (NPP)** if for every $(N,v) \in \Gamma$ every $x \in \sigma(N,v)$ satisfies $x_i = 0$ for every nullplayer $i \in N$. Here i is **nullplayer** if $v(S \cup \{i\}) = v(S)$ for $S \subseteq N$.

(5) σ is **covariant under strategic equivalence** (satisfies **COV**), if for $(N,v),(N,w) \in \Gamma$ with $w = \alpha v + \beta$ for some $\alpha > 0, \beta \in \mathbb{R}^N$

$$\sigma(N,w) = \alpha\sigma(N,v) + \beta$$

holds. The games v and w are called **strategically equivalent**.

(6) σ is **single valued** (satisfies **SIVA**), if $| \sigma(v) |= 1$ for $v \in \Gamma$.

(7) σ satisfies **nonemptiness (NE)**, if $\sigma(v) \neq \emptyset$ for $v \in \Gamma$.

(8) σ is **Pareto optimal** (satisfies **PO**), if $\sigma(v) \subseteq X(v)$ for $v \in \Gamma$.

(9) σ satisfies **reasonableness (on both sides) (REAS)**, if

(a)

$$x_i \geq \min\{v(S \cup \{i\}) - v(S) \mid S \subseteq N \setminus \{i\}\}$$

and

(b)

$$x_i \leq \max\{v(S \cup \{i\}) - v(S) \mid S \subseteq N \setminus \{i\}\}$$

for $i \in N, (N,v) \in \Gamma$, and $x \in \sigma(N,v)$.

Note that both equivalence and strategical equivalence commute with duality, i.e. $(\tau v)^* = \tau(v^*)$, $(\alpha v + \beta)^* = \alpha v^* + \beta$, where τ, α, β are chosen according to the definitions given above. With the help of assertion (9b) Milnor (1952) defined his notion of reasonableness.

It should be remarked (see Shapley (1953)) that the Shapley value φ (to be more precise the solution concept σ given by $\sigma(v) = \{\varphi(v)\}$) satisfies all above properties.

Some more notation will be needed. Let (N,v) be a game and $x \in \mathbb{R}^N$. The **excess of** a coalition $S \subseteq N$ **at** x is the real number

$$e(S,x,v) = e(S,x) = v(S) - x(S).$$

Let $\mu(x,v) = \mu(x)$ be the **maximal excess at** x, i.e. $\mu(x,v) = \max\{e(S,x) \mid S \subseteq N\}$. For **different** players $i,j \in N$ let

$$s_{ij}(x,v) = s_{ij}(x) = \max\{e(S,x) \mid i \in S \subseteq N \setminus \{j\}\}$$

denote the **maximal surplus of** i **over** j **at** x.

2 A Self Dual Modification of the Nucleolus

This section serves to define a self dual modification of the classical prenucleolus. Some well-known properties of this solution concept are recalled and an example is presented. For detailed proofs of all assertions in this section Sudhölter (1996a,b) is referred to.

The nucleolus of a game was introduced by Schmeidler (1966). Some corresponding definitions and results are recalled: Let $\vartheta : \bigcup_{n\in\mathbb{N}} \mathbb{R}^n \to \bigcup_{n\in\mathbb{N}} \mathbb{R}^n$ be defined by

$$\vartheta(x) = y \in \mathbb{R}^n \quad (x \in \mathbb{R}^n),$$

where y is the vector which arises from x by arranging the components of x in a nonincreasing order. The **nucleolus of v w.r.t** X, where $X \subseteq \mathbb{R}^N$, is the set

$$\mathcal{N}(X,v) = \{x \in X \mid \vartheta((e(S,x,v))_{S\subseteq N}) \leq_{lex} \vartheta((e(S,y,v))_{S\subseteq N}) \text{ for all } y \in X\}.$$

The **prenucleolus of** (N,v) is defined to be the nucleolus w.r.t. the set of feasible payoff vectors and denoted $\mathcal{PN}(v)$, i.e., $\mathcal{PN}(v) = \mathcal{N}(X^*(v),v)$. The prenucleolus of a game is a singleton and it is clearly Pareto optimal (see again Schmeidler (1966)). The unique element $\nu(v)$ of $\mathcal{PN}(v)$ is again called **prenucleolus (point)**.

For completeness reasons we recall that the nucleolus of (N,v) is the set $\mathcal{N}(X,v)$, where $X = \{x \in X(v) \mid x_i \geq v(\{i\})\}$ is the set of **imputations** of v. Maschler, Peleg and Shapley (1979) tried to give an intuitive meaning to the definition of the (pre)nucleolus by regarding the excess of a coalition as a measure of dissatisfaction which should be minimized. If the excess of a coalition can be decreased without increasing larger excesses, this process will also increase some kind of 'stability', they argued. Nevertheless, Maschler (1992) asked: "What is more 'stable', a situation in which a few coalitions of highest excess have it as low as possible, or one where such coalitions have a slightly higher excess, but the excesses of many other coalitions is substantially lowered?" Anyone, like the present author, who is not convinced by the first or latter, may try to search for a completely different solution concept. The concept which will be introduced in this paper constitutes an attempt to treat all coalitions equally w.r.t. excesses as far as this is possible. Therefore, instead of minimizing the highest excess, then minimizing the number of coalitions with highest excess, minimizing the second highest excess and so on - the highest difference of excesses is minimized, then the number of pairs of coalitions with highest difference of excesses is minimized... Here is the notation.

Definition 2.1 *Let (N,v) be a game. For each $x \in \mathbb{R}^N$ define $\tilde{\Theta}(x,v) = \vartheta((e(S,x,v) - e(T,x,v))_{(S,T)\in 2^N \times 2^N}) \in \mathbb{R}^{2^{2|N|}}$. The* **modified nucleolus of** *v is the set*

$$\Psi(v) = \{x \in X(v) \mid \tilde{\Theta}(x,v) \leq_{lex} \tilde{\Theta}(y,v)$$

Remark 2.2 *Let (N,v) be a game.*

(1) If x is any preimputation of the game v, then the following equality holds by definition and Pareto optimality:

$$e(T,x,v^*) = -e((N \setminus T),x,v).$$

With $\bar{\Theta}(y,v) = \vartheta((e(S,y,v)+e(T,y,v^*))_{(S,T)\in 2^N \times 2^N})$ for $y \in \mathbb{R}^N$ this equality directly implies for $x \in X(v)$ that $\bar{\Theta}(x,v) = \tilde{\Theta}(x,v)$ holds true. Note that x has to be Pareto optimal for this equation. Nevertheless the modified nucleolus can be redefined as

$$\Psi(v) = \{x \in X^*(v) \mid \bar{\Theta}(x,v) \leq_{lex} \bar{\Theta}(y,v) \text{ for all } y \in X^*(v)\}, \tag{2.1}$$

because Pareto optimality is, now, automatically satisfied. Indeed, this property can be verified by observing that for every nonvoid coalition both, the excess w.r.t. v and w.r.t. v^*, strictly decrease if all components of a feasible payoff vector can be strictly increased.

(2) The alternate definition of $\Psi(v)$ in the last assertion (see (2.1)) directly shows that Ψ is **self dual**, *i.e. $\Psi(v) = \Psi(v^*)$ holds. Note that Ψ shares this property with the Shapley value.*

In what follows two kinds of replicated games are defined. The first one will be used to present a property which allows an axiomatization of the modified nucleolus, which is the restriction of the prenucleolus of the second kind of replication.

Definition 2.3 *Let (N, v) be a game and $\bar{N} = N \times \{0,1\}$. We identify $N \times \{0\}$ with N and $N \times \{1\}$ with N^* in the canonical way, thus $\bar{N} = N \cup N^*$.*

(1) The game $(N \cup N^*, \bar{v})$, defined by

$$\bar{v}(S \cup T^*) = v(S) + v^*(T)$$

for all $S, T \subseteq N$ is the **dual replication of** v.

(2) The game $(N \cup N^*, \tilde{v})$, defined by

$$\tilde{v}(S \cup T^*) = \max\{v(S) + v^*(T), v(T) + v^*(S)\}$$

for all $S, T \subseteq N$ is the **dual cover of** v.

Sudhölter (1996a) proved the following result which shows a strong relation between the modified nucleolus and the prenucleolus of the dual cover of the game.

Theorem 2.4 *The modified nucleolus of a game (N, v) is the restriction of the prenucleolus of $(N \cup N^*, \tilde{v})$ to N; i.e. $\psi(v) = \nu(\tilde{v})_N$. Moreover, $\nu_i(\tilde{v}) = \nu_{i^*}(\tilde{v})$ for $i \in N$.*

In view of Theorem 2.4 the modified nucleolus of a game v is a singleton denoted by $\psi(v)$, i.e. $\{\psi(v)\} = \Psi(v)$. The unique point $\psi(v)$ of $\Psi(v)$ is again called **modified nucleolus (point)**.

Some properties of the modified nucleolus are presented in the following remark. For the necessary proofs Sudhölter (1996a,b) is referred to.

Remark 2.5 *Let (N, v) be a game.*

(1) If $\nu(v) = \nu(v^*)$, then $\psi(v) = \nu(v)$.

(2) If v is a constant-sum game (i.e. v coincides with v^*), then $\psi(v) = \nu(v)$.

(3) If v is convex (i.e. $v(S) + v(T) \leq v(S \cup T) + v(S \cap T)$ for $S, T \subseteq N$), then the modified nucleolus is contained in the core of v.

(4) The modified nucleolus satisfies REAS, COV, AN, NPP, ETP, and it respects desirability.

(5) The modified nucleolus of the dual replication $(N \cup N^*, \bar{v})$ arises from the modified nucleolus of (N, v) by replication, i.e. $\psi_i(\bar{v}) = \psi_i(v) = \psi_{i^*}(\bar{v})$ for $i \in N$ (written $\psi(\bar{v}) = (\psi(v), \psi(v)^*)$).

To illustrate the notion of the modified nucleolus its behavior on weighted majority games is sketched.

Example 2.6 *A game (N, v) is a* **weighted majority game**, *if there is a pair $(\lambda; m)$ satisfying*

(1) $\lambda \in \mathbb{R}_{>0}$, $m \in \mathbb{R}^N_{\geq 0}$, and $m(N) \geq \lambda$,

(2) $v(S) = \begin{cases} 1 & , \text{ if } m(S) \geq \lambda \\ 0 & , \text{ otherwise} \end{cases}$.

In this case $(\lambda; m)$ *is a* **representation** *of the game.*

For an arbitrary weighted majority constant-sum game (N, v) *Peleg (1968) showed that the nucleolus* $\nu = \nu(v)$ **induces** *a representation, i.e.* $(1 - \mu(\nu, v); \nu)$ *is a representation of* (N, v)*. By Remark 2.5 (2) the same property holds for the modified nucleolus. For general weighted majority games the nucleolus does not necessarily induce a representation (see, e.g., the glove game presented in the introduction which can be represented by* $(3; 2, 1, 1)$ *and possesses a nucleolus assigning 0 to players* **2** *and* **3***). In Sudhölter (1996b) the following assertion is proved.*

If (N, v) *is a weighted majority game and* ψ *is its modified nucleolus, then* $(1 - \mu(\psi, v); \psi)$ *is a representation of* (N, v).

For completeness reasons we present a proof of this assertion: Let $(\lambda; m)$ *be a representation of* (N, v) *which is* **normalized**, *i.e.* $m(N) = 1$ *(i.e. m is a preimputation of the game). Then*

$$0 \leq e(S, m, v) \leq 1 - \lambda \text{ for } S \in 2^N \text{ with } v(S) = 1$$

and

$$-\lambda < e(S, m, v) \leq 0 \text{ for } S \in 2^N \text{ with } v(S) = 0,$$

thus

$$e(S, m, v) - e(T, m, v) < 1 \text{ for } S, T \in 2^N.$$

By Remark 2.5 (4) $\psi = \psi(v) \geq 0$*. Let* x *be any preimputation of* v *satisfying* $x \geq 0$ *which does* **not** *induce a representation of* (N, v)*. Take* $S, T \in 2^N$ *with* $v(S) = 1$, $v(T) = 0$, *and* $x(S) \leq x(T)$*. Then*

$$e(S, x, v) - e(T, x, v) = 1 - x(S) + x(T) \geq 1 > \max_{S,T \in 2^N} e(S, m, v) - e(T, m, v),$$

thus $x \neq \psi$ *by definition. Additionally, this observation shows that the maximal excess at* ψ *is attained by some winning coalitions only, thus* $(1 - \mu(\psi, v); \psi)$ *is a representation of* v. *q.e.d.*

For a proof (which is more involved) showing that ψ *coincides with the normalized vector of weights of the unique minimal integer representation in the homogeneous case Sudhölter (1996b) is referred to.*

3 An Axiomatization of the Modified Nucleolus

In Sudhölter (1996a) two axiomatizations of the modified nucleolus are presented. We will present one of them.

First of all the characterizing axioms for the prenucleolus will be recalled.

Definition 3.1

(1) For a set U *let* $\Gamma_U = \{(N, v) \mid N \subseteq U\}$ *denote the set of games with player set contained in* U.

(2) Let (N, v) *be a game,* $x \in I\!R^N$, *and* $\bar{S}$ *be a nonvoid coalition of* N*. The game* $(\bar{S}, v^{\bar{S},x})$, *where*

$$v^{\bar{S},x}(S) = \begin{cases} v(N) - x(N \setminus \bar{S}), & \text{if } S = \bar{S} \\ 0, & \text{if } S = \emptyset \\ \max\{v(S \cup Q) - x(Q) \mid Q \subseteq N \setminus \bar{S}\}, & \text{otherwise} \end{cases},$$

is the **reduced game of** v *w.r.t.* x *and* $\bar{S}$.

(3) *A solution concept* σ *on a set* Γ *of games satisfies* **consistency** *(***CONS***) if* $(N,v) \in \Gamma, x \in \sigma(v), \emptyset \subset \bar{S} \subseteq N$ *implies* $(\bar{S}, v^{\bar{S},x}) \in \Gamma$ *and* $x_{\bar{S}} \in \sigma(\bar{S}, v^{\bar{S},x})$.

The notion of a reduced game was introduced by Davis and Maschler (1965). For the axiom CONS - also called **reduced game property** *- and for the following axiomatization of the prenucleolus Sobolev (1975) is referred to. Note that the condition* $(\bar{S}, v^{\bar{S},x}) \in \Gamma$ *in the definition of the reduced game can be dropped in Sobolev's result, because the considered set of games* (Γ_U) *is rich enough, i.e. each reduced game w.r.t. each feasible payoff vector automatically is an element of this set.*

Theorem 3.2 *(Sobolev) If* U *is an infinite set, then the prenucleolus is the unique solution concept on* Γ_U *satisfying SIVA, AN, COV, and CONS.*

For the definition of SIVA, AN, COV Section 1 is referred to. Moreover, Ψ *does not satisfy CONS on* Γ_U*, because it does not coincide with* ν*. In what follows it turns out that the modified nucleolus can be characterized by replacing the reduced game property and the anonymity by three additional axioms. Some notation is needed.*

Definition 3.3 *Let* (N,v) *be a game.*

(1) *For* $x \in \mathbb{R}^N$ *let* $\Lambda(x,v)$ *be defined by*

$$\Lambda(x,v) = \min\{v(T) - v^*(T) \mid \emptyset \subset T \subset N\} - \mu_0(x,v),$$

where $\mu_0(x,v) = max\{e(S,x,v) \mid \emptyset \subset S \subset N\}$ *denotes the maximal nontrivial excess at* x*. Here* $\min \emptyset = \infty$ *and* $\max \emptyset = -\infty$ *as usual and, in addition,* $\Lambda(x,v) = 0$ *for a 1-person game.*

(2) *The game* v *has the* **large excess difference property** *(satisfies* **LED***) w.r.t.* $x \in \mathbb{R}^N$*, if* $\Lambda(x,v) \geq 0$.

(3) *A solution concept* σ *on a set* Γ *of games satisfies* **large excess difference consistency** *(***LEDCONS***), if* $(S, v^{S,x}) \in \Gamma$ *and* $x_S \in \sigma(v^{S,x})$*, whenever* $(N,v) \in \Gamma, x \in \sigma(v)$*, and* v *satisfies LED w.r.t.* x.

In case a game (N,v) *satisfies LED w.r.t. a vector* x *the excess of a nontrivial coalition* S *(i.e.* $\emptyset \subset S \subset N$*) w.r.t.* v *weakly dominates the excess of* S *w.r.t. the dual game* v^**, even if this number is enlarged by the maximal excess of nontrivial coalitions w.r.t.* v*. Intuitively, the modified nucleolus is 'stable' against objections of coalitions* S *argueing that the own excess should be diminished if compared to the smaller excesses of further coalitions* T*. In case of LED 'stability' of* x *is checked as soon as 'stability' of excess differences of pairs* (S,T) *with* $T = \emptyset$ *or* $T = N$ *can be verified. To be more precise, the modified nucleolus and the prenucleolus coincide, whenever the game satisfies LED w.r.t. the latter (see Remark 3.5 (1)).*

An interpretation of LEDCONS will be given together with a verbal description of a further 'derived' game defined as follows with the help of the initial game, its dual, and a given payoff vector.

Definition 3.4 *Let* σ *be a solution concept on a set* Γ *of games, let* (N,v) *be a game and* $x \in \mathbb{R}^N$.

(1) Define a game (N, v^x) *by*

$$v^x(S) = \begin{cases} v(S) & , \text{ if } S \in \{\emptyset, N\} \\ \max\{v(S) + \mu + 2\mu^*, v^*(S) + \mu^* + 2\mu\} & , \text{ otherwise} \end{cases}$$

for $S \subseteq N$, *where* $\mu = \mu(x, v)$ *and* $\mu^* = \mu(x, v^*)$.

(2) σ *satisfies* **excess comparability** *(***EC***), if* $v \in \Gamma$, $x \in \sigma(v)$, *and* $v^x \in \Gamma$ *imply* $x \in \sigma(v^x)$.

The idea of the game v^x *is as follows. Assume that* x *is Pareto optimal, i.e.* x *constitutes a rule how to share* $v(N)$. *Moreover, assume that the players agree that this rule should take into account the worth* $v(S)$ *of each coalition* S *and the amount which* S *can be given, if the complement coalition* $N \setminus S$ *obtains its own worth* $v(N \setminus S)$. *Now the problem to compare these numbers* $v(S)$ *and* $v^*(S)$ *is solved here by adding constants to both,* $v(S)$ *and* $v^*(S)$, *such that the arising modified maximal excesses w.r.t.* v *and* v^* *coincide (as long as both initial maximal excesses are attained by nontrivial coalitions). Excess comparability now means that the solution* x *has not to be changed if the game* v *is replaced by* v^x, *i.e. by a game which contains* v *and its dual as totally symmetric ingredients in its definition such that the coalitions with maximal initial excesses possess coinciding new excesses (except if one maximal excess is attained by the empty and grand coalition only).*

If $x = \nu(\tilde{v})_N$ *is the restriction of the prenucleolus of the dual cover of the game, then* v^x *coincides - up to adding a constant to the worth of every nontrivial coalition - with the reduced game of the dual cover w.r.t. the initial player set and the prenucleolus, hence* $x = \nu(v^x)$ *in this case. Moreover,* v^x *satisfies LED w.r.t.* x, *hence* x *coincides with the modified nucleolus of* v^x. *Therefore* $\psi(v^{\psi(v)}) = \psi(v)$ *holds true. For these properties Remark 3.5 is referred to.*

The large excess difference property can be interpreted with the help of v^x *as follows. If* v *satisfies LED w.r.t. the Pareto optimal vector* x, *then* $\mu(x, v^*) = 0$. *Due to the definition of LED we obtain* $v(S) - v^*(S) - \mu(x, v) \geq 0$, *thus*

$$v(S) + 2\mu(x, v^*) + \mu(x, v) \geq v^*(S) + 2\mu(x, v) + \mu(x, v^*)$$

for $\emptyset \subset S \subset N$. *This motivates the notion of a shift game.*

The game (N, w) *is a* **shift** *game of the game* (N, v) *if there is a real number* $\alpha \in \mathbb{R}$ *such that*

$$w(S) = \begin{cases} v(S) + \alpha & , \text{ if } \emptyset \subset S \subset N \\ v(S) & , \text{ otherwise} \end{cases}.$$

In this case w *is the* α**-shift** *game of* v, *denoted* ${}^{\alpha}v$.

In this sense v^x *coincides with a shift game of* v *(provided* v *satisfies LED w.r.t.* x*) and, hence,* v *can be seen as the only significant ingredient of* v^x *in this case. If the coalitions agree to the 'comparability principle' (i.e. to the replacement of* v *by* v^x*), then each coalition should argue with its excess w.r.t. the original game instead of switching to the dual game* v^*.

Note that every reduced game w.r.t. x *of a game* (N, v) *which satisfies LED w.r.t. the feasible payoff vector* x *inherits this property, i.e.* $(S, v^{S,x})$ *satisfies LED w.r.t.* x_S. *(see Remark 3.5 (2)).*

For the following remark Sudhölter (1996a), Lemmata 4.5 and 4.8, is referred to.

Remark 3.5 *Let* (N, v) *be a game and* $x \in X^*(v)$ *be a feasible payoff vector.*

(1) If v satisfies LED w.r.t. the prenucleolus $\nu(v)$, then the prenucleolus coincides with the modified nucleolus ($\nu(v) = \psi(v)$).

(2) If v satisfies LED w.r.t. x, then every reduced game $(S, v^{S,x})$ satisfies LED w.r.t. the restricted vector x_S.

(3) If x is Pareto optimal, then v^x satisfies LED w.r.t. x.

(4) If $\nu = \nu(\tilde{v})$ is the prenucleolus of the dual cover $(N \cup N^*, \tilde{v})$, then the reduced game $(N, \tilde{v}^{N,\nu})$ is a shift game of v^{ν_N}.

(5) The prenucleolus of every shift game of v coincides with the prenucleolus of v.

A further axiom which requires, roughly speaking, that the solution concept of the dual replication arises from the solution concept of the initial game by replication (see Remark 2.5 (5)), implies self duality and will be used in the axiomatization.

Definition 3.6 *A solution concept σ on a set Γ of games satisfies the* **dual replication property (DRP)**, *if the following is true: If $v \in \Gamma, \tau : N \cup N^* \to \tilde{N}$ is a bijection such that $(\tilde{N}, w) \in \Gamma$, where $w = \tau\bar{v}, x \in \sigma(v)$, then $\tau(x, x^*) \in \sigma(w)$.*

This definition means that the replication of an element of the solution has to be a member of the solution of the dual replication of the game in case both, the game and its dual replication belong to the considered set of games. In order to get a strong instrument which can also be applied if dual replications of games do not belong to Γ we also demand the property just described in case there is a game which is only equivalent to the dual replication. It is straightforward (see Sudhölter (1996a)) to verify that both, the Shapley value and the modified nucleolus satisfy DRP.

Theorem 3.7 *Let U be an infinite set. Then the modified nucleolus is the unique solution concept on Γ_U satisfying SIVA, COV, LEDCONS, EC, and DRP.*

A proof of this theorem contained in Sudhölter (1996a). Nevertheless an outline of the proof is presented for completeness reasons:

The modified nucleolus satisfies the desired properties by Theorem 2.4, Remark 2.5, and Remark 3.5. To show uniqueness let σ be a solution concept which satisfies the desired properties. Lemmata 4.7 and 4.9 of Sudhölter (1996a) show that σ satisfies AN and PO. We proceed similarly to Sobolev's proof of Theorem 3.2. Let $(N, v) \in \Gamma_U$ be a game, $\{x\} = \sigma(v)$, and $y = \psi(v)$. As in the classical context we can assume $y = 0$ by COV. By the infinity assumption of the cardinality of U and AN we assume that the dual replication $(N \cup N^*, \bar{v})$ is a member of Γ_U. With $w = \bar{v}^{(x,x^*)}$ it can be shown that $w = {}^{\alpha}(\bar{v}^{(y,y^*)})$ for some nonnegative α (recall that the modified nucleolus minimizes sums of excesses w.r.t. v and v^*). The game w satisfies LED w.r.t. $(y, y^*) = \nu(w)$ by Remark 3.5. By DRP, EC, and SIVA it suffices to show that $(y, y^*) \in \sigma(w)$ holds true. This can be done by applying Sobolev's approach to w. He showed the existence of a game $(\tilde{N}, u) \in \Gamma_U$ with $N \cup N^* \subseteq \tilde{N}$ satisfying

(1) $u^{N \cup N^*, z} = w$ (where $z = 0 \in \mathbb{R}^{\tilde{N}}$),

(2) $u(S) \geq \min_{\emptyset \subset T \subset N} w(T)$ for $\emptyset \subset S \subset \tilde{N}$ and $u(\tilde{N}) = 0$, and

(3) u is transitive (i.e. u's symmetry group is transitive).

By AN and PO $z \in \sigma(u)$ can be concluded. The proof is finished by the observation that u satisfies LED w.r.t z. *q.e.d.*

It should be remarked that Sudhölter (1996a) contains examples which show that all properties in Theorem 3.7 (including the infinity assumption on the cardinality of the univers U of players) are logically independent.

4 Self Dual Modifications of the Prekernel

The (pre)kernel was introduced in Davis and Maschler (1965) and Maschler, Peleg, and Shapley (1979) respectively. According to the strong relationship between the prekernel, nucleolus, and least core the second paper is referred to and, in the homogeneous case, Peleg and Rosenmüller (1992). For the prenucleolus the corresponding modified solution concept is already defined, whereas the definition of the modified least core is straightforward (see Sudhölter (1996b)). The notion of the modified kernel is given as follows. Analogously to the prekernel the modified kernel will not only be used as an auxiliary solution concept but will be given an intuitive meaning with the help of an axiomatization. At first the definition of the prekernel is recalled. Let (N,v) be a game and $x \in \mathbb{R}^N$. The **prekernel** *of v is the set of* **balanced** *preimputations*

$$\mathcal{PK}(v) = \{x \in X(v) \mid s_{ij}(x,v) = s_{ji}(x,v) \text{ for } i,j \in N, i \neq j\}.$$

Definition 4.1 *Let (N,v) be a game, $x \in \mathbb{R}^N$, and $i,j \in N$ be different players of v.*

(1) Define two numbers

$$\tilde{s}_{ij}(x,v) = \max_{j \notin S \ni i} (e(S,x,v) + \mu(x,v^*), e(S,x,v^*) + \mu(x,v))$$

and

$$\bar{s}_{ij}(x,v) = \max_{i \in S, j \notin T} (e(S,x,v) + e(T,x,v^*), e(S,x,v^*) + e(T,x,v)).$$

Then $\tilde{s}_{ij}$ is the **maximal modified surplus of** i **over** j **at** x.

(2) The **modified kernel of** v *is the set*

$$\mathcal{MK}(v) = \{x \in X(v) \mid \tilde{s}_{ij}(x,v) = \tilde{s}_{ji}(x,v) \text{ for } i,j \in N, i \neq j\}$$

and the **proper modified kernel of** v *is the set*

$$\mathcal{MK}_0(v) = \{x \in \mathcal{MK}(v) \mid \bar{s}_{ij}(x,v) = \bar{s}_{ji}(x,v) \text{ for } i,j \in N, i \neq j\}.$$

The proper modified kernel is a subset of the modified kernel of the game and Example 4.4 shows that these concepts do not necessarily coincide.

There is a strong relationship between the prekernel of the dual cover of a game and the proper modified kernel of the game, implying nonemptiness.

Lemma 4.2 *Let (N,v) be a game. Then*

$$\mathcal{MK}_0(v) = \{x \in \mathbb{R}^N \mid (x,x^*) \in \mathcal{PK}(\tilde{v})\}.$$

Proof: *Let $x \in \mathbb{R}^N$ and $i,j \in N$ with $i \neq j$. By definition $s_{ij}((x,x^*),\tilde{v}) = \tilde{s}_{ij}(x,v)$ and $s_{ij^*}((x,x^*),\tilde{v}) = \bar{s}_{ij}(x,v)$ hold true. Hence $x \in \mathcal{MK}_0(v)$, iff $(x,x^*) \in \mathcal{PK}(\tilde{v})$.* q.e.d.

As a consequence of this lemma we obtain $\psi(v) \in \mathcal{MK}_0(v) \subseteq \mathcal{MK}(v)$ for each game v. The set $\{(x,x^*) \in \mathcal{PK}(\tilde{v})\} = \mathcal{SPK}(\tilde{v})$ could be called **symmetric** *prekernel of the dual cover $\tilde{v}$ of v.*

Remark 4.3

(1) $\mathcal{MK}(v) \supseteq \mathcal{MK}_0(v) \supseteq \mathcal{PK}(v) \cap \mathcal{PK}(v^*)$ holds true by definition. Moreover, both versions of the modified kernel coincide with the prekernel on constant-sum games.

(2) *The (proper) modified kernel satisfies reasonableness on both sides and respects desirability. A proof of these assertion is straightforward (see Sudhölter (1993)), because both, the game and its dual possess the same 'desirability structure' and the same maximal and minimal marginal contributions.*

(3) *Both modified kernels satisfy covariance, anonymity, the equal treatment property, and the nullplayer property.*

Example 4.4 *For glove games the proper modified kernel is a proper subset of the modified kernel. A game (N, v) is a* **glove game**, *if the player set can be partitioned into the sets R of 'right hand glove owners' and L of 'left hand glove owners' (i.e. $R \cup L = N$, $R \cap L = \emptyset$, $R \neq \emptyset \neq L$), whereas the coalitional function v counts the number of pairs of gloves owned by the coaltions (i.e. $v(S) = \min\{| R \cap S |, | L \cap S |\}$). Without loss of generality we may assume $r = | R | \leq | L | = l$. Moreover, we restrict our attention to the case $l \geq 2$, because for two-person games both modified kernels coincide with the Shapley value ($\mathcal{MK}$, $\mathcal{MK}_0$ and Ψ are* **Standard solutions**) *by NE, PO, COV, and ETP. We are going to show the following claims:*

(1) *The proper modified kernel of v is the singleton which treats the groups of left hand glove owners and right hand glove owners equally, i.e.*

$$\mathcal{MK}_0(v) = \{z^E\} \text{ where } z_i^E = \begin{cases} 1/2 & , \text{if } i \in R \\ r/2l & , \text{if } i \in L \end{cases}.$$

(2) *If $r < l$, then the modified kernel is the convex hull of the* **equal treatment vector** *z^E and the nucleolus z^R, defined by*

$$z_i^R = \begin{cases} 1 & , \text{if } i \in R \\ 0 & , \text{if } i \in L \end{cases}.$$

(3) *If $r = l$, then the modified kernel coincides with the core, i.e. with the convex hull of z^R and z^L. (Here z^L is defined analogously to z^R by $z_i^L = \begin{cases} 1 & , \text{if } i \in L \\ 0 & , \text{if } i \in R \end{cases}$.)*

Proof: *If $z \in \mathcal{MK}(v)$ and $i, j \in R$ or $i, j \in L$, then $z_i = z_j$ by ETP (see Remark 4.3 (3)). Moreover, $0 \leq z_i \leq 1$ for $i \in N$ by Remark 4.3 (2). With $z^\alpha \in \mathbb{R}^N$ defined by*

$$z^\alpha = \begin{cases} \alpha & , \text{if } i \in R \\ r(1-\alpha)/l & , \text{if } i \in L \end{cases}$$

Pareto optimality implies that

$$\mathcal{MK}(v) \subseteq \{z^\alpha \mid 0 \leq \alpha \leq 1\} = Z$$

holds true. For every $z^\alpha \in Z$ and $i, j \in R$ or $i, j \in L$ it is straightforward to verify

$$\begin{aligned} \tilde{s}_{ij}(x^\alpha, v) &= \tilde{s}_{ji}(x^\alpha, v) \\ \bar{s}_{ij}(x^\alpha, v) &= \bar{s}_{ji}(x^\alpha, v) \end{aligned} \quad \text{for } i, j \in R \text{ or } i, j \in L. \tag{4.1}$$

(1) *If $\alpha < 1/2$, then R is the unique coalition attaining $\mu(z^\alpha, v^*)$. In view of (2) we can assume that $r = l$ holds true. Then we have $\mu(z^\alpha, v) = e(S, z^\alpha, v)$ for every coalition S satisfying $\mid S \cap R \mid = \mid S \cap L \mid = 1$. This observation implies*

$$\bar{s}_{ij}(z^\alpha, v) = \mu(z^\alpha, v) + \mu(z^\alpha, v^*) > \bar{s}_{ij}(z^\alpha, v) \text{ for } i \in R,\ j \in L,$$

thus $z^\alpha \notin \mathcal{MK}_0(v)$.

If $\alpha > 1/2$ and $r = l$, the proof can be finished analogously by interchanging the rôles of R and L. If $\alpha > 1/2$ and $r < l$, then L is the unique coalition attaining $\mu(z^\alpha, v^*)$, whereas $\mu(z^\alpha, v)$ is attained by coalitions S satisfying $R \subseteq S$ and $\mid L \cap S \mid = r$. The observation

$$\bar{s}_{ij}(z^\alpha, v) = \mu(z^\alpha, v) + \mu(z^\alpha, v^*) > \bar{s}_{ji}(z^\alpha, v) \text{ for } i \in L,\ j \in R$$

finishes the proof of (1).

(2) *If $\alpha \geq 1/2$, then L attains $\mu(z^\alpha, v^*)$ and $R \cup T$ with $T \subset L$ such that $\mid T \mid = r$ attains $\mu(z^\alpha, v)$, thus*

$$\tilde{s}_{ij}(z^\alpha, v) = \tilde{s}_{ji}(z^\alpha, v) = \mu(z^\alpha, v) + \mu(z^\alpha, v^*) \text{ for } i \in R,\ j \in L.$$

This equality together with (4.1) implies that $z^\alpha \in \mathcal{MK}(v)$ holds true.

If $\alpha < 1/2$, then R is the unique coalition attaining $\mu(z^\alpha, v^*)$ and every coalition S attaining $\mu(z^\alpha, v)$ contains R, thus $\tilde{s}_{ij}(z^\alpha, v) = \mu(z^\alpha, v) + \mu(z^\alpha, v^*) > \tilde{s}_{ji}(z^\alpha, v)$ for $i \in R,\ j \in L$. This observation shows that z^α cannot be a member of the modified kernel in this case.

(3) *For $0 \leq \alpha \leq 1$ the maximal excess $\mu(z^\alpha, v)$ is attained by all coalitions S satisfying $\mid S \cap R \mid = 1 = \mid S \cap L \mid$, because z^α is a member of the core (recall that $r = l$ is assumed). Using the assumption $l > 1$, i.e. $r > 1$ is automatically satisfied by $r = l$, we come up with*

$$\tilde{s}_{ij}(z^\alpha, v) = \mu(z^\alpha, v) + \mu(z^\alpha, v^*) \text{ for } i, j \in N,$$

thus the proof is finished.

q.e.d.

Applied to the modified nucleolus this example shows that $\psi(v)$ assigns the same amount to both groups R and L. Glove games can be seen as **two-sided assignment games** *as discussed, e.g., in Shapley and Shubik (1972). It can be shown (see Sudhölter (1994)) that both sides of an assignment game are treated equally by the modified nucleolus in general.*

The following lemma is used to show that the (proper) modified kernel satisfies excess comparability as well as LEDCONS.

Lemma 4.5 *Let (N, v) be a game and $x \in X^*(v)$ be a feasible payoff vector. Assume v satisfies LED w.r.t. x. Then the following properties are valid.*

(1) *If $i, j \in N$ with $i \neq j$, then $s_{ij}(x, v) = s_{ji}(x, v)$ iff $\tilde{s}_{ij}(x, v) = \tilde{s}_{ji}(x, v)$.*

(2) *If $s_{ij}(x, v) = s_{ji}(x, v)$ for all $i, j \in N$ with $i \neq j$, then $\bar{s}_{ij}(x, v) = \bar{s}_{ji}(x, v)$ for all $i, j \in N$ with $i \neq j$.*

Proof: *Assume w.l.o.g.* $|N| \geq 2$ *(otherwise both assertions are trivially satisfied). Analogously to Remark 2.2 (1) it is obvious that*

$$e(S,x,v) = -e(N \setminus S, x, v^*) + v(N) - x(N) \tag{4.2}$$

for all $S \subseteq N$ *holds true. Using (4.2), LED and*

$$\Lambda(x,v) = \min\{\min\{e(S,x,v), e(T,x,v)\} - e(S,x,v) - e(T,x,v^*) \mid \emptyset \neq S, T \neq N\} \tag{4.3}$$

(for a proof of equation (4.3) see Sudhölter (1996a)) it can easily be seen that

$$e(S,x,v^*) \leq 0 \text{ for all } S \neq N, \tag{4.4}$$

thus

$$e(S,x,v) \geq v(N) - x(N) \geq 0 \text{ for all } S \neq \emptyset. \tag{4.5}$$

Therefore we come up with

$$\mu(x,v) = \mu_0(x,v), \mu(x,v^*) = v(N) - x(N). \tag{4.6}$$

Let $i, j \in N, i \neq j$ *and* $j \notin T \ni i$ *for some* $T \subseteq N$. *Then*

$$\begin{aligned} e(T,x,v) + \mu(x,v^*) &= e(T,x,v) + v(N) - x(N) \text{ (by (4.6))} \\ &\geq e(T,x,v) \text{ (by the feasibility of } x) \\ &\geq e(T,x,v^*) + \mu_0(x,v) \text{ (by (4.3))} \\ &= e(T,x,v^*) + \mu(x,v) \text{ (by 4.6))}, \end{aligned}$$

thus $\tilde{s}_{ij}(x,v) = s_{ij}(x,v) + v(N) - x(N)$; *hence the first assertion is established.*

In order to show the second one, observe that

$$\begin{aligned} &\bar{s}_{ij}(x,v) \\ &= \max_{i \in S, j \notin T}(e(S,x,v) + e(T,x,v^*), e(S,x,v^*) + e(T,x,v)) \\ &= \max\{e(S,x,v) \mid i \in S\} \cup \{e(T,x,v) + v(N) - x(N) \mid j \notin T\} \text{ (by (4.4))} \\ &\leq \mu(x,v) + v(N) - x(N) \text{ (by definition).} \end{aligned} \tag{4.7}$$

Take any coalition $S \subseteq N$ *with* $e(S,x,v) = \mu(x,v)$ *and* $\emptyset \neq S \neq N$ *- note that the existence of* S *is guaranteed by (4.6). If* $j \notin S$, *then* $\bar{s}_{ij}(x,v) = \mu(x,v) + v(N) - x(N)$ *(see (4.7)). If* $j \in S$, *then choose any* $k \in N \setminus S$. *Now, by assumption,* $s_{jk}(x,v) = \mu(x,v) = s_{kj}(x,v)$, *thus there is a coalition* $T \subseteq N$ *with* $j \notin T \ni k$ *and* $e(T,x,v) = \mu(x,v)$. *Again* $\bar{s}_{ij}(x,v) = \mu(x,v) + v(N) - x(N)$ *is concluded in view of (4.7).* *q.e.d.*

Note that Lemma 4.5 yields a relationship between the prekernel, the modified, and the proper modified kernel in case LED is satisfied. Indeed, under the assumptions of this lemma, the vector x *is a member of the prekernel of* v, *iff this is true for the modified kernel. Moreover, modified can be replaced by proper modified. These considerations together with consistency of the prekernel lead to*

Corollary 4.6 *The modified and proper modified kernel satisfy LEDCONS and EC on* Γ_U *for each set* U.

Proof: *For both modified solution concepts LEDCONS is directly implied by Lemma 4.5, Remark 3.5 (2), and consistency of the prekernel. By Lemma 4.5 and Remark 3.5 (3) it remains to show that the modified kernel satisfies EC. Let* $(N,v) \in \Gamma_U$, $x \in \mathcal{MK}(v)$ *and* $i,j \in N$. *The straightforward observations* $\mu(x,v^x) = 2\cdot(\mu(x,v)+\mu(x,v^*))$ *and* $\mu(x,(v^x)^*) = 0$ *imply*

$$\tilde{s}_{ij}(x,v^x) = \tilde{s}_{ij}(x,v) + \mu(x,v) + \mu(x,v^*), \tag{4.8}$$

thus the proof is finished. *q.e.d.*

5 An Axiomatization of the Modified Kernel

First of all Peleg's (1986) axiomatization of the prekernel is recalled. For a finite set N let $\Pi(N) = \{\{i,j\} \mid i,j \in N, i \neq j\}$ *denote the set of player pairs. A solution concept* σ *on a set* Γ *of games satisfies* **converse consistency** (**COCONS**), *if the following condition is satisfied:*

If $(N,v) \in \Gamma$, $x \in X(v)$, $(S,v^{S,x}) \in \Gamma$, *and* $x_S \in \sigma(S,v^{S,x})$ *for every* $S \in \Pi(N)$, *then* $x \in \sigma(N,v)$.

Theorem 5.1 *(Peleg) If U is a set, then the prekernel is the unique solution concept on* Γ_U *satisfying NE, PO, ETP, COV, CONS, and COCONS.*

In order to axiomatize $\mathcal{MK}$ *one further axiom is needed, which resembles COCONS and which finally leads to an analogon of Peleg's result.*

Definition 5.2 *A solution concept* σ *on a set* Γ *of games satisfies* **large excess difference converse consistency (LEDCOCONS)**, *if the following condition is satisfied:*

If $(N,v) \in \Gamma$, $x \in X(v)$, $(S,u^{S,x}) \in \Gamma$, *where* $u = v^x$, *and* $x_S \in \sigma(S,u^{S,x})$ *for every* $S \in \Pi(N)$, *then* $x \in \sigma(v)$.

LEDCOCONS is a modified converse consistency (COCONS) property in the sense of Peleg (1986). Indeed, if (N,v) *satisfies LED w.r.t.* x, *then* $v^x = u$ *coincides with* v *up to a nonnegative shift. Moreover, the reduced games* $u^{S,x}$ *coincide with* $v^{S,x}$ *up to a shift. For the general case COCONS is hardly comparable with the modified property. Nevertheless, at least together with EC both converse consistency properties are similar.*

Theorem 5.3 *Let U be a set. The modified kernel is the unique solution concept on* Γ_U *satisfying NE, COV, PO, ETP, LEDCONS, LEDCOCONS, and EC.*

Proof: *Clearly,* $\mathcal{MK}$ *satisfies NE, COV, PO, ETP, LEDCONS, and EC by Lemma 4.2, Remark 4.3 (3), definition, and Corollary 4.6. To verify LEDCOCONS, let* $(N,v) \in \Gamma_U$ *and* $x \in X(v)$ *such that* $x_S \in \mathcal{MK}(S,u^{S,x})$, *where* $u = x^x$, *for every* $S \in \Pi(N)$. *By Remark 3.5, (2) and (3), and Lemma 4.5 we conclude that* $x_S \in \mathcal{PK}(S,u^{S,x})$ *holds true for every* $S \in \Pi(N)$. *By COCONS of the prekernel* $x \in \mathcal{PK}(N,u)$. *Remark 3.5 (4) and Lemma 4.5 imply* $x \in \mathcal{MK}(N,u)$, *thus equation 4.8 (which is valid for every Pareto optimal* x) *shows that* $x \in \mathcal{MK}(N,v)$.

In order to show the uniqueness part let σ *be a solution concept on* Γ_U *which satisfies NE, COV, PO, ETP, LEDCONS, LEDCOCONS, and EC. Due to NE, COV, PO, and ETP, we have* $\sigma(N,v) = \mathcal{PK}(N,v) = \mathcal{MK}(N,v)$ *for all games* (N,v) *with* $N \subseteq U$ *and* $|N| = 2$ *as in the classical context (see Peleg (1986), Remark 4.4). From now on only games* $(N,v) \in \Gamma_U$ *satisfying* $|N| \geq 3$ *are considered.*

First we prove the inclusion $\mathcal{MK}(N,v) \subseteq \sigma(N,v)$. *Let* $x \in \mathcal{MK}(N,v)$. *Then* $x \in \mathcal{MK}(N,v^x)$, *because* $\mathcal{MK}$ *satisfies EC. Write* $u = v^x$. *In view of Corollary 4.6 (The modified kernel satisfies EC.) and Remark 3.5 (3) (The derived game* (N,u) *satisfies LED w.r.t.* x.) *we obtain* $x_S \in \mathcal{MK}(S,u^{S,x})$ *for every* $\emptyset \neq S \subseteq N$, *in particular for every coalition* S *with* $|S| = 2$. *For two-person games we already know that the solution concept* σ *coincides with the modified kernel, i.e.* $x_S \in \sigma(S,u^{S,x})$ *for* $S \subset N$ *with* $|S| = 2$. *We conclude* $x \in \sigma(N,v)$, *because* σ *satisfies LEDCOCONS. These considerations complete the proof of the inclusion* $\mathcal{MK}(N,v) \subseteq \sigma(N,v)$.

Secondly we prove the inverse inclusion $\sigma(N,v) \subseteq \mathcal{MK}(N,v)$. *Let* $x \in \sigma(N,v)$. *Then* $x \in \sigma(N,v^x)$, *because* σ *satisfies EC. Write* $u = v^x$. *In view of the assumption that* σ *satisfies EC and of Remark 3.5 (3) (The derived game* (N,u) *satisfies LED w.r.t.* x.) *we obtain* $x_S \in \sigma(S,u^{S,x})$ *for every* $\emptyset \neq S \subseteq N$, *in particular for every coalition* S *with* $|S| = 2$. *For two-person games we already know that the solution concept* σ *coincides with the modified kernel, i.e.* $x_S \in \mathcal{MK}(S,u^{S,x})$ *for* $S \subset N$ *with* $|S| = 2$. *We conclude* $x \in \mathcal{MK}(N,v)$, *because* $\mathcal{MK}$ *satisfies LEDCOCONS. These considerations complete the proof of the inclusion* $\sigma(N,v) \subseteq \mathcal{MK}(N,v)$. *q.e.d.*

Note that the universe U *of players in Theorem 5.3 may be finite or infinite as in the classical context (Theorem 5.1). For an axiomatization of the proper modified kernel Sudhölter (1993) is referred to. Peleg showed the logical independence of NE, COV, PO, ETP, CONS, and COCONS by defining six solution concepts which do not coincide with the prekernel satisfying all differing five of the preceding properties. Slightly modified, these examples also show the independence of the axioms of Theorem 5.3. Indeed, define* σ^i $(i \in \{1,\ldots,7\})$ *on* Γ_U *for each* $(N,v) \in \Gamma_U$ *by*

$\sigma^1(v) = \emptyset$,

$\sigma^2(v) = \{x \in \mathbb{R}^N \mid x_i = v(N)/\mid N \mid \text{ for } i \in N\}$,

$\sigma^3(v) = \{x \in X^*(v) \mid \tilde{s}_{ij}(x) = \tilde{s}_{ji}(x,v) \text{ for } i,j \in N, i \neq j\}$,

$\sigma^4(v) = X(v)$,

$\sigma^5(v) = \{x \in X(v) \mid v(\{i\}) - x_i = v(\{j\}) - x_j$ *equivalence relation defined by* $i \simeq_v j$, *if* $\max\{v(S\cup\{i\}), v^*(S\cup\{i\})\} - \max\{v(\{i\}), v^*(\{i\}\} = \max\{v(S\cup\{j\}), v^*(S\cup\{j\})\} - \max\{v(\{j\}), v^*(\{j\}\}$ *holds true for* $S \subseteq N \setminus \{i,j\}$,

$\sigma^6(v) = \Psi(v)$,

$$\sigma^7(v) = \begin{cases} \mathcal{MK}(v) & \text{, if } v \text{ satisfies LED w.r.t. } \varphi(v) \\ \mathcal{MK}(v) \cup \{\varphi(v)\} & \text{, otherwise.} \end{cases}$$

Following Peleg's approach (1986,1988/89) it is straightforward to verify that each σ^i *satisfies all axioms of Theorem 5.3 up to the i-th one. Clearly,* σ^1 *violates NE, if* $|U| \geq 1$. *The solution concepts* σ^2, σ^3, *and* σ^4 *violate COV, PO, and ETP, respectively, if* $|U| \geq 2$. *The modified kernel is contained in* $\sigma^5(v)$. *Both concepts only coincide in case* $|U| \leq 2$, *in which they cannot be distinguished from* σ^6 *and* σ^7. *The last two examples show the logical independence of LEDCOCONS and EC, respectively.*

In view of the just mentioned solution concepts Theorem 5.3 can be regarded as an axiomatization of the modified kernel.

In Sudhölter (1993) one crucial difference between the modified kernel and the proper modified kernel is observed. Indeed, in contrast to the modified kernel the proper modified kernel satisfies DRP. From the axiomatic viewpoint 'properness' seems to be less intuitive. The application to many examples (see, e.g., Example 4.4) indicates the converse statement.

References

[1] *Davis M and Maschler M (1965) The kernel of a cooperative game. Naval Research Logist. Quarterly 12, pp. 223-259*

[2] *Kopelowitz A (1967) Computation of the kernel of simple games and the nucleolus of n-person games. Research Program in Game Th. and Math. Economics, Dept. of Math., The Hebrew University of Jerusalem, RM 31*

[3] *Maschler M (1992) The bargaining set, kernel, and nucleolus: a survey. In: R.J. Aumann and S. Hart, eds., Handbook of Game Theory 1, Elsevier Science Publishers B.V., pp. 591-665*

[4] *Maschler M, Peleg B, and Shapley LS (1979) Geometric properties of the kernel, nucleolus, and related solution concepts. Math. of Operations Res. 4, pp.303-338*

[5] *Milnor JW (1952) Reasonable outcomes for n-person games. The Rand Corporation 916*

[6] *Ostmann A (1987) On the minimal representation of homogeneous games. Int. Journal of Game Theory 16, pp. 69-81*

[7] *Peleg B (1968) On weights of constant-sum majority games. SIAM Journal Appl. Math. 16, pp. 527-532*

[8] *Peleg B (1986) On the reduced game property and its converse. Int. Journal of Game Theory 15, pp. 187-200*

[9] *Peleg B (1988/89) Introduction to the theory of cooperative games. Research Memoranda No. 81-88, Center for Research in Math. Economics and Game Theory, The Hebrew University, Jerusalem, Israel*

[10] *Peleg B and Rosenmüller J (1992) The least core, nucleolus, and kernel of homogeneous weighted majority games. Games and Economic Behavior 4, pp. 588-605*

[11] *Rosenmüller J (1987) Homogeneous games: recursive structure and computation. Math. of Operations Research 12, pp. 309-330*

[12] *Sankaran JK (1992) On finding the nucleolus of an n-person cooperative game. Int. Journal of Game Theory 21*

[13] *Schmeidler D (1966) The nucleolus of a characteristic function game. Research Program in Game Th. and Math. Econ., The Hebrew University of Jerusalem, RM 23*

[14] *Shapley LS (1953) A value for n-person games. in: H. Kuhn and A.W. Tucker, eds., Contributions to the Theorie of Games II, Princeton University Press, pp. 307-317*

[15] *Shapley LS and Shubik M (1972) The assignment game I: the core. Int. Journal of Game Theory 2, pp. 111-130*

[16] *Sobolev AI (1975) The characterization of optimality principles in cooperative games by functional equations. In: N.N. Vorobiev, ed., Math. Methods Social Sci. 6, Academy of Sciences of the Lithunian SSR, Vilnius, pp. 95-151 (in Russian)*

[17] *Sudhölter P (1993) The modified nucleolus of a cooperative game. Habilitation Thesis, University of Bielefeld, 80 p*

[18] *Sudhölter P (1994) Solution concepts for c-convex, assignment, and M2-games. Working paper 232, Institute of Mathematical Economics, University of Bielefeld, 28 p*

[19] *Sudhölter P (1996a) The modified nucleolus: properties and axiomatizations. To appear in the Int. Journal of Game Theory, 34 p*

[20] *Sudhölter P (1996b) The modified nucleolus as canonical representation of weighted majority games. To appear in Mathematics of Operations Research, 30 p*

Peter Sudhölter
Institute of Mathematical Economics
University of Bielefeld, Postfach 100131
D-33501 Bielefeld, Germany

THE EGALITARIAN NONPAIRWISE-AVERAGED CONTRIBUTION (ENPAC-) VALUE FOR TU-GAMES

Theo Driessen and Yukihiko Funaki [1]

Abstract: *The paper introduces a new solution concept for transferable utility games called the Egalitarian Non-Pairwise-Averaged Contribution (*ENPAC-*) value. This solution arises from the egalitarian division of the surplus of the overall profits after each participant is conceded to get his pairwise-averaged contribution. Four interpretations of the ENPAC-value are presented. The second part of the paper provides sufficient conditions on the transferable utility game to guarantee that the ENPAC-value coincides with the well-known solution called prenucleolus. The main conditions require that the largest excesses at the ENPAC-value are attained at the $(n-2)$-person coalitions, whereas the excesses of $(n-2)$-person coalitions at the ENPAC-value do not differ.*

1 Notions and Summary

A *transferable utility game* (or cooperative game or coalitional game with side payments) is an ordered pair (N, v), where N is a finite set of *players* and $v : 2^N \to R$ is a *characteristic function*, defined on the power set 2^N of all subsets of N called *coalitions* (notation: $S \in 2^N$ or $S \subset N$). With every coalition S there is associated the real number $v(S)$ called the *worth* of S in the TU-game (N, v) and it represents the joint profits that the members of S can achieve due to their cooperative behaviour. It is a standard requirement that the empty set has no worth, i.e., $v(\emptyset) = 0$. The cardinality of coalition S is denoted by $|S|$ or, if no ambiguity is possible, by s. Particularly, we put $n := |N|$ and throughout the paper it is supposed that a game involves at least three players, so $n \geq 3$. For every coalition $S \in 2^N$, the corresponding expression $\Delta_S(N, v) := v(N) - v(N \backslash S)$ has to be interpreted as the *incremental return to scale for cooperation* by the members of S (with respect to the formation of the grand coalition in the TU-game (N, v)). Here $N \backslash S$ denotes the complement of coalition $S \in 2^N$, i.e., $N \backslash S := \{i \in N \mid i \notin S\}$. The set of all TU-games is denoted by **G**. The (one-point or multi-valued) solution concepts for TU-games are concerned with the essential problem how to divide the overall profits $v(N)$ of the grand coalition N among all the players of the TU-game (N, v). Generally speaking, a *value* (or one-point solution concept) on **G** is a function Φ that assigns a payoff vector $\Phi(N, v) = (\Phi^i(N, v))_{i \in N}$ to every TU-game (N, v). The corresponding payoff vector $\Phi(N, v)$ is called an *allocation* if it meets the efficiency principle $\sum_{i \in N} \Phi^i(N, v) = v(N)$. The value $\Phi^i(N, v)$ of player i in the TU-game (N, v) represents an assessment by i of his gains from participating in the game.

[1] The research for this paper was done during the sabbatical leave period (From April 1992 till September 1993) at the Department of Applied Mathematics, University of Twente, Enschede, The Netherlands. This author gratefully acknowledges funding from Toyo University.

In this paper we aim to discuss a new value for TU-games called the Egalitarian Non-Pairwise-Averaged Contribution (**ENPAC-**) value. This new value refers to the egalitarian division of the surplus of the overall profits after each participant is conceded to get his pairwise-averaged contribution. In order to justify the new value in different ways, we present, in Section 2, four interpretations of the ENPAC-value (see Definition 2.1, Remarks 2.2-2.3, Proposition 2.4). Example 2.6 illustrates the determination of the ENPAC-value in the context of a bankruptcy situation.
Particularly, Proposition 2.4 states that the ENPAC-value is the unique solution of a system of linear equations requiring that the so-called excesses of coalitions of size $n-2$ with respect to the (variable) payoff vectors do not differ. Because of this interpretation of the ENPAC-value, it is worthwhile to compare the ENPAC-value with the so-called prenucleolus, a well-known value which is strongly based on the notion of excess. In fact, in Section 3 we are interested in the coincidence of the ENPAC-value and the prenucleolus. Theorem 3.1 provides two main conditions on the ENPAC-value of a TU-game which are sufficient for the ENPAC-value to agree with the prenucleolus concept. These two main conditions require that the largest excesses at the ENPAC-value are attained at the $(n-2)$-person coalitions, whereas the excesses of $(n-2)$-person coalitions at the ENPAC-value do not differ. For the sake of computational matters, an equivalent formulation of the first condition is provided in Section 4 (see Theorem 4.1). In a general setting we examine, in the last part of Section 3, the set of allocations for which the largest excesses are attained at the $(n-2)$-person coalitions. Theorem 3.4 states that, under the given circumstances, the intersection of this set with the prekernel is either empty or a singleton consisting of the ENPAC-value. Moreover, the nonemptiness of the intersection of this set with the prekernel guarantees the coincidence of the ENPAC-value and the prenucleolus. The technical proof of the second main Theorem 3.4 is given in Section 5, whereas the illustrative proof of the first main Theorem 3.1 is given in Section 3 itself.
Four concluding remarks are listed in Section 6 and two of them are concerned with the individual rationality of the ENPAC-value (see Remark 6.2) and the relationship of the ENPAC-value with the core (see Remark 6.1). In Remark 6.3 we compare the obtained results concerning the ENPAC-value with similar results concerning another value called the Egalitarian Non-Separable Contribution (**ENSC-**) value. Suggestions for further research are mentioned in the last two Remarks 6.3 and 6.4.

2 The Egalitarian Non-Pairwise-Averaged Contribution (ENPAC-) value for TU-games

An allocation rule with firm roots in the multipurpose water resource project evaluation literature is that no user should be charged less than the separable cost of including the user in the joint project. For instance, the well-known separable costs remaining benefits (SCRB-) method is a widely used approach in the water resources field (cf. [17]). Within the game theoretic framework concerning profits instead of costs, the former allocation rule states that the payoff to any player in a game is at most his separable contribution in the game. The *separable contribution* of a player (with respect to the formation of the grand coalition in the TU-game) is defined to be the individually incremental return to scale for cooperation, so $SC^i(N,v) := v(N) - v(N\backslash\{i\})$, where (N,v) is a game and $i \in N$. Subsequently, the egalitarian division of the surplus of the overall profits, the size of $v(N) - \sum_{j\in N} SC^j(N,v)$, gives rise to the so-called *Egalitarian Non-Separable Contribution* (**ENSC-**) value. In other words, with every TU-game (N,v) there is associated the allocation

$\mathbf{ENSC(N,v)} = (ENSC^i(N,v))_{i\in N}$ given by

$$\mathbf{ENSC^i(N,v)} := SC^i(N,v) + \frac{1}{n}\left[v(N) - \sum_{j\in N} SC^j(N,v)\right] \qquad \text{for all } i \in N. \qquad (21)$$

The ENSC-value has been studied by many researchers (cf. [1], [2], [3], [4], [5], [6], [7], [8], [9], [17]). Our main objective here is to introduce a new value for TU-games which is basically composed of pairwise contributions with respect to the formation of the grand coalition. That is, the standard notion of contributions dedicated to single players is replaced by a similar notion dedicated to pairs of players. For instance, we have in mind that single players are desperately searching for partners to communicate with and that such a communication happens to occur by means of a dialogue (in which not more than two persons are involved to avoid noise). Like the ENSC-value, the determination of the new value involves two stages. In the first step we describe the pairwise-averaged contribution of a player (with respect to the formation of the grand coalition in a TU-game (N,v)). For notation sake, for any pair $\{i,j\}$ of players, let $\Delta_{ij}(N,v) := v(N) - v(N\backslash\{i,j\})$ represent their *pairwise-incremental return to scale for cooperation*. The two members of a pair of players are distinguished in the sense that one player is to be considered a *leader* and the other a *follower*. Player i's *pairwise-averaged contribution* is obtained by averaging, over all followers j, $j \in N\backslash\{i\}$, the amount what is left of the overall pairwise-incremental return $\Delta_{ij}(N,v)$ (due to the pair $\{i,j\}$ inclusive of leader i) after follower j is conceded to get the per-capita worth of the complementary coalition $N\backslash\{i,j\}$. More precisely, in order to evaluate follower j's part of the overall pairwise-incremental return $\Delta_{ij}(N,v)$, leader i makes no distinction between the randomly chosen follower j and the alternative followers contained in the complementary coalition $N\backslash\{i,j\}$, of which the worth per person is given by $\frac{v(N\backslash\{i,j\})}{n-2}$. According to this reasoning, follower j is expected to receive the amount the size of $\frac{v(N\backslash\{i,j\})}{n-2}$ and consequently, leader i's part of the overall pairwise-incremental return $\Delta_{ij}(N,v)$ is expected to be the amount the size of $\Delta_{ij}(N,v) - \frac{v(N\backslash\{i,j\})}{n-2}$.
In the second step the resulting allocation to player i is obtained by the egalitarian division of the surplus of the overall profits after each player is conceded to get his pairwise-averaged contribution. That is, the remaining amount of the overall profits is equally allocated to all the players of the game.

Definition 2.1 *The* pairwise-averaged contribution *of a player in a TU-game is given by*

$$PAC^i(N,v) := \frac{1}{n-1}\sum_{j\in N\backslash\{i\}}\left[\Delta_{ij}(N,v) - \frac{v(N\backslash\{i,j\})}{n-2}\right] \qquad \textit{or equivalently,} \qquad (22)$$

$$PAC^i(N,v) = v(N) - \frac{1}{n-2}\sum_{j\in N\backslash\{i\}} v(N\backslash\{i,j\}) \qquad (23)$$

where (N,v) is a game, $i \in N$, and $\Delta_{ij}(N,v) := v(N) - v(N\backslash\{i,j\})$ for all $i,j \in N$, $i \neq j$ (which denotes the pairwise-incremental return of the pair $\{i,j\}$).
The Egalitarian Non-Pairwise-Averaged Contribution *(***ENPAC-***) value assigns to every game (N,v) the allocation* $\mathbf{ENPAC(N,v)} = (ENPAC^i(N,v))_{i\in N}$ *given by*

$$\mathbf{ENPAC^i(N,v)} := PAC^i(N,v) + \frac{1}{n}\left[v(N) - \sum_{j\in N} PAC^j(N,v)\right] \qquad \textit{for all } i \in N. \qquad (24)$$

Before continuing our study of the ENPAC-value in Sections 3, 4, and 5, we attempt to put forward, besides the above-mentioned interpretation on the basis of formulae (22) and (24), three additional motivations for the introduction of the ENPAC-value. Firstly, we illustrate that the ENPAC-value is a particular example out of a large class of values with common properties. Several other examples out of this large class of values are well-known (e.g. the Shapley value) and well-studied in the context of the game theoretic literature on values for TU-games. Secondly, we provide one striking relationship between, on the one hand, the Shapley value and on the other, the ENPAC- and ENSC-values. More precisely, the Shapley value of a game is presented as the sum of various terms, of which two separate terms are determined by the ENPAC- and ENSC-values. Thirdly, we discuss an interesting relationship between the ENPAC-value and a well-known concept called excess.

Remark 2.2 *Let $\mathcal{N} = \{1, 2, 3, \ldots\}$ denote the set of natural numbers (exclusive of 0). Let's consider a value Φ on $\mathbf{G}$ such that, for every game (N, v), the corresponding allocation $\Phi(N, v) = (\Phi^i(N, v))_{i \in N}$ is of the form*

$$\Phi^i(N,v) = \sum_{S \subset N \backslash \{i\}} p^n_{|S|} \left[b^n_{|S|+1} v(S \cup \{i\}) - b^n_{|S|} v(S) \right] \qquad \text{for all } i \in N, \tag{25}$$

where the underlying collection of constants $\{b^k_s : k \in \mathcal{N} \backslash \{1\}, s = 1, 2, \ldots, k\}$ is arbitrary, provided that $b^k_k := 1$ for all $k \in N \backslash \{1\}$, and the suitably defined collection of constants $\{p^k_s : k \in \mathcal{N} \backslash \{1\}, s = 0, 1, \ldots, k-1\}$ such that

$$p^k_s := k^{-1} \binom{k-1}{s}^{-1} \qquad \text{whenever } 0 \le s \le k-1.$$

Given a player $i \in N$, the collection $\{p^n_{|S|} : S \subset N \backslash \{i\}\}$ can be regarded as a probability distribution over the collection of coalitions not containing player i (inclusive of the empty set). This probability distribution arises from the belief that the coalition, to which player i joins, is equally likely to be of any size s, $0 \le s \le n-1$, and that all such coalitions of the same size s are equally likely. If, for each $S \subset N \backslash \{i\}$, $p^n_{|S|}$ is seen as the probability that player i joins coalition S and the weighted *marginal contribution $b^n_{|S|+1} v(S \cup \{i\}) - b^n_{|S|} v(S)$ is paid to player i for joining coalition S, then the value $\Phi^i(N, v)$ of player i in the game (N, v), as given by formula (25), is simply the expected payoff to player i in the game (N, v). In case of unit weights for all coalitions (i.e., $b^k_s = 1$ whenever $1 \le s \le k$), then the resulting value of the form (25) agrees with the well-known Shapley value (cf. [12], [15], [16]). In case of zero weights for all coalitions different from the grand coalition (i.e., $b^k_s = 0$ whenever $1 \le s \le k-1$), then the resulting value of the form (25) agrees with the most simplest egalitarian value Φ on $\mathbf{G}$ given by $\Phi^i(N, v) = \frac{v(N)}{n}$ for every game (N, v) and player $i \in N$. Secondly, it turns out that the ENSC-value is of the form (25) with corresponding constants $b^n_{n-1} = n-1$ and $b^n_s = 0$ whenever $1 \le s \le n-2$. Thirdly, it turns out that the ENPAC-value is of the form (25) with corresponding constants $b^n_{n-2} = n-1$ and $b^n_s = 0$ whenever $1 \le s \le n-1$, $s \ne n-2$. In summary, the ENPAC- and ENSC-values can be interpreted as two particular examples out of a large class of values of the form (25), so that for all coalitions, except for those of size $n-2$ and $n-1$ respectively, the corresponding weights are chosen to be zero.*

For the sake of completeness we recall two other interesting values of the form (25). In case $b^n_s = (s+1)^{-1}$ whenever $1 \le s \le n-1$, then the so-called solidarity value arises (cf. [10]). In case $b^n_s = \frac{s}{2^{n-2}} \binom{n-1}{s}$ whenever $1 \le s \le n-1$, then the so-called least square prenucleolus arises (cf. [13]). A reader with interest in an axiomatic characterization of values of the form (25) should consult [6].

Remark 2.3 *For every TU-game (N,v), define the* average worth *of coalitions of size h, $h \in \{1,2,\ldots,n\}$, by*

$$v_h := \binom{n}{h}^{-1} \sum_{S\in\Gamma_h} v(S), \qquad \textit{where} \qquad \Gamma_h := \{S \mid S \subset N, |S| = h\}.$$

Similarly, for every TU-game (N,v) and player $i \in N$, define the average worth *of coalitions of size h, $h \in \{1,2,\ldots,n-1\}$, not containing player i, by*

$$v_h^i := \binom{n-1}{h}^{-1} \sum_{S\in\Gamma_h^i} v(S), \qquad \textit{where} \qquad \Gamma_h^i := \{S \mid S \subset N, |S| = h, i \notin S\}.$$

For convenience' sake, put $v_n^i := 0$ for all $i \in N$. In [1] it is shown that the Shapley value is a weighted sum, over all possible sizes, of differences between the average worth. More precisely, $Sh^i(N,v) = \sum_{h=1}^n h^{-1}(v_h - v_h^i)$, where (N,v) is a game and $i \in N$. Further, it is established that the two components in the latter sum, corresponding to the sizes $n-2$ and $n-1$ respectively, determine almost fully the ENPAC- and ENSC-values. More precisely, $\mathbf{ENPAC^i(N,v)} = \frac{v(N)}{n} + \frac{n-1}{n-2}(v_{n-2} - v_{n-2}^i)$ and $\mathbf{ENSC^i(N,v)} = \frac{v(N)}{n} + (v_{n-1} - v_{n-1}^i)$, where (N,v) is a game and $i \in N$. With these particular descriptions of the three values at hand, the collinearity property between, on the one hand, the Shapley value and on the other, the ENPAC-, ENSC- and CIS-values is investigated for a suitably defined class of TU-games. Here the CIS-value represents the egalitarian value Φ on $\mathbf{G}$ given by $\Phi^i(N,v) = \frac{v(N)}{n} + (n-1)(v_1 - v_1^i)$, or equivalently, $\Phi^i(N,v) = v(\{i\}) + \frac{1}{n}\left[v(N) - \sum_{j\in N} v(\{j\})\right]$ for every game (N,v) and player $i \in N$. By the way, the ENPAC-value has been called, in [1], the Egalitarian Non-Averaged Contribution (ENAC-) value. In the present paper we subjoin the prefix "pairwise (P)" to emphasize the importance of "pairs of players" within the notion of the pairwise-averaged contribution of a player (as given in Definition 2.1).

Let's conclude, besides the three above-mentioned interpretations of the ENPAC-value, with a fourth motivation for the introduction and study of the ENPAC-value. According to the next proposition, the ENPAC-value is the unique solution of a system of linear equations requiring that the so-called excesses of coalitions of size $n-2$ with respect to the (variable) payoff vectors do not differ. As usual, the *excess* of coalition $S \subset N$ with respect to a payoff vector $x = (x_i)_{i\in N}$ is given by $e^v(S,x) = v(S) - \sum_{j\in S} x_j$, where (N,v) is a game. For notation sake, given $x = (x_i)_{i\in N}$, we put $x(S) := \sum_{j\in S} x_j$ for all $S \subset N$, $S \neq \emptyset$, and $x(\emptyset) := 0$. The idea of excess plays a prominent part in the definitions of many solution concepts for TU-games like the prenucleolus and the prekernel (see Section 3).

Proposition 2.4 *Let (N,v) be a TU-game. Consider the following system of linear equations in the unknowns $x = (x_i)_{i\in N}$ and c:*

$$x(N) = v(N) \qquad \textit{and} \qquad e^v(N\backslash\{i,j\},x) = c \qquad \textit{for all } i,j \in N,\ i \neq j. \tag{26}$$

If the system (26) is solvable, then the solution for x is unique and coincides with the ENPAC-value, that is $x = \mathbf{ENPAC(N,v)}$ as well as $c = \frac{n-2}{n(n-1)}\left[v(N) - \sum_{k\in N} PAC^k(N,v)\right]$.

Proof.

Suppose that a vector $x = (x_i)_{i\in N}$ satisfies $x(N) = v(N)$ and $v(N\backslash\{i,j\}) - x(N\backslash\{i,j\}) = c$ for all $i, j \in N$, $i \neq j$, some $c \in R$. In particular, we have

$$x_i + x_j = v(N) - x(N\backslash\{i,j\}) = v(N) - v(N\backslash\{i,j\}) + c \qquad \text{for all } i,j \in N,\ i \neq j.$$

For a fixed $i \in N$, summing up the relevant equalities over j, $j \in N\backslash\{i\}$, yields the equality

$$(n-1)x_i + x(N\backslash\{i\}) = (n-1)v(N) - \sum_{j\in N\backslash\{i\}} v(N\backslash\{i,j\}) + (n-1)c \quad \text{or equivalently,}$$

$$(n-2)x_i = (n-2)v(N) - \sum_{j\in N\backslash\{i\}} v(N\backslash\{i,j\}) + (n-1)c. \qquad \text{By (23), it follows that}$$

$$x_i = PAC^i(N,v) + \frac{(n-1)c}{n-2} \qquad \text{for all } i \in N.$$

Summing up the latter equalities over i, $i \in N$, yields the equality

$$v(N) = \sum_{i\in N} PAC^i(N,v) + \frac{n(n-1)c}{n-2}, \quad \text{thus} \quad c = \frac{n-2}{n(n-1)}\left[v(N) - \sum_{k\in N} PAC^k(N,v)\right].$$

Now we conclude that for all $i \in N$ 0pt

$$x_i = PAC^i(N,v) + \frac{(n-1)c}{n-2} = PAC^i(N,v) + \frac{1}{n}\left[v(N) - \sum_{k\in N} PAC^k(N,v)\right] = \mathbf{ENPAC^i(N,v)}.$$

Hence, the solution x of the system (26) is unique and equals $\mathbf{ENPAC(N,v)}$. This completes the proof of Proposition 2.4. □

The next two examples illustrate the behaviour of the ENPAC-value for three-person TU-games as well as bankruptcy situations.

Example 2.5 *Consider any three-person TU-game (N,v). Put $\alpha := v(N) - v(\{1\}) - v(\{2\}) - v(\{3\})$. It turns out that, by straightforward calculations (cf. formulae (23) and (24)), the pairwise-averaged contribution of a player equals his individual worth plus the surplus α, while the egalitarian non-pairwise-averaged contribution of a player equals his individual worth plus the egalitarian fraction of the surplus α. That is, $PAC^i(N,v) = v(\{i\}) + \alpha$ and $\mathbf{ENPAC^i(N,v)} = v(\{i\}) + \frac{\alpha}{3}$ for all $i \in N$.*

Example 2.6 *We illustrate the determination of the ENPAC-value in the context of a bankruptcy situation. The data are given by the estate E of a bankrupt firm and the claims $d_1, d_2, \ldots, d_n$ of n creditors satisfying $0 < E < \sum_{j=1}^n d_j$. Let $\Delta := \sum_{j=1}^n d_j - E$ denote the surplus of claims in comparison with the estate. With every bankruptcy situation there is associated a TU-game (N,v) so that the player set N consists of the n creditors and the characteristic function $v: 2^N \to R$ is given by $v(S) = \max\left[0, E - \sum_{j\in N\backslash S} d_j\right]$ for all $S \subset N$ (cf. [11]). Henceforth we suppose that the estate is sufficient to meet every pair of claims, i.e., $E \geq d_i + d_j$ for all $i,j \in N$, $i \neq j$. In other words, the so-called bankruptcy TU-game (N,v) satisfies $v(N\backslash\{i,j\}) = E - d_i - d_j$ for all $i,j \in N$, $i \neq j$. Under these circumstances, it turns out that, by straightforward calculations (cf. formulae (23) and (24)), the pairwise-averaged contribution of a creditor equals his claim plus a fraction of the surplus of claims, while the egalitarian non-pairwise-averaged contribution of a creditor equals his claim minus the egalitarian fraction of the surplus of claims. That is, $PAC^i(N,v) = d_i + \frac{\Delta}{n-2}$*

and $\mathbf{ENPAC^i(N,v)} = d_i - \frac{\Delta}{n}$ *for all* $i \in N$*. Thus the ENPAC-value charges the surplus of claims equally to the creditors, on the understanding that every creditor is already paid his claim. Finally we remark that, under the given circumstances, the ENPAC-value agrees with the ENSC-value, although the separable contribution of a creditor is different from his pairwise-averaged contribution (here* $SC^i(N,v) = d_i$ *for all* $i \in N$*).*

3 Coincidence of the ENPAC-value and the prenucleolus

The aim of the section is to present conditions which are sufficient for the ENPAC-value to agree with the prenucleolus concept. The definition of the prenucleolus is as follows. Let (N,v) be a TU-game and denote the corresponding set of allocations by $A(N,v) := \{x = (x_i)_{i\in N} \mid x(N) = v(N)\}$. With every allocation $x \in A(N,v)$ there is associated the complaint vector $\theta(x) \in R^{2^n}$ whose components are the excesses $e^v(S,x)$, $S \in 2^N$, arranged in nonincreasing order. The idea behind the prenucleolus is to select, among all allocations, those which minimize the complaint function $\theta(x)$ in the lexicographic order $\leq_L$ on R^{2^n}. Formally, the *prenucleolus* is defined by $\mathcal{N}^*(N,v) := \{x \in A(N,v) \mid \theta(x) \leq_L \theta(y) \text{ for all } y \in A(N,v)\}$, where (N,v) is a TU-game. It is well-known that the prenucleolus is a singleton (cf. [14]). The unique point in the prenucleolus is denoted by $\eta^*(N,v)$. Notice that the prenucleolus concept relies very much on the essential notion of excess in contradistinction to the ENPAC-value. Nevertheless, both solution concepts appear to coincide for a suitably defined class of TU-games. For that purpose, we concentrate on the levels of excesses over all coalitions at the ENPAC-value itself.

Theorem 3.1 *Let* (N,v) *be a TU-game. Then* $\eta^*(N,v) = \mathbf{ENPAC(N,v)}$ *whenever the following two conditions hold:*

- *The largest excesses at the ENPAC-value are attained at the* $(n-2)$*-person coalitions, i.e.,*

$$e^v(S,z) \leq e^v(N\backslash\{i,j\},z) \quad \text{for all } i,j \in N,\ i \neq j, \text{ and all } S \subset N \text{ such that } \emptyset \neq S \subset N\backslash\{i,j\} \text{ or } S = N\backslash\{i\}, \text{ where } z = \mathbf{ENPAC(N,v)}. \tag{31}$$

- *The excesses of* $(n-2)$*-person coalitions at the ENPAC-value do not differ, i.e., there exists some* $c \in R$ *so that*

$$e^v(N\backslash\{i,j\},z) = c \quad \text{for all } i,j \in N,\ i \neq j, \text{ where } z = \mathbf{ENPAC(N,v)}. \tag{32}$$

Remark 3.2 *Let* (N,v) *be a TU-game and recall that, for any pair* $\{i,j\}$ *of players, the expression* $\Delta_{ij}(N,v) := v(N) - v(N\backslash\{i,j\})$ *represents their pairwise-incremental return to scale for cooperation. In Section 5 we establish a reformulation of condition (32) in terms of differences between these pairwise-incremental returns. More precisely, it turns out that (32) is fully equivalent to the following condition: 0pt*

$$\text{For all } i,j \in N,\ i \neq j, \quad \Delta_{jk}(N,v) - \Delta_{ik}(N,v) \text{ is the same for all } k \in N\backslash\{i,j\}. \tag{33}$$

In words, every player achieves the same gain (or loss) with respect to pairwise-incremental returns by shifting from one partner i to another partner j. Obviously, every three-person TU-game satisfies condition (33), whereas a four-person TU-game (N,v) *satisfies condition (33) if and only if* $v(\{1,2\}) + v(\{3,4\}) = v(\{1,3\}) + v(\{2,4\}) = v(\{1,4\}) + v(\{2,3\})$*. In the framework of the bankruptcy TU-game* (N,v)*, as presented in Example 2.6, condition (33) holds because the resulting expression* $\Delta_{jk}(N,v) - \Delta_{ik}(N,v) = d_j - d_i$ *is indeed independent of player* $k \in N\backslash\{i,j\}$*.*

Proof of Theorem 3.1.
Put $x := \eta^*(N,v)$ and $z :=$ **ENPAC(N, v)**. Suppose that both conditions (31) and (32) hold. We prove $x = z$. Concerning the lexicographic comparison between the two complaint vectors $\theta(x)$ and $\theta(z)$, we may ignore the (zero) excesses of N and $\emptyset$ since $e^v(S,x) = e^v(S,z) = 0$ whenever $S = N$ or $S = \emptyset$. Notice that $x = \eta^*(N,v)$ implies $\theta(x) \leq_L \theta(z)$, so we have $\theta_1(x) \leq \theta_1(z)$. By assumption of the two conditions (31) and (32), we obtain $\theta_1(z) = c$ (i.e., the largest excess over all coalitions at z equals c). From $\theta_1(x) \leq \theta_1(z) = c$, we deduce that

$$v(N\backslash\{i,j\}) - x(N\backslash\{i,j\}) = e^v(N\backslash\{i,j\},x) \leq \theta_1(x) \leq \theta_1(z) = c \quad \text{and consequently,}$$

$$x_i + x_j = v(N) - x(N\backslash\{i,j\}) \leq v(N) - v(N\backslash\{i,j\}) + c \qquad \text{for all } i,j \in N,\ i \neq j.$$

For a fixed $i \in N$, summing up the relevant inequalities over j, $j \in N\backslash\{i\}$, yields the inequality

$$(n-1)x_i + x(N\backslash\{i\}) \leq (n-1)v(N) - \sum_{j \in N\backslash\{i\}} v(N\backslash\{i,j\}) + (n-1)c \quad \text{or equivalently,}$$

$$(n-2)x_i \leq (n-2)v(N) - \sum_{j \in N\backslash\{i\}} v(N\backslash\{i,j\}) + (n-1)c. \qquad \text{By (23), it follows that}$$

$$x_i \leq PAC^i(N,v) + \frac{(n-1)c}{n-2} \qquad \text{for all } i \in N.$$

By assumption of (32), the ENPAC-value solves the system (26) listed in Proposition 2.4 and hence, the result listed in Proposition 2.4 states that $c = \frac{n-2}{n(n-1)}\left[v(N) - \sum_{k \in N} PAC^k(N,v)\right]$. Together with $x_i \leq PAC^i(N,v) + \frac{(n-1)c}{n-2}$ for all $i \in N$, this implies that $x_i \leq$ $\mathbf{ENPAC^i(N,v)}$ $= z_i$ for all $i \in N$. From both $x_i \leq z_i$ for all $i \in N$ and $x(N) = v(N) = z(N)$, we conclude that $x_i = z_i$ for all $i \in N$. Therefore, $x = z$ as was to be shown. This completes the proof of Theorem 3.1. □

In the remainder of the section our main goal is to examine the set of allocations for which the largest excesses are attained at the $(n-2)$-person coalitions. The next auxiliary lemma provides a suitable description of this set in terms of individually and coalitionally incremental returns to scale for cooperation in subgames. Given a TU-game (N,v) and a coalition $T \subset N$, we write (T,v) for the *subgame* obtained by restricting the characteristic function v to subsets of T only (i.e., to 2^T). In the context of subgames of a TU-game (N,v), we recall that, by definition, $\Delta_S(T,v) = v(T) - v(T\backslash S)$ whenever $S \subset T \subset N$.

Lemma 3.3 *Let (N,v) be a TU-game and $x = (x_i)_{i \in N}$. The following two statements are equivalent. Opt*

(i) $\quad e^v(S,x) \leq e^v(N\backslash\{i,j\},x)$ *for all $i,j \in N$, $i \neq j$, and all $S \subset N$ such that $\emptyset \neq S \subset N\backslash\{i,j\}$ or $S = N\backslash\{i\}$.* (34)

(ii) $\quad x_i \geq \max_{j \in N\backslash\{i\}} \Delta_i(N\backslash\{j\},v)$ *for all $i \in N$;* (35)

$\quad x(S) \leq \min_{i,j \in N\backslash S,\, i\neq j} \Delta_S(N\backslash\{i,j\},v)$ *for all $S \subset N$ with $1 \leq |S| \leq n-3$.* (36)

The proof of Lemma 3.3 is straightforward and is left to the reader. For every TU-game (N,v), define the set $U(N,v) := \{x \in A(N,v) \mid x \text{ subject to (35) and (36)}\}$. In words, according to an allocation out of the set $U(N,v)$, a player receives at least his largest separable

contribution (with respect to the formation of the grand coalition in any suitably defined $(n-1)$-person subgame) and moreover, the payoff to a coalition containing at most $n-3$ players is not greater than the smallest incremental return to scale for cooperation by members of the coalition (with respect to the formation of the grand coalition in any suitably defined $(n-2)$-person subgame). Notice that the set $U(N,v)$ may be empty.
In the context of excess, we say the $(n-2)$-person coalitions are *effective* at the allocation x if condition (34) holds. Lemma 3.3 expresses that an allocation x satisfies the effectiveness condition (34) if and only if $x \in U(N,v)$. In view of Theorem 3.1, the coincidence $\eta^*(N,v) = \mathbf{ENPAC(N,v)}$ is valid whenever $\mathbf{ENPAC(N,v)} \in U(N,v)$ and (32) holds.
Generally speaking it is known that the prenucleolus is contained in the so-called prekernel (cf. [14]). Formally, the *prekernel* of a TU-game (N,v) is defined by $\mathcal{K}^*(N,v) := \{x \in A(N,v) \mid s^v_{ij}(x) = s^v_{ji}(x) \text{ for all } i,j \in N, i \neq j\}$, where the maximum surplus of player i over another player j at an allocation x is given by $s^v_{ij}(x) := \max[e^v(S,x) \mid S \subset N, i \in S, j \notin S]$.
The last part of the section concerns the relationship between the ENPAC-value and the intersection of the set $U(N,v)$ with the prekernel. The main theorem states that the intersection of the set $U(N,v)$ with the prekernel is either empty or a singleton consisting of the ENPAC-value. Moreover, the nonemptiness of the intersection of the set $U(N,v)$ with the prekernel guarantees the coincidence of the ENPAC-value and the prenucleolus.

Theorem 3.4 *Let (N,v) be a TU-game so that (32) (or equivalently, (33)) holds.*

(i) $U(N,v) \cap \mathcal{K}^*(N,v) \subset \{\mathbf{ENPAC(N,v)}\}$.

(ii) $U(N,v) \cap \mathcal{K}^*(N,v) = \{\mathbf{ENPAC(N,v)}\}$ *iff* $\mathbf{ENPAC(N,v)} \in U(N,v)$.

(iii) $\eta^*(N,v) \in U(N,v)$ *iff* $\mathbf{ENPAC(N,v)} \in U(N,v)$.

(iv) If $\mathbf{ENPAC(N,v)} \in U(N,v)$, *then* $\eta^*(N,v) = \mathbf{ENPAC(N,v)}$.

Proof.
For the technical proof of the statements (i) and (ii), we refer to Section 5. The statement (iv) as well as the "if" part of statement (iii) are due to Lemma 3.3 and Theorem 3.1. To prove the "only if" part of statement (iii), suppose $\eta^*(N,v) \in U(N,v)$. We always have $\eta^*(N,v) \in \mathcal{K}^*(N,v)$. From $\eta^*(N,v) \in U(N,v) \cap \mathcal{K}^*(N,v)$ and the statement (i), we conclude that $\mathbf{ENPAC(N,v)} = \eta^*(N,v) \in U(N,v)$. This completes the proof of the statements (iii) and (iv). □

Example 3.5 *In the framework of the bankruptcy TU-game (N,v), as presented in Example 2.6, recall that condition (33) holds and the ENPAC-value is determined by $\mathbf{ENPAC^i(N,v)} = d_i - \frac{\Delta}{n}$ for all $i \in N$. The ENPAC-value, however, does not satisfy condition (35) since*

$$\max_{j \in N\setminus\{i\}} \Delta_i(N\setminus\{j\},v) = \max_{j \in N\setminus\{i\}} [(E-d_j)-(E-d_i-d_j)] = d_i > \mathbf{ENPAC^i(N,v)}$$

for all $i \in N$. By definition of the set $U(N,v)$, we arrive at $z := \mathbf{ENPAC(N,v)} \notin U(N,v)$. Finally we remark that the effectiveness condition (34) for the ENPAC-value turns out to fail because of $e^v(N\setminus\{i\},z) = -\frac{\Delta}{n} > -\frac{2\Delta}{n} = e^v(N\setminus\{i,j\},z)$ for all $i,j \in N, i \neq j$.

4 Condition (31) reconsidered

¿From the computational viewpoint, condition (31) has the serious drawback that the worth of *various* $(n-2)$-person coalitions appear within *each* inequality to be verified. The purpose of the section is to illustrate that the verification of the essential condition (31) can be carried out ever so much faster by using the notion of the so-called *gap function* $g^v : 2^N \rightarrow R$ corresponding to the characteristic function $v : 2^N \rightarrow R$. With every coalition $S \subset N$ there is associated the real number $g^v(S) := \sum_{j \in S} PAC^j(N,v) - v(S)$ called the *gap* of S in the TU-game (N,v) and it represents the surplus of the pairwise-averaged contributions of members of S in comparison with the worth of S. According to the next theorem, condition (31) is fully equivalent to the requirement that, for every coalition, the corresponding gap is not less than a fraction of the gap of the grand coalition, where the fraction is linearly dependent on the size of the coalition. Evidently, an advantage of the latter requirement is that the relevant inequalities involve (the gap of) one coalition, besides (the gap of) the grand coalition, at a time. As a matter of fact, the latter requirement elucidates that not the TU-game (N,v) itself is utmost important, but its gap function g^v is the only important tool needed for the verification of condition (31).

Theorem 4.1 *Let (N,v) be a TU-game so that (32) holds and define its corresponding gap function $g^v : 2^N \rightarrow R$ to be $g^v(S) := \sum_{j \in S} PAC^j(N,v) - v(S)$ for all $S \subset N$. Then condition (31) is equivalent to the following condition:*
Opt

$$g^v(S) \geq \left[\frac{n-2}{n(n-1)} + \frac{|S|}{n}\right] g^v(N) \qquad \text{for all } S \subset N \text{ with } 1 \leq |S| \leq n-3 \text{ or } |S| = n-1. \tag{41}$$

Proof.
Put $z :=$ **ENPAC(N, v)**. For every coalition S, the excess of S at z satisfies
Opt

$$e^v(S,z) = v(S) - \sum_{j \in S} PAC^j(N,v) - \frac{|S|}{n}\left[v(N) - \sum_{k \in N} PAC^k(N,v)\right] = -g^v(S) + \frac{|S|}{n} g^v(N).$$

Moreover, from (32) and Proposition 2.4, we derive that 0pt

$$e^v(N \backslash \{i,j\}, z) = \frac{n-2}{n(n-1)}\left[v(N) - \sum_{k \in N} PAC^k(N,v)\right] = -\frac{n-2}{n(n-1)} g^v(N)$$

for all $i, j \in N$, $i \neq j$. Now it follows that condition (31) is fully equivalent to 0pt

$$-g^v(S) + \frac{|S|}{n} g^v(N) \leq -\frac{n-2}{n(n-1)} g^v(N) \quad \text{or equivalently,} \quad g^v(S) \geq \left[\frac{n-2}{n(n-1)} + \frac{|S|}{n}\right] g^v(N)$$

for all $S \subset N$ with $1 \leq |S| \leq n-3$ or $|S| = n-1$. Concerning $(n-2)$-person coalitions, the inequalities in (31) are trivially satisfied as equalities and consequently, the same holds in (41). That is, by asumption of (32) and by definition of the gap function g^v, it holds that $g^v(S) = \frac{n-2}{n-1} g^v(N)$ for all $S \subset N$ with $|S| = n-2$. This completes the proof of Theorem 4.1. □

Example 4.2 *For every three-person TU-game (N,v), condition (41) reduces to $g^v(\{i,j\}) \geq \frac{5}{6} g^v(N)$ for all $i, j \in N$, $i \neq j$. For every four-person TU-game (N,v), condition (41)* *reduces to* $g^v(\{i\}) \geq \frac{5}{12}g^v(N)$ *and* $g^v(N\backslash\{i\}) \geq \frac{11}{12}g^v(N)$ *for all* $i \in N$. *For instance, consider the numerical four-person TU-game* (N,v) *given by* $v(\{i\}) = 0$ *for all* $i \in N$, $v(\{1,2\}) = 2$, $v(\{1,3\}) = v(\{2,3\}) = 3$, $v(\{1,4\}) = v(\{2,4\}) = 4$, $v(\{3,4\}) = v(\{1,2,3\}) = 5$, $v(\{1,2,4\}) = v(\{1,3,4\}) = v(\{2,3,4\}) = 6$, *and* $v(N) = 11$. *By (23),* $PAC^i(N,v) = 5,5,6,7$ *for* $i = 1,2,3,4$ *respectively and hence, by (24),* $\mathbf{ENPAC^i(N,v)} = 2,2,3,4$ *for* $i = 1,2,3,4$ *respectively. Notice that the TU-game* (N,v) *satisfies (33) because of* $v(\{1,2\}) + v(\{3,4\}) = v(\{1,3\}) + v(\{2,4\}) = v(\{1,4\}) + v(\{2,3\})$. *In order to check (41), we calculate the gap function* g^v *so that* $g^v(N) = 12$, $g^v(\{i\}) = 5,5,6,7$ *and* $g^v(N\backslash\{i\}) = 12,12,11,11$ *for* $i = 1,2,3,4$ *respectively. Obviously,* $g^v(\{i\}) \geq \frac{5}{12}g^v(N)$ *and* $g^v(N\backslash\{i\}) \geq \frac{11}{12}g^v(N)$ *for all* $i \in N$. *We conclude that both conditions (33) and (41) (or equivalently, (31) and (32)) hold and therefore, by Theorem 3.1, the prenucleolus coincides with the ENPAC-value. It is left to the reader to check that, by straightforward calculations (cf. formulae (35) and (36)), the set* $U(N,v)$ *for this numerical four-person TU-game* (N,v) *is a singleton consisting of the allocation* $(2,2,3,4)$. *Obviously,* $\mathbf{ENPAC(N,v)} = (2,2,3,4) \in U(N,v)$ *which causes the coincidence of the prenucleolus and the ENPAC-value.*

5 Technical proof of Theorem 3.4

In order to prove Theorem 3.4, we start with two preliminary lemmata which are interesting on their own.

Lemma 5.1 *Let* (N,v) *be a TU-game. The following three statements are equivalent. Opt*

(i) $e^v(N\backslash\{i,j\}, \mathbf{ENPAC(N,v)}) = c$ *for all* $i,j \in N$, $i \neq j$, *some* $c \in R$.

(ii) $PAC^i(N,v) + PAC^j(N,v) - \Delta_{ij}(N,v) = \bar{c}$ *for all* $i,j \in N$, $i \neq j$, *some* $\bar{c} \in R$.

(iii)
$$\Delta_{jk}(N,v) - \Delta_{ik}(N,v) = \Delta_{j\ell}(N,v) - \Delta_{i\ell}(N,v) := \delta_{ij}(N,v) \quad \text{for all } i,j \in N \text{ and } k,\ell \in N\backslash\{i,j\}. \tag{51}$$

Lemma 5.2 *Let* (N,v) *be a TU-game so that (51) (or equivalently, (32)) holds. Then*

$$\mathbf{ENPAC^i(N,v)} = \frac{1}{n}\left[v(N) - \sum_{j\in N}\delta_{ij}(N,v)\right] \qquad \text{for all } i \in N.$$

Given two players $i,j \in N$, condition (51) expresses that every other player achieves the same gain (or loss), the size of $\delta_{ij}(N,v)$, with respect to pairwise-incremental returns by shifting from partner i to partner j. Moreover, Lemma 5.2 states that, under the above-mentioned circumstances, the deviation of the ENPAC-value of a player i in comparison with the most simplest egalitarian division the size of $\frac{v(N)}{n}$ is computable as the averaged sum over all gains $\delta_{ij}(N,v)$, $j \in N$ (where $\delta_{ii}(N,v) = 0$).

Proof of Lemma 5.1.

Put $z := \mathbf{ENPAC(N,v)}$. Generally speaking, it holds that for all $i,j \in N$, $i \neq j$,

$$\begin{aligned}
e^v(N\backslash\{i,j\}, z) &= v(N\backslash\{i,j\}) - z(N\backslash\{i,j\}) \\
&= v(N\backslash\{i,j\}) - v(N) + z_i + z_j = z_i + z_j - \Delta_{ij}(N,v) \\
&= PAC^i(N,v) + PAC^j(N,v) - \Delta_{ij}(N,v) + \frac{2}{n}\left[v(N) - \sum_{k\in N} PAC^k(N,v)\right].
\end{aligned}$$

This proves the equivalence $(i) \iff (ii)$. Next we establish the equivalence $(ii) \iff (iii)$. Suppose that the statement (ii) holds. Let $i,j \in N$ and $k,\ell \in N\backslash\{i,j\}$. By assumption of (ii), we have

$$PAC^i(N,v) + PAC^k(N,v) - \Delta_{ik}(N,v) = PAC^j(N,v) + PAC^k(N,v) - \Delta_{jk}(N,v),$$

$$PAC^i(N,v) + PAC^\ell(N,v) - \Delta_{i\ell}(N,v) = PAC^j(N,v) + PAC^\ell(N,v) - \Delta_{j\ell}(N,v).$$ Thus,

$$\Delta_{jk}(N,v) - \Delta_{ik}(N,v) = PAC^j(N,v) - PAC^i(N,v) = \Delta_{j\ell}(N,v) - \Delta_{i\ell}(N,v).$$

So, (ii) implies (iii). To prove the converse implication, suppose that the statement (iii) holds. We derive from (23) that for all $i,j \in N$, $i \neq j$,

$$\begin{aligned} PAC^j(N,v) - PAC^i(N,v) &= \frac{1}{n-2}\Big[\sum_{k\in N\backslash\{i\}} v(N\backslash\{i,k\}) - \sum_{k\in N\backslash\{j\}} v(N\backslash\{j,k\})\Big] \\ &= \frac{1}{n-2}\sum_{k\in N\backslash\{i,j\}} [v(N\backslash\{i,k\}) - v(N\backslash\{j,k\})] \\ &= \frac{1}{n-2}\sum_{k\in N\backslash\{i,j\}} [\Delta_{jk}(N,v) - \Delta_{ik}(N,v)] = \delta_{ij}(N,v) \end{aligned}$$

where the very last equality follows by assumption of (iii). So far, we conclude that condition (51) yields

$$PAC^j(N,v) - PAC^i(N,v) = \delta_{ij}(N,v) \qquad \text{for all } i,j \in N. \tag{52}$$

¿From (52) we derive that for all $i,j,k,\ell \in N$ with $i \neq j$, $k \neq \ell$, $j \neq k$,

$$PAC^k(N,v) = PAC^i(N,v) + \delta_{ik}(N,v) = PAC^i(N,v) + \Delta_{jk}(N,v) - \Delta_{ij}(N,v),$$

$$PAC^\ell(N,v) = PAC^j(N,v) + \delta_{j\ell}(N,v) = PAC^j(N,v) + \Delta_{k\ell}(N,v) - \Delta_{jk}(N,v),$$

and next, summing up both equalities yields the equality

$$PAC^k(N,v) + PAC^\ell(N,v) - \Delta_{k\ell}(N,v) = PAC^i(N,v) + PAC^j(N,v) - \Delta_{ij}(N,v).$$

So, (iii) implies (ii). This completes the proof of the equivalence involving the three statements. □

Proof of Lemma 5.2.

Put $z := \mathbf{ENPAC(N,v)}$. As already shown in the proof of Lemma 5.1, the assumption of condition (51) yields (52), that is $z_j - z_i = PAC^j(N,v) - PAC^i(N,v) = \delta_{ij}(N,v)$ for all $i,j \in N$, $i \neq j$. For a fixed $i \in N$, summing up the relevant equalities over j, $j \in N\backslash\{i\}$, yields the equality 0pt

$$z(N\backslash\{i\}) - (n-1)z_i = \sum_{j\in N\backslash\{i\}} \delta_{ij}(N,v) \quad \text{or equivalently,} \quad nz_i = v(N) - \sum_{j\in N\backslash\{i\}} \delta_{ij}(N,v).$$

Hence, $\mathbf{ENPAC^i(N,v)} = z_i = \frac{1}{n}\Big[v(N) - \sum_{j\in N}\delta_{ij}(N,v)\Big]$ for all $i \in N$. This completes the proof of Lemma 5.2. □

Proof of Theorem 3.4.

Let (N,v) be a TU-game so that (51) (or equivalently, (32)) holds. As already indicated in Section 3, it remains to prove the statements (i) and (ii). The proof proceeds in three stages.

Stage one. Suppose $x \in U(N,v)$ and let $i,j \in N$, $i \neq j$. Our goal is to compare the maximum surplus $s_{ij}^v(x)$ with $s_{ji}^v(x)$. By Lemma 3.3, the effectiveness condition (34) holds for $x \in U(N,v)$ and thus, $s_{ij}^v(x) = e^v(N\backslash\{j,k\},x)$, where $k \in N\backslash\{i,j\}$ is chosen so that $e^v(N\backslash\{j,k\},x) \geq e^v(N\backslash\{j,\ell\},x)$ for all $\ell \in N\backslash\{i,j\}$. In order to determine $s_{ji}^v(x)$ in a similar way, we notice that for all $\ell \in N\backslash\{i,j\}$

$$\begin{aligned} e^v(N\backslash\{i,\ell\},x) &= v(N\backslash\{i,\ell\}) - x(N\backslash\{i,\ell\}) \\ &= v(N\backslash\{i,\ell\}) - x(N\backslash\{j,\ell\}) + x_i - x_j \\ &= v(N\backslash\{j,\ell\}) + \delta_{ij}(N,v) - x(N\backslash\{j,\ell\}) + x_i - x_j \\ &= e^v(N\backslash\{j,\ell\},x) + \delta_{ij}(N,v) + x_i - x_j \end{aligned}$$

where the third equality follows by assumption of (51). From this and the effectiveness condition (34) for $x \in U(N,v)$, we deduce that $s_{ji}^v(x) = e^v(N\backslash\{j,k\},x) + \delta_{ij}(N,v) + x_i - x_j$. In summary, we conclude that every $x \in U(N,v)$ satisfies $s_{ji}^v(x) = s_{ij}^v(x) + \delta_{ij}(N,v) + x_i - x_j$ for all $i,j \in N$, $i \neq j$.

Stage two. Now we are in a position to prove the statement (i) of Theorem 3.4. Suppose $U(N,v) \cap \mathcal{K}^*(N,v) \neq \emptyset$, say $x \in U(N,v) \cap \mathcal{K}^*(N,v)$. Let $i,j \in N$, $i \neq j$. By stage one, $x \in U(N,v)$ implies $s_{ji}^v(x) = s_{ij}^v(x) + \delta_{ij}(N,v) + x_i - x_j$, whereas $s_{ij}^v(x) = s_{ji}^v(x)$ because of $x \in \mathcal{K}^*(N,v)$. From this, it follows that $x_j - x_i = \delta_{ij}(N,v)$ for all $i,j \in N$, $i \neq j$. For a fixed $i \in N$, summing up the latter equalities over j, $j \in N\backslash\{i\}$, yields the equality $x(N\backslash\{i\}) - (n-1)x_i = \sum_{j \in N\backslash\{i\}} \delta_{ij}(N,v)$ or equivalently, $nx_i = v(N) - \sum_{j \in N\backslash\{i\}} \delta_{ij}(N,v)$. We arrive at $x_i = \frac{1}{n}\left[v(N) - \sum_{j \in N} \delta_{ij}(N,v)\right] = \mathbf{ENPAC^i(N,v)}$ for all $i \in N$, where the last equality follows from Lemma 5.2. We conclude that $x = \mathbf{ENPAC(N,v)}$ whenever $x \in U(N,v) \cap \mathcal{K}^*(N,v)$. This proves the statement (i).

Stage three. Now we establish the statement (ii) of Theorem 3.4. The "only if" part is trivial. In order to prove the "if" part, suppose $z := \mathbf{ENPAC(N,v)} \in U(N,v)$. In view of statement (i), it suffices to show that $z \in \mathcal{K}^*(N,v)$. Let $i,j \in N$, $i \neq j$. By stage one, $z \in U(N,v)$ implies $s_{ji}^v(z) = s_{ij}^v(z) + \delta_{ij}(N,v) + z_i - z_j$. Further, as already shown in the proof of Lemma 5.1, the assumption of condition (51) yields (52), that is $z_j - z_i = PAC^j(N,v) - PAC^i(N,v) = \delta_{ij}(N,v)$. Now we conclude that $s_{ji}^v(z) = s_{ij}^v(z) + \delta_{ij}(N,v) + z_i - z_j = s_{ij}^v(z)$ for all $i,j \in N$, $i \neq j$. So, $z \in \mathcal{K}^*(N,v)$ as was to be shown. This proves the statement (ii) and as such, the proof of Theorem 3.4 is completed. □

6 Concluding remarks

Remark 6.1 *As usual, the solution concept called* core *of a TU-game (N,v) is defined by $CORE(N,v) := \{x = (x_i)_{i \in N} \mid x(N) = v(N)$ and $x(S) \geq v(S)$ for all $S \in 2^N\}$. In other words, an allocation x belongs to the core if and only if all excesses $e^v(S,x)$, $S \in 2^N$, are less than or equal to zero. Clearly, the separable contributions $SC^i(N,v)$, $i \in N$, of players provide an* upper bound *for the core in the sense that $x_i \leq SC^i(N,v)$ for all $i \in N$ and all $x \in CORE(N,v)$ (simply because $x_i = v(N) - x(N\backslash\{i\}) \leq v(N) - v(N\backslash\{i\}) = SC^i(N,v)$). Now we claim that the pairwise-averaged contributions $PAC^i(N,v)$, $i \in N$, of players provide another upper bound for the core in the sense that $x_i \leq PAC^i(N,v)$ for all $i \in N$ and all $x \in CORE(N,v)$ (due to the fact that $x_i + x_j = v(N) - x(N\backslash\{i,j\}) \leq v(N) - v(N\backslash\{i,j\})$ for all $i,j \in N$, $i \neq j$, and so, the middle part of the proof of Theorem 3.1 is applicable once again by ignoring the role of the constant c). In the framework of Theorem 3.1, notice that the largest excess at the ENPAC-value the size of $c = \frac{n-2}{n(n-1)}\left[v(N) - \sum_{k \in N} PAC^k(N,v)\right]$ is* *less than or equal to zero as soon as* $v(N) - \sum_{k \in N} PAC^k(N,v) \leq 0$. *We conclude that the ENPAC-value belongs to the core for TU-games whenever (31) and (32) hold together with the weak condition* $\sum_{k \in N} PAC^k(N,v) \geq v(N)$.

Remark 6.2 *The construction of the ENPAC-value on the basis of pairwise-averaged contributions of players has been carried out so that the value satisfies the efficiency principle. It is, however, not guaranteed that the ENPAC-value meets the individual rationality principle (i.e.,* $\mathbf{ENPAC^i(N,v)} \geq v(\{i\})$ *for all* $i \in N$*) since the pairwise-averaged contribution of some player may be lowered too much by the egalitarian division of the surplus of the overall profits. Our aim is to introduce a slightly adapted version of the ENPAC-value which does meet the individual rationality principle.*
For that purpose we replace the notion of the surplus of the overall profits by some (yet unknown) amount, on the understanding that the pairwise-averaged contribution of every player is lowered by the variable amount as long as the individual rationality principle is not violated. The variable amount itself is fully determined by the efficiency condition for the solution. More precisely, the adapted ENPAC-value of a TU-game (N,v) *is given by the allocation* $z = (z_i)_{i \in N}$, $z_i := \max\left[v(\{i\}),\ PAC^i(N,v) - \lambda^v\right]$ *for all* $i \in N$, *where* $\lambda^v \in R$ *is (uniquely) determined by* $z(N) = v(N)$. *To show that this solution is well-defined, let* (N,v) *be a TU-game so that* $v(N) \geq \sum_{j \in N} v(\{j\})$ *and consider the corresponding function* $f : R \rightarrow R$ *given by* $f(\lambda) := \sum_{j \in N} \max\left[v(\{j\}),\ PAC^j(N,v) - \lambda\right]$ *for all* $\lambda \in R$. *Put* $\bar{\lambda} := \max_{j \in N}\left[PAC^j(N,v) - v(\{j\})\right]$. *Obviously, the function f is continuous and strictly decreasing on* $(-\infty,\ \bar{\lambda}]$ *such that* $\lim_{\lambda \to -\infty} f(\lambda) = +\infty$ *and* $f(\bar{\lambda}) = \sum_{j \in N} v(\{j\})$. *In view of the latter properties of f and the assumption* $v(N) \geq \sum_{j \in N} v(\{j\})$, *we conclude that there exists a unique* $\lambda^v \in (-\infty,\ \bar{\lambda}]$ *satisfying* $f(\lambda^v) = v(N)$. *Hence, the adapted ENPAC-value is well-defined. A further study of the adapted ENPAC-value is beyond the scope of the present paper since we put the emphasis on the original ENPAC-value. Evidently, both values coincide if and only if the original ENPAC-value is individually rational.*

Remark 6.3 *In Section 3 we examined the set of allocations for which the largest excesses are attained at the* $(n-2)$*-person coalitions (cf. Lemma 3.3 and particularly, conditions (35)-(36)) and the relationship of this set* $U(N,v)$ *with the ENPAC-value (cf. Theorem 3.4). Let us compare our results concerning the ENPAC-value with similar results concerning the ENSC-value as presented in [4]. In accordance with condition (34), we say the largest excesses at* $x = (x_i)_{i \in N}$ *are attained at the* $(n-1)$*-person coalitions whenever* $e^v(S,x) \leq e^v(N\backslash\{i\},x)$ *for all* $i \in N$ *and all* $\emptyset \neq S \subset N\backslash\{i\}$, *where* (N,v) *is a TU-game. It turns out that the latter condition is fully equivalent to the following condition (cf. Proposition 4.1 in [4]):*

$$x(S) \leq \min_{j \in N\backslash S} \Delta_S(N\backslash\{j\}, v) \qquad \textit{for all } S \subset N \textit{ with } 1 \leq |S| \leq n-2. \tag{61}$$

In words, the payoff to a coalition containing at most $n-2$ *players is not greater than the smallest incremental return to scale for cooperation by members of the coalition (with respect to the formation of the grand coalition in any suitably defined* $(n-1)$*-person subgame). Obviously, condition (61) with reference to* $(n-1)$*-person subgames is of the same form as condition (36) with reference to* $(n-2)$*-person subgames. In similarity to our Theorem 3.1, we recall that Theorem 3.1 in [4] states that the prenucleolus coincides with the ENSC-value whenever the ENSC-value satisfies (61). In similarity to our Theorem 4.1 and condition (41), we also recall that Proposition 3.2 in [4] provides a reformulation of (61) applied to the ENSC-value so that* $h^v(S) \geq \frac{(|S|+1)}{n} h^v(N)$ *for all* $S \subset N$ *with* $1 \leq |S| \leq n-2$, *where* $h^v(S) := \sum_{j \in S} SC^j(N,v) - v(S)$ *for all* $S \subset N$. *Obviously, the latter condition concerning the gap function with reference to the surplus of the separable contributions of players is of*

the same form as condition (41) concerning the gap function with reference to the surplus of the pairwise-averaged contributions of players.
For every TU-game (N,v), define the set $\bar{U}(N,v) := \{x \in A(N,v) \mid x$ subject to (61)$\}$. By Theorems 4.6 and 4.7 in [4], a similar version of our Theorem 3.4 holds in the framework of the ENSC-value in the sense that the set $U(N,v)$ and the ENPAC-value, as listed in Theorem 3.4, should be replaced by the set $\bar{U}(N,v)$ and the ENSC-value respectively. In other words, the intersection of the set $\bar{U}(N,v)$ with the prekernel is either empty or a singleton consisting of the ENSC-value. Moreover, the nonemptiness of the intersection of the set $\bar{U}(N,v)$ with the prekernel guarantees the coincidence of the ENSC-value and the prenucleolus.
It is still an open problem to develop a similar theory about the (yet unknown) value for which the largest excesses are attained at the k-person coalitions and the excesses of those k-person coalitions do not differ (where k, $1 \leq k \leq n-3$, is fixed). The case $k = n-2$ was treated throughout the present paper and yields the ENPAC-value. The case $k = n-1$ was treated in [4] and yields the ENSC-value. The remaining cases k, $1 \leq k \leq n-3$, are yet unexplored as far as the authors are aware of.

Remark 6.4 *Let us consider once again the values Φ on $\mathbf{G}$ of the form (25) as presented in Remark 2.2. As already mentioned, the ENSC-value arises in case the underlying collection of constants $\{b_s^k : k \in \mathcal{N}\backslash\{1\}, s = 1,2,\ldots,k\}$ is given by $b_{n-1}^n = n-1$ and $b_s^n = 0$ whenever $1 \leq s \leq n-2$, whereas the ENPAC-value arises in case $b_{n-2}^n = n-1$ and $b_s^n = 0$ whenever $1 \leq s \leq n-1$, $s \neq n-2$. In view of this similarity, it is natural to ask which value arises in case $b_k^n = n-1$ and $b_s^n = 0$ whenever $1 \leq s \leq n-1$, $s \neq k$, (where k, $1 \leq k \leq n-3$, is fixed). The answer to this question is yet unknown.*
Let us also consider once again the values Φ on $\mathbf{G}$ of the form $\Phi^i(N,v) = \frac{v(N)}{n} + a_h^n(v_h - v_h^i)$, where (N,v) is a game, $i \in N$, $h \in \{1,2,\ldots,n\}$. As already indicated in Remark 2.3, the ENSC-value arises in case $h = n-1$ and $a_h^n = 1$, while the ENPAC-value arises in case $h = n-2$ and $a_h^n = \frac{n-1}{n-2}$. It is a natural question which value arises in case h, $1 \leq h \leq n-3$, is fixed, with an appropriately defined number a_h^n. The answer to this question is yet unknown. Moreover, it is of interest to figure out whether these open problems are related or not to the open problem stated at the end of the previous remark.

References

[1] Dragan, I., Driessen, T.S.H., and Y. Funaki, (1996), *Collinearity between the Shapley value and the egalitarian division rules for cooperative games.* OR Spektrum **18**, 97-105.

[2] Driessen, T.S.H., (1985), *Properties of* 1*-convex n-person games.* OR Spektrum **7**, 19–26.

[3] Driessen, T.S.H., (1988), *Cooperative Games, Solutions, and Applications.* Kluwer Academic Publishers, Dordrecht, The Netherlands.

[4] Driessen, T.S.H., and Y. Funaki, (1991), *Coincidence of and collinearity between game theoretic solutions.* OR Spektrum **13**, 15–30.

[5] Driessen, T.S.H., and Y. Funaki, (1993), *Reduced game properties of egalitarian division rules for cooperative games.* Memorandum No. 1136, Department of Applied Mathematics, University of Twente, Enschede, The Netherlands.

[6] Driessen, T.S.H., Radzik, T., and R. Wanink, (1996), *Potential and consistency: a uniform approach to values for TU-games.* Memorandum No. 1323, Department of Applied Mathematics, University of Twente, Enschede, The Netherlands.

[7] Funaki, Y., (1986), *Upper and lower bounds of the kernel and nucleolus.* International Journal of Game Theory **15**, 121–129.

[8] Legros, P., (1986), *Allocating joint costs by means of the nucleolus.* International Journal of Game Theory **15**, 109–119.

[9] Moulin, H., (1985), *The separability axiom and equal-sharing methods.* Journal of Economic Theory **36**, 120–148.

[10] Nowak, A.S., and T. Radzik, (1994), *A solidarity value for n-person transferable utility games.* International Journal of Game Theory **23**, 43–48.

[11] O'Neill, B., (1982), *A problem of rights arbitration from the Talmud.* Mathematical Social Sciences **2**, 345–371.

[12] Roth, A.E. (editor), (1988), *The Shapley value: Essays in honor of Lloyd S. Shapley.* Cambridge University Press, Cambridge, U.S.A.

[13] Ruiz, L.M., Valenciano, F., and J.M. Zarzuelo, (1996), *The least square prenucleolus and the least square nucleolus: two values for TU-games based on the excess vector.* International Journal of Game Theory **25**, 113–134.

[14] Schmeidler, D., (1969), *The nucleolus of a characteristic function game.* SIAM Journal of Applied Mathematics **17**, 1163–1170.

[15] Shapley, L.S., (1953), *A value for n-person games.* Annals of Mathematics Study **28**, 307–317 (Princeton University Press). Also in [12], 31–40.

[16] Weber, R.J., (1988), *Probabilistic values for games.* In: [12], 101–119.

[17] Young, H.P., Okada, N., and T. Hashimoto, (1982), *Cost allocation in water resources development.* Water Resources Research **18**, 463–475.

Dr. Theo S.H. Driessen
Department of Applied Mathematics
University of Twente, P.O. Box 217
7500 AE Enschede, The Netherlands

Dr. Yukihiko Funaki
Faculty of Economics
Toyo University, Hakusan
Bunkyo-ku, Tokyo 112, Japan.

CONSISTENCY PROPERTIES OF THE NONTRANSFERABLE COOPERATIVE GAME SOLUTIONS

Elena Yanovskaya [1]

Abstract: *We consider solutions of NTU cooperative games defined with help of an excess function – the ε-core, the prenucleolus, the prekernel. It is shown that both the prenucleolus and the prekernel don't possess the reduced game property and the converse reduced game property for all excess functions satisfying the Kalai's (Kalai (1978)) conditions. The ε-core may possess these properties or not in dependence on excess functions. Axiomatic characterizations of the ε-core for arbitrary fixed ε and of the collection of ε-cores for all ε and for a particular excess function are given.*

1 Introduction

There are solutions of cooperative games with nontransferable utilities (NTU games) which are direct generalizations of the corresponding games with transferable utilities (TU games). These solutions are the core, the ε-core, the (pre)nucleolus and the (pre)kernel. Only the core has a good axiomatic characterization *(Peleg (1985))*. As for the ε-core, the (pre)nucleolus and the (pre)-kernel, they depend on excess functions assigning to each NTU game $\Gamma = \langle N, v \rangle$, and to its payoff vector $x \in v(N)$ an excess vector $e_v(x, S)$, $S \subset N$ whose components are negative utility functions of the payoff vectors. Some natural properties of the excess functions were given by Kalai *(Kalai (1978))*. Notice that for TU games the Schmeidler excess function $e_v(x,S) = v(S) - x(S)$ is usually considered. Up to present game researchers try to find a universal excess function defining the solutions possessing good properties *(Maschler (1992))*. In this paper three above mentioned solutions: the ε-core, the prenucleolus and the prekernel are examined from this point of view.

It is well-known that the main properties used for the characterization of the prenucleolus and the prekernel are the consistency or the reduced game property (RGP) and the converse consistency – the converse reduced game property (CRGP) (only for the prekernel) in sense of Davis – Maschler definition of the reduced game. Unfortunately, it turns out that these properties fail in the corresponding NTU game solutions.

It is shown that both the prenucleolus and the prekernel don't possess RGP and CRGP for all excess functions satisfying the Kalai's conditions. The ε-core may possess these properties or not in dependence on excess functions. Axiomatic characterizations of the ε-core for arbitrary fixed ε and the collection of ε-cores for all ε and for a particular excess function are given.

The paper is organized as follows. In Section 2 we recall basic nucleolus and kernel concepts and give the corresponding definitions for NTU games. In Section 3 we give examples

[1] The research for this paper was supported by the Russian Science Foundation (project 95-01-00118) and by the ACE091-R02 project grant from the European Community.

of NTU game showing that for a particular excess function the prenucleolus and the prekernel don't possess the reduced game property and the converse reduced game pproperty. Further we prove that for any excess function satisfying Kalai's conditions there is an NTU game, whose prenucleolus and prekernel don't possess the reduced game property. Moreover, another example shows that the prenucleolus may not be consistent even for the composition of games, i.e when we consider the reduced game property only for reduced games on the game–components, when an original game is the composition of several NTU games.

In Section 4 the excess function, equal to the maximal coordinate distance from a vector to the Pareto boundary of the corresponding characteristic function set, is considered. It is shown that for this excess function the ε-core possesses both RGP and CRGP. Thus, for an axiomatic characterization of the ε-core it suffices to characterize it only for two-person games. One system of axioms, similar to one for the characterization of the core, characterizes the ε-core for an arbitrary but fixed ε. Another system of axioms characterizes the collection of ε-cores for all values of ε's. It contains a new Axiom Independence of scale transformations of players' utilities, which are not independent one from another as usual, but depend on both coordinates at once. The independence of axioms in both systems are proven.

2 Basic definitions and notation.

Let N be an arbitrary finite set. For each vector $x \in \Re^N$ we denote by x^i the vector $x^i \in \Re^{N\setminus i}$ which is obtained from x by deleting its i-th component x_i. For each $S \subset N$ the vectors x_S, x^S are the vectors from $\Re^S$, $\Re^{N\setminus S}$ respectively, whose components are the corresponding components of x. The vector $x||y_S$ denotes the vector x whose components from S are changed by y_S..

By $\succ_{lex}$, $\succ_{lexmin}$ we denote the relations of lexicographical and lexmin dominance in $\Re^N$ respectively:

$$x, y \in \Re^N, \quad x \succ_{lex} y \quad \stackrel{\text{def}}{=} \quad x_k > y_k \quad \text{for some } k \in N, x_i = y_i \quad \text{for all } i < k,$$

$$x \succ_{lexmin} y \quad \stackrel{\text{def}}{=} \quad \tau x_k > \tau y_k \text{ for some } k \in N,$$

$$\tau x_i = \tau y_i \text{ for all } i < k,$$

where τx is the vector whose components are the components of x but disposed in a weakly increasing manner: $(\tau x)_1 \geq (\tau x)_2 \geq \ldots (\tau x)_n$, $n = |N|$. Let m, n be arbitrary positive numbers, $f : \Re^m \to \Re^n$, $A \subset \Re^m$ be arbitrary mapping and set respectively. Denote by

$$\arg\min_{y \in A} \text{lex } f(x) = \{x \in A | \; f(y) \succ_{lex} f(x) \text{ for all } y \in A\},$$

$$\arg\max_{y \in A} \text{lexmin } f(y) = \{x \in A | \; f(x) \succ_{lexmin} f(y) \text{ for all } y \in A\}$$

the sets of lexicographically minimal and lexmin maximal vectors for the set A respectively.

For any $S \subset \{1, ..., m\}$, $x_S \in \Re^S$ denote by $A|_{x_S}$ the section of A by the hyperplanes $y_i = x_i, \quad i \in S$:

$$A|_{x_S} = \{y \in A \,|\, y_S = x_S\}.$$

We shall consider nontransferable (NTU) cooperative games. Each NTU game Γ is defined by a pair $\Gamma = \langle N, v \rangle$, where N is a finite set of players, $v : 2^N \setminus \emptyset \rightarrow \Re^N$ is a characteristic function, such that the sets $v(S)$ are $|S|$ - dimensional cylinders in $\Re^N$: $x \in v(S) \Rightarrow (x||y_{N\setminus S}) \in v(S)$ for any vector $y_{N\setminus S} \in \Re^{N\setminus S}$. For brevity, we shall denote by $v(i) = v\{(i)\}$.

We denote by $\partial v(S)$ the Pareto optimal boundary of $v(S)$:

$$x \in \partial v(S) \implies \text{there is no } y \in v(S) \text{ such that } y_i \geq x_i \text{ for all } i \in S \text{ and } y_i > x_i \text{ for some } i \in S.$$

Let the sets $v(S)$ satisfy the following conditions:

1_v. $v(S)$ are closed for all $S \subset N$;

2_v. $v(S)$ are upper bounded, i.e. for any S and $a \in \Re^S$ the set $v(S) \cap \{y \in \Re^S |\ y_i \geq a_i, i \in S\}$ is bounded;

3_v. $v(S)$ are comprehensive, i.e. $x \in v(S) \Rightarrow y \in v(S)$ for each $y \leq x$.

These properties are usually supposed to be hold in the definition of NTU cooperative games. If $A \subset \Re^N$, then by compr$A$ we shall denote the comprehensive hull of A :

$$\text{compr}A = \{x \in \Re^N |\text{ there exists } y \in A,\ y \geq x\}.$$

Then a set $A \subset \Re^N$ is comprehensive iff $A = \text{compr}A$.

A solution for a class $\mathcal{G}$ of NTU games is a mapping σ assigning to each game $\Gamma \in \mathcal{G}$ a subset of its payoff vectors $\sigma(\Gamma) \subset v(N)$.

In the sequel we denote by $\mathcal{G}_{NTU}$ the class of NTU games possessing the properties $1_v - 3_v$. For such games the characteristic function values for one-element coalitions are completely defined by the numbers $v(i) = \max\{x \in \Re^1 | x \in v(\{i\})\}$.

For each game $\Gamma = \langle N, v \rangle \in \mathcal{G}_{NTU}$ we denote by $e_v : \Re^N \times 2^{N\setminus\emptyset} \rightarrow \Re^1$ an excess function satisfying the Kalai's conditions (*Kalai (1978)*):

1_e. *Independence of Irrelevant Coalitions.* $x, y \in \Re^N$, $x_i = y_i$ for all $i \in S \subset N$, $v_1(S) = v_2(S)$, then $e_{v_1}(x, S) = e_{v_2}(y, S)$.
2_e. *Normalization condition.* $x \in \partial v(S) \Rightarrow e_v(x, S) = 0$.
3_e. *Monotonicity.* $x, y \in \Re^N$, $x_i < y_i$ for all $i \in S \Rightarrow e_v(x, S) > e_v(y, S)$
4_e. *Continuity.* The function $e_v(x, S)$ is continuous jointly in x and v.

We give here one more condition on excess functions, implying the anonymity of solutions defined with its help;

5_e. *Anonymity.* For any coalition $S \subset N$, $x \in \Re^N$ and permutation $\pi : N \rightarrow N$ $e_v(x, S) = e_v(\pi x, \pi S)$, where $\pi x_i = x_{\pi i}$.

By $\bar{e}_v(x) = \{e_v(x, S)\}_{S \subset (\neq) N}$ we denote the excess vector, corresponding to the payoff vector x, whose components are the corresponding values of the excess function e_v. The

prenucleolus of Γ is defined by

$$PN(\Gamma) = \{x \in v(N) | -\bar{e}_v(x) \succeq_{lexmin} -\bar{e}_v(y)_S \text{ for all } y \in v(N)\},$$

where $\succeq_{lexmin} = \succ_{lexmin} \cup \approx_{lexmin}$, $\approx_{lexmin}$ is the equivalence relation, corresponding to the relation $\succ_{lexmin}$. A *transfer* from a player i to a player$j$ is a vector $t^{ij} \in \Re^N$ such that $t^{ij}_k = 0$ for all $k \neq i, j$, and $t^{ij}_i t^{ij}_j \leq 0$. The *prekernel* of Γ is defined by

$$PK(\Gamma) = \{x \in \partial v(N) | \quad - \bar{e}_v(x) \succeq_{lexmin} -\bar{e}_v(x + t^{ij})$$

for all transfers t^{ij}, $i, j \in N$ such that $x + t^{ij} \in v(N)$.

These definitions are similar to the Kalai's definitions of the nucleolus and the kernel, and by following his way, it is not dificult to show that for the excess functions satisfying the conditions $1_e - 4_e$ the prenucleolus and the prekernel are non-empty on the class $\mathcal{G}_{NTU}$, for each $\Gamma \in \mathcal{G}_{NTU}$ the prenucleolus is finite-valued and $PN(\Gamma) \subset PK(\Gamma)$. It is obvious that for the excess functions satisfying the condition 5_e both the prenucleolus and the prekernel are anonymous. However, they don't possess the reduced game property and the converse reduced game property for the Davis–Mashler definition of the reduced game. We shall prove this fact in the next Section and here we give the definitions of these properties.

Let $\mathcal{G}$ be a class of NTU games, $\Gamma \in \mathcal{G}$, $S \subset N, x$ be an arbitrary coalition and a payoff vector of Γ respectively.

A reduced game of Γ on the player set S and with respect to the payoff vector x is the game $\Gamma^x_S = \langle S, v^x_S \rangle$, where the characteristic function v^x_S is defined by

$$v^x_S(T) = \begin{cases} v(N)|_{x_{N \setminus S}}, & \text{if } T = S, \\ \bigcup_{Q \subset N \setminus S} v(T \cup Q)|_{x_Q} \cap \Re^S & \text{otherwise}, \end{cases}$$

A solution σ for a class $\mathcal{G}$ of NTU games possesses *the reduced game property* if for each game $\Gamma = \langle N, v \rangle \in \mathcal{G}$, its coalition $S \subset N$, and payoff vector $x \in \sigma(\Gamma)$ the reduced game $\Gamma^x_S \in \mathcal{G}$ and $x_S \in \sigma(\Gamma^x_S)$.

A solution σ for a class $\mathcal{G}$ of NTU games possesses *the converse reduced game property* if from the relations $\Gamma = \langle N, v \rangle \in \mathcal{G}$, $x \in v(N)$, $x_S \in \sigma(\Gamma^x_S)$, $\Gamma^x_S \in \mathcal{G}$ for any two-person coalition $S \subset N$, it follows that $x \in \sigma(\Gamma)$.

In the following Section we show that the prekernel and the prenucleolus for the class $\mathcal{G}_{NTU}$ are not consistent, i.e. they don't possess the reduced game property and its converse, and moreover, they even don't possess a weaker version of these properties.

3 Inconsistency of the nucleoli and the prekernels.

To begin with we give some examples of NTU games showing that for particular excess functions the correponding prenucleoli and prekernels don't possess the reduced game property and the converse reduced game property.

For any vector x we denote by x^i the same vector without its $i-$th component x_i.

Consider the excess function defined by:

$$e_v(x,S) = \max_{i \in S} e_v(x,i,S), \tag{1}$$

where

$e_v(x,i,S) = \mathrm{v}(S)|_{x^i} - x_i,$

$\mathrm{v}(S)|_{x^i} = \max\{y_i \mid (x^i, y_i) \in v(S)\} = \max\{y_i \mid y_i \in v(S)|_{x^i}\}.$

If $(x^i, y_i) \notin v(S)$ for all $y \in \Re^1$ then we put $e_v(x,i,S) = -\infty$.

Example 1. Let $N = \{1,2,3\}, \rho > 0, b > a > 0$ be arbitrary numbers, π be the permutation $\pi : \{1,2,3\} \to \{1,2,3\}$ such that $\pi(1) = 2, \pi(2) = 1, \pi(3) = 3,$. Consider the following vectors:

$$A = (a,a,0), \quad B = (0, a+\rho, b), \quad K = (-\rho, a+\rho, 0).$$

With help of these vectors we define the values of a characteristic function v for $S \subset \{1,2,3\}, \ S \neq N$ as follows:

$v(1,2) = \mathrm{compr}\{A^3, K^3, (\pi K)^3\} \times \Re^1,$

$v(2,3) = \mathrm{compr} B^1 \times \Re^1,$

$v(1,3) = \pi v(2,3),$

$-(a+\rho) < \mathrm{v}(3) < \mathrm{v}(1) < \mathrm{v}(2) < -\rho$

where $a, \rho > 0$ are arbitrary numbers. Consider two vectors

$$\begin{aligned} C &= (0, a+\rho, b+\rho), \\ D &= \pi C = (a+\rho, 0, b+\rho). \end{aligned}$$

The corresponding vectors $\tau\bar{e}_v(C), \tau\bar{e}_v(D)$ are equal to:

$$\begin{aligned} \tau\bar{e}_v(C) &= (\mathrm{v}(3) - (b+\rho), \ , \mathrm{v}(2) - (a+\rho), \ \mathrm{v}(1), -\rho, -\rho, -\rho) \\ \tau\bar{e}_v(D) &= (\mathrm{v}(3) - (b+\rho), \ \mathrm{v}(1) - (a+\rho), \ , \mathrm{v}(2), -\rho, -\rho, -\rho) \end{aligned}$$

and

$$-\bar{e}_v(C) \succ_{lexmin} -\bar{e}_v(D).$$

Denote by Γ the three-person game with the characteristic function $v(S)$ for the coalitions $S \subset (1,2,3), S \neq (1,2,3)$ and with $v(1,2,3) = \mathrm{compr}\{C, D\}$, and by $\Gamma'_\varepsilon == \langle (1,2,3,4), v'_\varepsilon \rangle$ the four-person game whose characteristic function v'_ε is defined by the following way:

$$\begin{aligned} \mathrm{v}(4) &= 1, \\ v'_\varepsilon(S) &= v(S) \text{ for all } S \subset (\neq)\{1,2,3\} \text{ except for } S = (1,2), \\ v'_\varepsilon(S \cup 4) &= v'(S) \cap v'(4) \text{ for all } S \subset \{1,2,3\} \\ v'_\varepsilon(1,2) &= \mathrm{compr}\{(-\rho, a+\rho), (a,a), (a+\rho, -\rho-\varepsilon)\} \times \Re^2, \\ v'_\varepsilon(1,2,3,4) &= \mathrm{compr}\{(C,1),(D,1)\}, \end{aligned}$$

where v is the characteristic function of Γ, ε is an arbitrary positive number, less than ρ and such that $\rho + \varepsilon < -\mathrm{v}(1)$. The only candidates for the prenucleolus of Γ' are the vectors $C' = (C,1)$, $D' = (D,1)$. The corresponding vectors $\bar{e}_{v'}(C')$ and $\bar{e}_{v'}(D')$ have maximal components equal to 0 on the coalitions $\{4\}, (1,2,3)$, $(1,2,4)$. The second value component is equal to $-\rho$ and is attained on the coalitions (1,3), (2,3), (1,3,4), (1,2,4). For the coalition (1,2) we have

$$e_{v'_\varepsilon}(C', (1,2)) = -\rho,$$

$$e_{v'_\varepsilon}(D', (1,2)) = -\rho - \varepsilon.$$

The other components of the excess vectors are less then $-\rho-\varepsilon$, and therefore $-\bar{e}_{v'_\varepsilon}(D') \succ_{lexmin} -\bar{e}_{v'_\varepsilon}(C')$ in Γ'_ε.

However, it is easy to check that the reduced game of Γ' on the player set $\{1,2,3\}$ with respect to any vector x with $x_4 = 1$ coincides with the game Γ and $C = C'|_{x_4=1} \succ_{lexmin} D'|_{x_4=1} = D$.

Remark 1. The excess function (1), considered in Example 1, is not continuous in payoff vector $x \in \Re^N$ and hence it doesn't satisfy one of the Kalai's conditions. However, on the sets $v(S)$ the functions $e_v(x, S)$ are continuous.

Remark 2. Evidently, the prekernel of Γ'_ε coincides with its prenucleolus. Therefore, Example 1 also shows that the prekernel defined by the excess function (1) doesn't possess the reduced game property.

This example also shows that the prekernel and, therefore, the prenucleolus defined with help of the excess function (1), don't possess the converse reduced game property. Indeed, as it has been already shown, the maximal components of the excess vectors $\bar{e}_v(C), \bar{e}_v(D)$ of Γ are equal to $-\rho$ and they are attained in all two-element coalitions. Hence, in any reduced game of Γ on a two-player set with respect to C or D, the corresponding excess vectors of C^i and $D^i, i = 1,2,3$ are equal to $(-\rho, -\rho)$. Therefore, the vectors $C^i, D^i, i = 1,2,3$ constitute the prenucleoli (and the prekernels) of the corresponding reduced games. However, the vector D belongs neither to the prenucleolus nor to the prekernel of Γ. Similarly to Example 1 it is possible to give examples of games whose prenucleoli and prekernels don't possess the reduced game property and the converse reduced game property for other excess functions as well. This statement we formulate as the following Proposition:

Proposition 1 *For any excess function e, satisfying the conditions $1_e - 5_e$ and considered as a mapping associated with each game $\Gamma = \langle N, v\rangle \in \mathcal{G}_{NTU}$ and its payoff vector $x \in v(N)$ the excess vector $\bar{e}_v(x) = \{e(x,S)\}_{S \subset N}$, there is a four-person game $\Gamma(e) \in \mathcal{G}_{NTU}$ such that the prenucleolus $PN(\Gamma(e))$ and the prekernel $PK(\Gamma(e))$ don't possess the reduced game property with respect to the excess function e.*

P r o o f. Let e be an arbitrary excess function, satisfying the Kalai's conditions. Its dependence on a characteristic function v we shall denote by its lower index: $e = e_v$. First as in Example 1 we define a three-person game in which the players 1 and 2 have "almost" equal treatments and the characteristic function value of the big coalition $N = \{1,2,3\}$ is the comprehensive hull of arbitrary two vectors C and D, symmetric with respect to the plyers 1 and 2: $D = \pi C,\ \pi : N \to N,\ \pi 1 = 2, \pi 2 = 1, \pi 3 = 3$. Let $v(1,2) = \pi v(1,2),\ v(2,3) = \pi v(1,3)$ be arbitrary sets satisfying the conditions $1_v - 3_v$ and such that $C, D \notin v(S)$ for all two-person coalitions. Let the v(1), v(2), v(3) be arbitrary numbers satisfying the inequalities

$$\text{v}(3) < \text{v}(1) < \text{v}(2) \tag{2}$$

and

$$e_v(C,S) > e_v(C,1) > e_v(C,2) > e_v(C,3)$$

$$e_v(D,S) > e_v(D,2) > e_v(D,1) > e_v(D,3) \tag{3}$$

for all two-person coalitions S. Notice that for such coalitions S $e_v(C,S) = e_v(D,S)$. Then $PN(\Gamma(e)) = PK(\Gamma(e)) = C$. Now by the same way as in Example 1 we define a four-person game $\Gamma'(e) = \langle\{1,2,3,4\}, v'_e\rangle$ as follows:

$$\begin{array}{rcl} v'(4) & & \text{is defined arbitrarily,} \\ v'(S) & = & v(S) \text{ for all } S \subset \{1,2,3\}, \\ v'(S \cup 4) & = & v'(S) \cap v'(4) \text{ for all } S \subset \{1,2,3\} \\ & & \quad \text{except for } S = (1,2), \\ v'(1,2,4) \subset v'(1,2) \cap v'(4) & & \text{is an arbitrary set such that} \\ e_{v'}(C',(1,2,4)) & > & e_{v'}(D',(1,2,4)) \\ & & \quad \text{where } C' = (C, \mathrm{v}(4)),\ D' = (D, \mathrm{v}(4)), \\ v'(1,2,3,4) & = & \mathrm{compr}\{C', D'\}. \end{array} \tag{4}$$

A characteristic function value $v'(1,2,4)$, satisfying the conditions (4) always exists, because of the monotonicity of excess functions in payoff vectors and the continuity in characteristic functions.

Therefore, by symmetry of vectors C and D and characterstic function values in players 1,2 except for $v'(1,2,4)$ and by the following from (4) inequality

$$e_{v'}(C',(1,2,4)) > e_{v'}(D',(1,2,4))$$

we obtain that

$$-\bar{e}_{(v')}(D') \succ_{lexmin} -\bar{e}_{v'}(C')$$

and $PN(\Gamma(e)) = D'$. As in Example 1 the reduced game $\Gamma^x_{N\setminus 4} = \langle N \setminus 4, v^x_{N\setminus 4}\rangle$ of $\Gamma'(e)$ on the player set $\{1,2,3\}$ with respect to any vector x with $x_4 = \mathrm{v}(4)$ coincides with the game Γ in which $-\bar{e}_v(C) \succ_{lexmin} -\bar{e}_v(D)$.

It is obvious, that $PN(\Gamma(e)) = PK(\Gamma(e))$ and, therefore, the Proposition has been proven. ∎

Proposition 2 *For any excess function e, satisfying the the conditions $1_e - 5_e$ there is a game $\Gamma(e) \in \mathcal{G}_{NTU}$ such that the prekernel of $\Gamma(e)$ doesn't possess the converse reduced game property.*

P r o o f. Let e be an arbitrary excess function, satisfying the conditions of the Proposition. We define a three-person game $\Gamma(e)$ analogous to the game Γ in Example 1 by the following way: Let $C, D \in \Re^3$ be arbitrary vectors such that $C_1 = D_2, C_2 = D_1, C_3 = D_3$ and let the characteristic function values $v(S), |S| = 2$ be symmetric with respect to the players 1,2 and therefore they satisfy the equalities

$$e_v(C,S) = e_v(D,S) = \theta_1$$

for all two-person coalitions S. As in the proof of the Proposition 1 the numbers $\mathrm{v}(i), i \in \{1,2,3\}$ are supposed arbitrary but satisfying the inequalities (2) and (3). Such numbers exist for any excess function e because of its continuity and monotonicity. Then as in Example 2 we obtain that for a three-person game $\Gamma(e)$ with an arbitrary characteristic function satisfying the given above conditions and with $v(1,2,3) = \mathrm{compr}\{C, D\}$ and for all its reduced games on two-person player sets and with respect C and D the corresponding pairs of components of C and D are the prenucleoli coinciding with the prekernels of the reduced games, but only $C = PN(\Gamma(e)) = PK(\Gamma(e))$. ∎

Consider now the composition of NTU games. Let $\Gamma_1 = \langle N_1, v_1\rangle, \Gamma_2 = \langle N_2, v_2\rangle \in \mathcal{G}_{NTU}$. *The composed game* $\Gamma = \Gamma_1 \otimes \Gamma_2$ is the the game $\Gamma = \langle N_1 \cup N_2, v\rangle$, where $v(S) = v_1(S \cap N_1) \cap v_2(S \cap N_2)$.[2] It is evident that the reduced game of the composed game on the player set of any component game with respect to arbitrary payoff vector coincides with the corresponding game–component. The reduced game property of a solution fulfilling only when reducing on the player sets of game–components, can be considered as a generalization of "the dummy property" of a solution. Of course, it is much weaker than the reduced game property. However, in the following example we show that the prenucleolus may not possess it.

Example 2. Consider the following excess functions for NTU games (*Kalai (1975)*):

$$e_v(x, S) = \sup\{t|\ x + \frac{t}{|S|}e_N \in v(S)\ \}, \tag{5}$$

where e_N is the unit vector in $\Re^N$. Define two games $\Gamma_1 = \langle N_1, v_1\rangle$, $\Gamma_2 = \langle N_2, v_2\rangle \in \mathcal{G}_{NTU}$:

$|N_1| = |N_2| = n$, where n is an arbitrary even integer,

$$\begin{aligned}
v_1(N_1) &= \{x^1 \in \Re^n \mid \sum_{i \in N_1} x_i^1 \le a_{N_1}\},\\
v_1(N_1 \setminus 1) &= \{x^1 \in \Re^n \mid \sum_{i \in N_1 \setminus 1} x_i^1 \le a_1\},\\
&\quad \text{where } a_{N_1} < a_1 + \mathrm{v}_1(1),\\
v_1(S) &= \bigcap_{i \in S} v_1(i) \text{ for other } S \subset N_1,\\
\mathrm{v}_1(i) &\quad \text{are arbitrary numbers for } i \ne 1\\
&\quad \text{such that } \sum_{i \in N_1 \setminus 1} \mathrm{v}_1(i) < a_1.
\end{aligned}$$

Denote by $b = \frac{a_{N_1} + a_1 - \mathrm{v}_1(1)}{2}$. Define the vector $x \in \partial v_1(N_1)$ by:

$$\begin{aligned}
x_1 &= b - a_1 + \mathrm{v}_1(1) = a_{N_1} - b,\\
x_i &= \frac{b - \sum_{j \in N_1 \setminus 1} \mathrm{v}(j)}{n-1} + \mathrm{v}_1(i),\ \forall i \in N_1 \setminus 1.
\end{aligned}$$

Then

[2]This definition is not analogous to the definition of the composition of TU games, where $v(S) = v(S \cap N_1) + v(S \cap N_2)$. In this case the reduced games of on the sets N_1, N_2 may not coincide with game–components.

$$
\begin{aligned}
e_{v_1}(x,1) &= e_{v_1}(x, N_1\setminus 1) = a_1 - b = \theta_1 > 0,\\
e_{v_1}(x,S) &= \frac{\sum_{j\in N_1\setminus 1} \mathrm{v}_1(j) - b}{n-1}|S| < \theta_1,\ \ \forall S\subset N_1\setminus 1, \qquad (6)\\
e_{v_1}(x,S) &= \min\{(\mathrm{v}_1(1) - x_1)|S|, \frac{\sum_{j\in N_1\setminus 1} \mathrm{v}_1(j) - b}{n-1}|S|\} < \theta_1\\
& \forall S \subset N_1,\ 1\in S.
\end{aligned}
$$

Moreover for any $x' \in \mathcal{S}_1(\Gamma_1) = \arg\min_{x\in v_1(N_1)} \max_{S\in N_1} e_{v_1}(x,S)$

$$
e_{v_1}(x',1) = e_{v_1}(x', N_1\setminus 1) = \theta_1. \qquad (7)
$$

The equalities (6) and (7) imply

$$
x_{PN} \in PN(\Gamma_1) \implies \max_{S\subset N_1} e_{v_1}(x_{PN}, S) = \theta_1
$$

and the maximum is attained only on the coalitions 1 and $N_1 \setminus 1$. Define the game $\Gamma_2 = \langle N_2, v_2\rangle$. Let $N_2 = S_1 \cup S_2$ be a partition of N_2 such that $|S_1| = |S_2| = n/2$, v_2 be a characteristic function, satisfy the following conditions:

$$
\begin{aligned}
v_2(N_2) &= v_1(N_1),\\
v_2(S_k) &= \{x\in \Re^n |\ \textstyle\sum_{i\in S_k} x_i \le a\},\ \ k=1,2,\\
v_2(S) &= \textstyle\bigcap_{i\in S} v_2(i) \text{ for other } S\subset N_2,
\end{aligned}
$$

where $a, \mathrm{v}_2(i)$ are arbitrary numbers satisfying the equality and inequalities

$$
\begin{aligned}
a &= \frac{a_1 + \mathrm{v}_1(1)}{a}\\
\sum_{j\in S_1} \mathrm{v}_2(j) &= \sum_{j\in S_2} \mathrm{v}_2(j) < \frac{a_1 + a_{N_1} + \mathrm{v}_1(1)}{4.} \qquad (8)
\end{aligned}
$$

By the same way as in the case of the game Γ_1 and by using the conditions (8) We obtain that

$$
y \in \mathcal{S}_1(\Gamma_2) \Rightarrow e_{v_2}(y, S_1) = e_{v_2}(y, S_2) = \theta_1,
$$

and for $y_{PN} \in PN(\Gamma_2)$

$$
e_{v_2}(y_{PN}, S) < \theta_1 \text{ for other coalitions } S \subset N_2.
$$

Consider the composed game $\Gamma = \Gamma_1 \otimes \Gamma_2 = \langle N = N_1 \cup N_2, v\rangle$ and its payoff vector $z = (x_{PN}, y_{PN})$. It is evident that $\max_{S\subset N} e_v(z,S)$ is attained only on coalitions from the collection $\{1, N_1\setminus 1, S_i, 1\cup S_i, (N_1\setminus 1)\cup S_i, i=1,2\}$. To check all the possibilities it suffices only to find $e_v(z, (N_1\setminus 1)\cup S_1)$, $e_v(z, 1\cup S_1)$. We have

$$e_v(z,(N_1\setminus 1)\cup S_1) = \max\{t|\ z+\tfrac{2t}{3n-2}\in v_1(N_1\setminus 1)\cap v_2(S_1)\}$$
$$= \min\{\tfrac{\theta_1(n-1+n/2)}{n-1},\ \tfrac{\theta_1(n-1+n/2)}{n/2}\} =$$
$$= \tfrac{\theta_1(3n-2)}{2(n-1)}. \tag{9}$$

$$e_v(z,1\cup S_1) = \max\{t|\ z+\tfrac{t}{n/2+1}\in v(1)\cap v(S_1)\}$$
$$= \min\{\theta_1(n/2+1), \tfrac{\theta_1(n/2+1)}{n/2}\} = \tfrac{n+2}{n}\theta_1. \tag{10}$$

Then by comparing the equalities (9) and (10) we obtain that for $n \geq 3$ $\max_{S\subset N} e_v(z,S) = \theta_1' = \frac{3n-2}{2(n-1)}\theta_1 > \theta_1$ and it is attained only on the coalitions $(N_1\setminus 1)\bigcup S_1$, $(N_1\setminus 1)\bigcup S_2$. Consider the vector $z_\varepsilon \in v(N)$, whose components are equal to:

$$\begin{aligned} z_{\varepsilon j} &= y_j \text{ for } j\in N_2, \\ z_{\varepsilon 1} &= x_1-\varepsilon(n-1), \\ z_{\varepsilon k} &= x_k+\varepsilon,\ \ k\in N_1\setminus 1, \end{aligned}$$

where ε is chosen by such a manner that $z_\varepsilon \in v_1(N_1\setminus 1)$, i.e. $x(N_1\setminus 1)+\varepsilon(n-1)\leq a_1$. It is always possible, because $x(N_1\setminus 1) < a_1$. Then by monotonicity of the excess function e_v the following inequalities hold for enough small values of ε :

$$\begin{aligned} e_v(z_\varepsilon,(N_1\setminus 1)\cup S_i) &< e_v(z_\varepsilon,(N_1\setminus 1)\cup S_i), \\ e_v(z_\varepsilon,S) &< \theta_1' \text{ for other } S\subset N. \end{aligned}$$

Thus, $z \notin \mathcal{S}_1\Gamma)$, and $z \notin PN(\Gamma)$. Therefore, if $w = (w_1,w_2) \in PN(\Gamma)$, then either $w_{N_1} \notin PN(\Gamma_1)$, or $w_{N_2} \notin PN(\Gamma_2)$.

4 Epsilon-cores in NTU games.

It is well-known that the core has RGP both in TU and NTU games *(Peleg (1985))*. However, its generalization – the ε-cores for arbitrary $\varepsilon\in\Re^1$ have RGP in TU games, but in general don't possess it in NTU games. The last fact is connected with the dependence of ε-cores on excess functions. In fact, let $\Gamma = \langle N,v\rangle \in \mathcal{G}_{NTU}$. *The ε-core* of Γ is defined by

$$\varepsilon\text{–}c(\Gamma) = \{x\in\partial v(N)|e(x,S)\leq\varepsilon \text{ for all } S\subset(\neq)N\}$$

for an excess function e, satisfying the conditions $1_e - 3_e$. [3] Notice that the continuity condition 4_e is not necessary for the definition of ε-cores and in the sequel we shall not suppose the continuity of excess functions.

[3] This definition does not coincide with Kalai's ε–core (*Kalai (1978)*). The last one is contained in the set of individually rational payoff vectors. Our definition is consistent with the one of the ε–core for TU games in (*Maschler M., Peleg B., Shapley L (1979)*).

For some excess functions the ε−core is inconsistent. For example, consider the excess function defined by

$$e(x, S) = \max\{t | x + te_N \in v(S)\}.$$

Define a game $\Gamma = \langle N, v\rangle \in \mathcal{G}_{NTU}$ as follows: let S be a coalition with even set of players and the set $v(S)$ have the spherical Pareto boundary:

$$\partial v(S) = \{x \in \Re^N_+ | \sum_{i \in S} x_i^2 = 1\},$$

other coalitions $K \neq S$ are inessential: $v(K) = \bigcap_{i\in K} v(i)$, $v(i) = 0$, $i \in N$. The set $v(N)$ is arbitrary, but has the property that the vector y with

$$y_i = \begin{cases} \frac{\alpha}{\sqrt{|S|}}, & \text{if } i \in S, \\ 0, & \text{if } i \notin S \end{cases}$$

belongs to $\partial v(N)$ for some $\alpha \in (0, 1)$.

Then

$$e_v(y, S) = \frac{1-\alpha}{\sqrt{|S|}},$$

and $e_v(y, K) \leq 0$ for other coalitions K. Therefore, $y \in (\frac{1-\alpha}{\sqrt{|S|}})$−core (Γ).

Now consider a reduced game $\Gamma^y_T = \langle T, v^y_T\rangle$ on the player set $T = N \setminus S_1$, where $S_1 \subset S$, $|S_1| = |S|/2$, and with respect to y. By the definition of reduced games

$$v^y_T(S \cap T) \supset v(S)|_{y_{S\setminus T}}$$

and, therefore,

$$e_{v^y_T}(y, S \cap T) \geq \max\{t \,|\, y_{S\cap T} + te_N \in v(S)|_{y_{S\setminus T} = t_{S\cap T}}\}. \tag{11}$$

¿From the definition of the set $v(S)$ it follows that

$$\sum_{i \in S\cap T} (y_i + t_{S\cap T})^2 = 1 - \sum_{i\in S\setminus T} y_i^2$$

and $t_{S\cap T} = \frac{\sqrt{2-\alpha^2}-\alpha}{\sqrt{|S|}}$. By rewriting the inequality (11) we obtain

$$e_{v^y_T}(y, S\cap T) \geq \sqrt{\frac{2-\alpha^2}{|S|}} - \frac{\alpha}{\sqrt{|S|}} > \frac{1-\alpha}{\sqrt{|S|}} = e_v(y, S).$$

Thus, the vector y_T doesn't belong to the $(\frac{1-\alpha}{\sqrt{|S|}})$−core of the reduced game Γ^y_T.

However, it turned out that for the excess function, defined in (1), the corresponding ε-core possesses RGP and its converse – the converse reduced game property. Moreover, we give its axiomatic characterization, reminding the Peleg's axiomatic characterization of the core of the NTU games. Let $\Gamma = \langle N, v\rangle \in \mathcal{G}_{NTU}$. In the rest part of the Section we denote by e the excess function defined in (1).

For each $x \in v(N)$ the vector $\bar{s}(X) \in \Re^N$ is defined by

$$\bar{s}(x) = (s_i(x))_{i\in N},$$

$$s_i(x) = \max_{\substack{S \subset (\neq) N \\ S \ni i}} e(x, S) \quad (12)$$

Notice that $s_i(x) > -\infty$ for any $i \in N$, $x \in v(N)$. The vector function $\bar{s}(x)$ can be considered as a mapping assigning to each game $\Gamma \in \mathcal{G}_{NTU}$ a real vector whose dimension is equal to the number of players. So, in the sequel we shall use the notations $\bar{s}(x)(\Gamma)$, $s_i(x)(\Gamma)$ if it is necessary to indicate the game into consideration.

The equalities (12) imply

$$\max_{S \subset (\neq) N} e(x, S) = \max_{i \in N} s_i(x)$$

and if for some $x \in \varepsilon - c(\Gamma)$ and $S \subset N \quad e(x, S) = \varepsilon$ then there is $i \in S$ such that $s_i(x) = \varepsilon$. Therefore, we can equivalently define the ε-core of Γ by

$$\varepsilon - c(\Gamma) = \{x \in v(N) | s_i(x) \leq \varepsilon \text{ for all } i \in N\}. \quad (13)$$

Proposition 3 *For the class $\mathcal{G}_{NTU}$ the ε-core possesses RGP.*

P r o o f. First notice that any reduced game of an arbitrary game $\Gamma \in \mathcal{G}_{NTU}$ also belongs to the class $\mathcal{G}_{NTU}$. The definition (10) of the functions $s_i(x)$ implies that their values may only decrease with reducing of a game: if $\Gamma = \langle N, v \rangle \in \mathcal{G}_{NTU}$, $x \in v(N)$, Γ^x_S is the reduced game on the player set S and with respect to x, then for all $i \in S$

$$s_i(x)(\Gamma^x_S) \leq s_i(x)(\Gamma).$$

Now the proof follows from the definition (13) and the last inequalities. ∎

It turns out that the ε-core also possesses the converse reduced game property (CRGP).

Remind that a solution σ for a class $\mathcal{G}$ of cooperative games possesses *the converse reduced game property* if from the relations $\Gamma =< N, v >\in \mathcal{G}, x \in v(N)$, $(x_i, x_j) \in \sigma(\Gamma^x_{ij})$, $i, j \in N$, $\Gamma^x_{ij} \in \mathcal{G}$ for all two-person reduced games Γ^x_{ij} it follows that $x \in \sigma(\Gamma)$.

Proposition 4 *The ε-core possesses CRGP.*

P r o o f. Let $\Gamma = \langle N, v \rangle$ be an arbitrary game from $\mathcal{G}_{NTU}$, , $x \in v(N)$ be a payoff vector such that $(x_i, x_j) \in \varepsilon - c(\Gamma^x_{ij})$ for all $i, j \in N$. Then

$$v^x_{ij}(i) - x_i = \max_{S \ni i, S \not\ni j} v(S) \mid_{x^i} - x_i \leq \varepsilon$$

for all $i, j \in N$.

Therefore, the definition (10) of the vector $\bar{s}(x)$ implies $s_i(x) \leq \varepsilon$ for all $i \in N$ and $x \in \varepsilon{-}c(\Gamma)$. ∎

Notice that any cooperative game solution, possessing both RGP and CRGP, is completely defined by its values for two-person games. In fact, let σ be such a solution for a class $\mathcal{G}$ of (TU or NTU) cooperative games. Then for any $\Gamma = \langle N, v \rangle \in \mathcal{G}$

$$\sigma(\Gamma) = \{x \in v(N) | (x_i, x_j) \in \sigma(\Gamma^x_{ij}) \forall i, j \in N\}.$$

Thus, in order to obtain an axiomatic characterization of the ε -core for the class $\mathcal{G}_{NTU}$ with help of RGP and CRGP we have only to obtain an axiomatic characterization for the class $\mathcal{G}_2 \subset \mathcal{G}_{NTU}$ of all two-person games. It is our immediate purpose.

The excess function (1) take the following values on one-element coalitions:

$$e_v(x,i) = \mathrm{v}(i) - x_i,\ i \in N,$$

and, therefore, the ε-core of a two-person game $\Gamma \in \mathcal{G}_2$ is the set

$$\varepsilon - c(\Gamma) = \{(x_1,x_2) \in \partial_w v(1,2) \mid x_i \geq \mathrm{v}(i) - \varepsilon,\ i = 1,2.\}$$

Let σ be a solution for the class $\mathcal{G}_2$. We give now some properties of σ in form of axioms. For any coalition S we denote by $\partial_w v(S)$ the weak Pareto boundary of the set $v(S):\ \partial_w v(S) = \{x \in v(S) \mid y_i > x_i$ for all $i \in S \rightarrow y \notin v(S).\}$

Efficiency (Paretc optimality)(PO). $\sigma(\Gamma) \subset \partial_w v(1,2)$ for any $\Gamma \in \mathcal{G}_2$;
if $x \in \sigma(\Gamma),\ y \geq x,\ y \in \partial_w v(S)$, then $y \in \sigma(\Gamma)$.

Anonymity (ANO). For any $\Gamma = \langle(1,2),v\rangle \in \mathcal{G}_2$ and the permutation $\pi\{1,2\} = \{2,1\}$ $\sigma(\pi\Gamma) = \pi(\sigma(\Gamma))$, where $\pi\Gamma = \langle(1,2),\pi v\rangle,\ \pi v(i) = v(j),\ i,j = 1,2, i \neq j,\ \pi v(1,2) = \{(x_1,x_2) \in \Re^2 | (x_2,x_1) \in v(1,2)\}$.

Symmetry (SYM). If a game $\Gamma = \langle(1,2),v\rangle \in \mathcal{G}_2^\circ$ is symmetric, i.e.

$$(x_1,x_2) \in v(1,2) \iff (x_2,x_1) \in v(1,2),$$

then

$$(y,y) \in \partial_w v(1,2) \Rightarrow (y,y) \in \sigma(\Gamma).$$

Weak Covariance (WCOV). For any $\Gamma = \langle(1,2),v\rangle \in \mathcal{G}_2, \beta \in \Re^N$ $\sigma(\Gamma+\beta) = \sigma(\Gamma) + \beta$, where $\Gamma + \beta = <(1,2), v+\beta>$.

Independence of Ordinal Transformations with a Fixed ε-difference (IORD$_\varepsilon$.) Let $\Gamma = \langle(1,2),v\rangle \in \mathcal{G}_2$ and $\varphi_i : \Re^1 \rightarrow \Re^1,\ i = 1,2$ be arbitrary monotonically non-decreasing functions such that $\varphi_i(\mathrm{v}(i)) - \varphi_i(\mathrm{v}(i) - \varepsilon) = \varepsilon,\ i = 1,2.$ for some ε. Then $\varphi\sigma(\Gamma) = \sigma(\varphi(\Gamma))$, where $\varphi = (\varphi_1,\varphi_2),\ \varphi\Gamma = \langle(1,2),\varphi v\rangle,\ \varphi v(i) = \varphi_i v(i)$,

$$\varphi v(1,2) = \mathrm{compr}\{x \in \Re^2 | x = (\varphi_1(y_1),\varphi_2(y_2)),\ y = (y_1,y_2) \in v(1,2)\}.$$

ε-Individual Rationality (IR$_\varepsilon$). If $x \in \sigma(\Gamma),\ \Gamma = \langle(1,2,v\rangle \in \mathcal{G}_2$, then

$$\sigma(\Gamma) \subset IR_\varepsilon(\Gamma) = \{x \in v(1,2) \mid x_i \geq \mathrm{v}(i) - \varepsilon,\ i = 1,2.\}$$

Closedness(CL). The sets $\sigma(\Gamma)$ are closed for any $\Gamma \in \mathcal{G}_2$.

Give now a little discussion of the axioms. The second part of Axiom Efficiency follows from the first part, demanding the weak Pareto optimality of a solution set. Axiom Anonymity doesn't need in any explanations.

Axiom Symmetry (SYM) can be interpreted as a generalization of Axiom Equal Treatment Property see e.g. (*Peleg (1985)*) for set-valued solutions.

Axiom Independence of Ordinal Transformations with a fixed ε-difference characterizes a scale of measurement of players' utilities. For characterizing of TU game solutions the interval scales with a common unit are used as a rule. The solutions satisfying Axion Independence of the corresponding transformations are called covariant. However, for TU games the ε-core for $\varepsilon \neq 0$ is not covariant, it satisfies only a weak its version – Weak Covariance (WCOV) – which means that the players' payoffs are measured in the translation scale. It is clear that for NTU games Axiom Weak Covariance follows from Axiom IORD$_\varepsilon$. Axiom WCOV permits to consider only 0-reduced games, i.e. games with zero numbers v(1), v(2). For such normalized games (or games in 0-reduced form) Axiom IORD$_\varepsilon$ states that players' payoffs are measured in independent ordinal scales in each orthant of $\Re^2$, bounded by the lines: $x_i = -\varepsilon, i = 1,2$. Notice that the functions φ_1, φ_2 are not supposed continuous. For discontinuous functions φ_1, φ_2 the set $\varphi v(1,2)$ may not be comprehensive, so it is necessary to take the comprehensive hull in the definition of the set $\varphi v(1,2)$.

In the statement of Axiom IORD$_\varepsilon$ we use an external with respect to a game number ε. We use it also in the axiom ε-Individual Rationality, which extends the usual Axiom Individual Rationality (with $v(i)$ instead of $v(i) - \varepsilon$.)

The last axiom – Closedness (CL) – has a merely technical character.

We don't use Axiom Non-emptiness, because for each game $\Gamma \in \mathcal{G}_2$ the ε-cores exist for some ε's and are empty for others. So, to avoid the trivial solution $\sigma(\Gamma) = \emptyset$ for any $\Gamma \in \mathcal{G}_2$ in the sequel we shall refine the definition of cooperative game solutions and by a solution σ for a class $\mathcal{G}$ of cooperative games we shall consider mappings from $\mathcal{G}$ to the corresponding payoff vector sets, such that $\sigma(\Gamma) \neq \emptyset$ for some $\Gamma \in \mathcal{G}$.

The following Theorem characterizes the ε-core for an arbitrary fixed number $\varepsilon \geq 0$ and the ε-core together with the core for $\varepsilon \leq 0$.

Theorem 1 *If $\varepsilon \geq 0$, then there is a unique solution for the class $\mathcal{G}_2$, satisfying Axioms PO, SYM, IORD$_\varepsilon$, IR$_\varepsilon$ and CL. It is the ε-core. If $\varepsilon < 0$, then there are two solutions for the class $\mathcal{G}_2$, satisfying these Axioms and Axiom ANO: they are the core and the ε-core.*

P r o o f. It is not difficult to check that the ε- core for arbitrary ε and the core for $\varepsilon \leq 0$ satisfy all the Axioms stated in the Theorem.

Let now σ be any solution satisfying all the Axioms in the statement of the Theorem. As it has been already noticed, Axiom IORD$_\varepsilon$ permits to consider only normalized games. We denote the class of all normalized games (i.e. in 0-reduced form) from $\mathcal{G}_2$ by $\mathcal{G}_2^\circ$.

For any $\Gamma = \langle(1,2), v\rangle \in \mathcal{G}_2^\circ$ denote by SymΓ the game Sym$\Gamma = \langle(1,2), \mathrm{Sym}\, v(1,2)\rangle \in \mathcal{G}_2^\circ$, where

$$\mathrm{Sym}\, v(1,2) = \{(x_1,x_2) \,|\, (x_1,x_2) \text{ or } (x_2,x_1) \in v(1,2)\}.$$

Evidently, the game Sym$\,\Gamma$ is symmetric for any $\Gamma \in \mathcal{G}_2^\circ$ and by axiom SYM

$$(a,a) \in \partial_w \mathrm{Sym}\, v(1,2) \;\rightarrow\; (a,a) \in \sigma(\mathrm{Sym}\Gamma). \tag{14}$$

Let us show that there are non-decreasing functions φ_1, φ_2 such that Sym$\Gamma = \varphi\Gamma$. In fact, define them by

$$\varphi_1(x) = \begin{cases} x & \text{for all } x \leq a, \\ x & \text{if } x > a \text{ and there is } y \text{ such that } (x,y) \in \partial_w \mathrm{Sym}\, v(1,2) \\ x_1 > x, & \text{if } x > a \text{ and } (x,y) \in \partial_w v(1,2) \setminus \partial_w \mathrm{Sym}\, v(1,2) \\ & \begin{cases} (y,x_1) \in & \partial_w \mathrm{Sym}\, v(1,2) \text{ and such a } x_1 \text{ is unique}, \\ (y,x_1) \in & \partial \mathrm{Sym}\, v(1,2) \text{ otherwise}, \end{cases} \end{cases}$$

$$\varphi_2(y) = \begin{cases} y & \text{for all } y \geq a, \\ y & \text{if } x < a \text{ and there is } x \text{ such that } (x,y) \in \partial_w \mathrm{Sym}\, v(1,2) \\ y_1 > y, & \text{if } y < a \text{ and } (x,y) \in \partial_w v(1,2) \setminus \partial_w \mathrm{Sym}\, v(1,2) \\ & \begin{cases} (y_1, x) \in & \partial_w \mathrm{Sym}\, v(1,2) \text{ and such a } y_1 \text{ is unique,} \\ (y_1, x) \in & \partial \mathrm{Sym}\, v(1,2) \text{ otherwise.} \end{cases} \end{cases}$$

It is evident that $\varphi\Gamma = \mathrm{Sym}\,\Gamma$, and by Axiom IORD$_\varepsilon$ $(a,a) \in \sigma(\Gamma)$.

Let $(x_1, x_2) \in \partial_w v(1,2)$, $x_i > \max\{0, -\varepsilon\}, i = 1,2$. Then there are nondecreasing functions $\varphi_i : \Re^1 \to \Re^1$, such that

$$\varphi_i(-\varepsilon) = \varepsilon,\ \varphi_i(0) = 0,\ \ \varphi_1(x_1) = \varphi_2(x_2) = a.$$

The vector $(a,a) \in \partial_w \varphi v(1,2)$, therefore, $(a,a) \in \sigma(\varphi\Gamma)$, where $\varphi = (\varphi_1, \varphi_2)$, and by Axiom IORD$_\varepsilon$ $(x_1,x_2) \in \sigma(\Gamma)$. By Axiom CL the intersection of the core and the ε-core is contained in the solution:

$$\{(x_1,x_2) \mid x_i \geq \max\{0,-\varepsilon\}\} = c(\Gamma) \cap (\varepsilon{-}c(\Gamma)) \subset \sigma(\Gamma). \tag{15}$$

The relation (15) and axiom IR$_\varepsilon$ prove the statement of the Theorem for $\varepsilon \geq 0$.

If $\varepsilon < 0$, then by the same way we can prove that if a vector $(y_1,y_2) \in \sigma(\Gamma) \cap (\varepsilon{-}c(\Gamma) \setminus c(\Gamma))$ then

$$(\varepsilon{-}c(\Gamma)) \setminus c(\Gamma) \subset \sigma(\Gamma). \tag{16}$$

Now the relations (15), (16) and Axiom IR$_\varepsilon$ prove the Theorem for $\varepsilon < 0$. ∎

Theorem 2 *The Axioms used in Theorem 1 are independent.*

P r o o f. We shall give examples of solutions satisfying all the axioms except any single one. Let $\Gamma = \langle(1,2), v\rangle$ be any game from $\mathcal{G}_2$. Denote by

$$k = \max\{t \mid (t,t) \in v(1,2)\}.$$

If Axiom IORD$_\varepsilon$ is supposed to be held, then it suffices only to define the solution for the class $\mathcal{G}_2^0$. Let $\Gamma = \langle(1,2), v\rangle \in \mathcal{G}_2^0$. Denote by

$$I_i^\varepsilon = \max\{x_i \mid (x_i, -\varepsilon) \in v(1,2)\},\ i = 1,2.$$

Without PO.

$$\sigma_1(\Gamma) = IR_\varepsilon(\Gamma).$$

Without ANO. In the example we have to suppose that $\varepsilon < 0$.

$$\sigma_2(\Gamma) = c(\Gamma) \cup \{(x_1,x_2) \in \partial v(1,2) \mid \varepsilon \leq x_1 \leq 0\}.$$

Without SYM.

$$\sigma_3(\Gamma) = \{(I_1^\varepsilon, -\varepsilon), (-\varepsilon, I_2^\varepsilon)\}.$$

Without IORD$_\varepsilon$.

$$\sigma_4(\Gamma) = \{(x_1,x_2) \in \partial_w v(1,2) \cap IR_\varepsilon \mid x_1 \text{ or } x_2 = k\},$$

for all $\Gamma \in \mathcal{G}_2$.

Without IR_ε.

$$\sigma_5(\Gamma) = \partial_w v(1,2).$$

Without CL. $\sigma_6(\Gamma)$ is the relative interior of the ε-core.

∎

Theorems 1 and 2 completely characterize the ε-core for a fixed ε . The number ε turns out external to a class of cooperative games. Therefore it would be interesting to characterize the collection of ε-cores for all ε's. For this purpose it would be necessary to change Axioms IORD_ε and ε IR_ε in the statement of Theorems 1 and 2 because the number ε is used only in those Axioms.

Axiom IORD_ε describes the scale of measurent of players' payoffs being independent ordinal scales with the fixed difference equal to ε between the points $(\text{v}(1), \text{v}(2))$ and $(\text{v}(1) - \varepsilon, \text{v}(2) - \varepsilon)$. For the normalized games it means that the solution is invariant with respect to ordinal transformations with fixed two points: (0,0) and $(-\varepsilon. -\varepsilon)$. Therefore, if we intend to characterize the collection of all ε-cores we should consider scales not changing all the points of the diagonal of $\Re^2$. However, only the absolute scale, defined by the identical transformations, doesn't change all such points. Nevertheless, it is not necessary to consider the transformations of players' payoffs separately. For example, covariant solutions suppose that there are interpersonal comparisons of players' payoffs.

Thus, for two-person games it would be useful to consider transformations of payoff vectors $\varphi : \Re^2 \to \Re^2$, mapping each game $\Gamma \in \mathcal{G}_2$ to another game $\varphi\Gamma = \langle (1,2), \varphi v\rangle \in \mathcal{G}_2$, where $\varphi\text{v}(i) = \varphi(\text{v}(1), \text{v}(2))_i$, $i = 1,2$, $\varphi v(1,2) = \{(y_1, y_2) | (y_1, y_2) = \varphi(x_1, x_2), (x_1, x_2) \in v(1,2)\}$.

We use such an approach to formulate a new Independence Axiom.

Let $\varphi_1, \varphi_2 : \Re^1_+ \to \Re^1_+$ be arbitrary monotonically non decreasing functions such that $\varphi_i(0) = 0$, $i = 1,2$ such that

$$\varphi_i(t + a) - \varphi_i(t) \geq a \text{ for any } a > 0. \tag{17}$$

Define the function $\varphi : \Re^2 \to \Re^2$ by

$$\varphi(x_1, x_2) = \begin{cases} (x_1, & \text{v}(2) + x_1 - \text{v}(1) + \varphi_1(x_2 - \text{v}(2) - x_1 + \text{v}(1))) \\ & \text{if } x_1 - \text{v}(1) \leq x_2 - \text{v}(2), \\ (\text{v}(1) & +x_1 - \text{v}(2) + \varphi_2(x_1 - \text{v}(1) - x_2 + \text{v}(2)), x_2) \\ & \text{if } x_1 - \text{v}(1) \geq x_2 - \text{v}(2). \end{cases} \tag{18}$$

Lemma 1 *For any $\Gamma \in \mathcal{G}_2$ the game $\varphi\Gamma = \langle (1,2), \varphi v\rangle$ also belongs to the class $\mathcal{G}_2$.*

P r o o f. We have to prove that the set $\varphi v(1,2)$ is comprehensive, i.e. if $x = (x_1, x_2)$, $y = (y_1, y_2) \in \partial v(1,2)$, then $\varphi(x_1, x_2) \ngeq (\nleq) \varphi(y_1, y_2)$.

For simplicity consider only normalized games – the proof is the same for the general case.

As any transformation φ defined in (16) doesn't change the minimal components of vectors it suffices to consider the vectors x, y from the same halfplane bounded by the diagonal of $\Re^2$. Thus, suppose that $x_1 < x_2, y_1 < y_2$, $x_1 > y_1, y_2 > x_2$.

By the condition (17) on the function φ_1 we have

$$\varphi_1(y_2 - y_1) - \varphi_1(x_2 - x_2) \geq y_2 - y_1 - (x_2 - x_1) = y_2 - x_2 + x_1 - y_1 > x_1 - y_1$$

and ,therefore, $y_1 + \varphi_1(y_2 - y_1) > x_1 + \varphi_1(x_2 - x_1)$. ∎

We give now two new Independence Axioms:

Independence of Transformations not changing the Maximal Excess (IMAXE). For any function $\varphi : \Re^2 \to \Re^2$, defined in (15) and (16), $\sigma(\varphi\Gamma) = \varphi(\sigma(\Gamma))$ for any $\Gamma \in \mathcal{G}_2$. It is clear that $\varphi(x_1, x_2) = (x_1, x_2)$ for all $(x_1, x_2) \in \Re^2$ such that $x_1 - \mathrm{v}(1) = x_2 - \mathrm{v}(2)$ and the ε-core satisfies the Axiom for an arbitrary ε.

Antiindependence (AI). If for some $\Gamma = \langle(1,2), v\rangle \in \mathcal{G}_2^0$ $(x, y) \in \partial_w v(1,2) \setminus \sigma(\Gamma)$, then $(x, y) \notin \sigma(\Gamma')$ for all $\Gamma' \in \mathcal{G}_2^0$. Usual Axiom Independence of Irrelevant Alternatives and its versions, applied to the class of NTU games into consideration establishes a connection between the solution sets of different games. Antiindependence Axiom establishes a similar connection between the vectors not belonging to the solution sets. Axiom ε-Individual Rationality has no sense for our purpose, because there is a fixed ε in its statement. We replace it by Axiom Boundedness:

Boundedness (BOUND). For any $\Gamma \in \mathcal{G}_2$ the set $\sigma(\Gamma)$ is either empty or bounded.

Theorem 3 *If a solution σ for the class $\mathcal{G}_2$ satisfies Axioms PO, ANO, WCOV, IMAXE, AI, CL and BOUND then there exists a number ε such that $\sigma(\Gamma) = \varepsilon\text{-}c(\Gamma)$ for all $\Gamma \in \mathcal{G}_2$.*

P r o o f. Notice that Axiom WCOV doesn't follow from Axiom IMAXE as it follows from $\mathrm{IORD}_\varepsilon$ and therefore, it is essential in the statement of the Theorem.

For an arbitrary ε the ε-core satisfies all the Axioms.

Let now σ be any solution for the class $\mathcal{G}_2$, satisfying the Axioms. Then Axiom WCOV implies that without loss of generality it suffices only to consider the class $\mathcal{G}_2^0$ of normalized games, Axiom PO implies that $\sigma(\Gamma) \subset \partial_w v(1,2)$ for any $\Gamma = \langle(1,2), v\rangle \in \mathcal{G}_2$.

Let $\Gamma = \langle(1,2), v\rangle$ be an arbitrary normalized game, $x = (x_1, x_2) \in \sigma(\Gamma)$, and without loss of generality suppose that $x_1 < x_2$. Let $y = (y_1, y_2) \in \partial v_w(1,2)$ be an arbitrary vector such that $y_1 < y_2$ and $y_1 > x_1$. We are going to prove that $y \in \sigma(\Gamma)$.

Define a function φ_1, satisfying (17) as follows:

$$\begin{aligned} \varphi_1(0) &= 0, \\ \varphi_1(x_2 - x_1) &= x_2 - x_1, \\ \varphi_1(z_2 - z_1) &= x_2 - z_1 \text{ for all } z \in \partial_w v(1,2) \\ &\quad\ \text{such that } x_1 \le z_1 \le y_1. \end{aligned}$$

For other values of the arguments let the function $\varphi_1$be arbitrary, but satisfy (17). Then for all functions φ_2, satisfying (17) and for the corresponding transformations $\varphi = (\varphi_1, \varphi_2)$ we have by Axiom IMAXE $\varphi x \in \sigma\varphi\Gamma$ and $\varphi y \ge \varphi x$. Hence by Axiom PO $\varphi y \in \sigma(\varphi\Gamma)$ and by Axiom IMAXE $y \in \sigma(\Gamma)$.

The similar proof shows that if $x \in \sigma(\Gamma)$, $x_1 > x_2$, then any vector $y = (y_1, y_2) \in \partial_w v(1,2)$ such that $y_1 > y_2$ also belongs to the solution. Analogously to the proof of Theorem 1 it is possible to show that there is a transformation φ defined in (18) such that $\varphi\Gamma = \mathrm{Sym}\,\Gamma$. Therefore, by Axioms ANO and CL

$$(x_i, x_j) \in \sigma(\Gamma),\ x_i < x_j \ \to\ \text{there is } (y_i, y_j) \in \sigma(\Gamma),\ y_i > y_j = x_i,\ i, j = 1, 2. \tag{19}$$

The relations (19) together with Axioms CL and BOUND imply that $\sigma(\Gamma) = \varepsilon\text{-}c(\Gamma)$ for some ε.

It remains to prove that the number ε is the same for all $\Gamma \in \mathcal{G}_2^0$. Let Γ be an arbitrary game from the class $\mathcal{G}_2^0$. Then $\sigma(\Gamma) = \varepsilon\text{-}c(\Gamma)$ for some $\varepsilon = \varepsilon_\Gamma$. Then any vector (x, y) with $\min\{x, y\} < -\varepsilon$ doesn't belong to $\sigma(\Gamma)$ and by Axiom AI $(x, y) \notin \sigma(\Gamma')$ for all $\Gamma' \in \mathcal{G}_2^0$. Hence, $\varepsilon_{\Gamma'} \leq \varepsilon_\Gamma$. As the game Γ was chosen arbitrarily, we have $\varepsilon_\Gamma = \varepsilon_{\Gamma'}$ for all Γ, $\Gamma' \in \mathcal{G}_2^0$. ∎

The last result shows that the axioms used in Theorem 3 are independent

Theorem 4 *The Axioms in the statement of Theorem 3 are independent.*

P r o o f. As usual we give examples of solutions $\tau_1, \ldots, \tau_7$ not equal the ε-core and satisfying all the axioms except anyone. If Axiom WCOV is present in the corresponding systems of axioms, it suffices to consider only the class $\mathcal{G}_2^0$.

Without PO.

$$\tau_1(\Gamma) = IR(\Gamma) \text{ for all } \Gamma = \langle(1,2), v\rangle \in \mathcal{G}_2^0.$$

Without ANO.

$$\tau_2 = \sigma_2 \text{ for some } \varepsilon < 0,$$

where the solution σ_2 was defined in the proof of Theorem 2.

Without WCOV.

$$\tau_5(\Gamma) = (\mathrm{v}_1 + \mathrm{v}_2)\text{-}c(\Gamma) \text{ for all } \Gamma = \langle(1,2), v\rangle \in \mathcal{G}_2.$$

Without IMAXE.

$$\tau_4(\Gamma) = \varepsilon\text{-}c(\Gamma) \cap \{(x_1, x_2) \,|\, x_1 + x_2 \geq a\}$$

for some numbers ε, a and for all $\Gamma \in \mathcal{G}_2^0$.

Without AI.

$$\tau_3(\Gamma) = \varepsilon_\Gamma\text{-}c(\Gamma) \text{ for all } \Gamma = \langle(1,2), v\rangle \in \mathcal{G}_2^\circ,$$

where the number ε_Γ may depend on $k = \{t \,|\, (t,t) \in v(1,2)\}$.

Without CL. τ_6 is the relative interior of the ε-core for some ε.

Without BOUND. $\tau_7(\Gamma) = \partial_w v(1,2)$.

∎

References

Dutta B.(1990) The Egalitarian Solution and the Reduced Game Properties in Convex Games. *International Journal of Game Theory*, **19**, 153-169.

Kalai E.(1978) Excess Functions for Cooperative Games without Sidepaymentts. *SIAM Journal of Appl. Math.* **20**, 1, 60-71.

Maschler M.(1992) The bargaining set, kernel and nucleolus in: Aumann R., Hart S. eds. *Handbook of Game Theory*, v.1, Elsevier Sci.Publishers, pp.591-663.

Peleg B.(1985) An axiomatization of the core of cooperative games without side payments, *Journal of Math. Economics*, **14**, 203-214.

Elena Yanovskaya
Insitute for Economics and Mathematics
Russian Academy of Sciences
Tchaikovsky st.1, 191187 St.Petersburg
Russia

REDUCED GAME PROPERTIES OF EGALITARIAN DIVISION RULES FOR TU-GAMES

T.S.H. Driessen and Y. Funaki [1]

Abstract: *The egalitarian non-individual contribution (ENIC-)value represents the equal division of the surplus of the total profits, given that each player is already allocated some kind of a yet unspecified individual contribution. Four particular versions, the CIS-, ENSC-, ENPAC-, ENBC-values, are also considered by choosing the individual worth, the separable contribution, the pairwise-averaged contribution and the Banzhaf contribution as the notion of individual contribution. Axiomatic characterizations of the ENIC-value in general and the four particular ENIC values are provided on the class of cooperative games with a fixed player set as well as a variable player set. The latter axiomatization involves a consistency axiom in terms of reduced games.*

1 Introduction

The paper focuses on a uniform treatment of a special type of one-point solutions for cooperative games, called the egalitarian non-individual contribution (ENIC-)value. The value in question is special in the sense that it is constructed in two stages: first and foremost, each player in the game is allocated some kind of a yet unspecified individual contribution and subsequently, the surplus of the total profits is equally divided among all the players. In addition to the general setting, the unspecified notion of individual contribution will be chosen to be one of the following four particular notions:

- the individual worth of the player to participate in the game as a solitary player. The corresponding ENIC-value is known as the center of the imputation set, called CIS-value.
- the so-called separable (or marginal) contribution of the player to participate in the game as a member of the player set. The corresponding ENIC-value goes by the name of egalitarian non-separable contribution (ENSC-)value and has widely been discussed in the field of multipurpose water resource projects (cf. Young et al. [23]), social choice problems (cf. Moulin [15]) and cost allocation problems in combination with cooperative game theory (cf. Dragan et al.[2]; Driessen [3],[4]; Funaki [9]; Legros [14]; Driessen and Funaki [6]).
- the so-called pairwise averaged contribution of the player in the game which is obtainable as some average of marginal contributions of pairs of players including the player in question. The corresponding ENIC-value is called the egalitarian non-pairwise-averaged contribution (ENPAC-)value and has been studied in detail in Driessen and Funaki [7].

[1]The research for this paper was done during the sabbatical leave period (April 1992 till September 1993) of the second author at the Department of Applied Mathematics, University of Twente, Enschede, the Netherlands.

- the so-called Banzhaf contribution of the player in the game which is interpretable as the average of all the player's marginal contributions to participate in the game anyhow. The corresponding ENIC-value is called the egalitarian non-Banzhaf contribution (ENBC-) value and coincides with the least square prenucleolus which is studied in Ruiz et al.[17]. The notion of Banzhaf contribution itself is closely related to the well-known Shapley value as well as the Banzhaf index for simple games.

The main goal of the paper is to provide an axiomatic characterization of the ENIC-value in general and the four particular ENIC-values as well. In fact, our goal is twofold because the player set may be considered as fixed or variable. Involving a fixed player set, the main Theorem 3.1 states that the ENIC-value is fully characterized by the classical axiom of relative invariance under strategic equivalence (RISE) and a modification of the classical axiom of equal treatment of substitutes in the game (ETP). Here the modified axiom requires equal value payoffs for players with identically individual contributions. Involving a variable player set, the main Theorem 4.3 states that the ENIC-value is fully characterized by RISE, ETP and a so-called consistency axiom in terms of reduced games. The notion of a reduced game has been used in axiomatizations of many other game theoretic solutions (cf. Funaki [10]; Funaki and Yamato [11]; Sobolev [21], [22]; Peleg [18], [19]; Hart and Mas-Colell [13]) and is elucidated at the beginning of Section 4. Concerning the axiomatization of the ENIC-value in general, the reduced game is yet unspecified except for the player set, whereas the axiomatizations of the four particular ENIC-values do involve specified reduced games (cf. Corollaries 4.4 - 4.6). At the same time, Theorem 4.3 gives a new method to find several reduced games to characterize the ENIC-values. If we find a reduced game which provides *weakly* consistency of the notion IC, the consistency with respect to the reduced game characterizes the ENIC-value.

The organization of the paper is as follows. In Section 2 we introduce the general form of the egalitarian non-individual contribution value and four particular versions of it. Moreover, two basic properties of one-point solutions for cooperative games are mentioned. Section 3 provides an axiomatic characterization of the ENIC-value on the class of games with a fixed player set, while Section 4 provides it on the class of all games. Concluding remarks are listed in Section 5.

2 The class of egalitarian non-individual contribution (ENIC-) values

Let (N, v) be a *cooperative n-person game in characteristic function form* with a finite set N of n *players*, $n \geq 1$, and the *characteristic function* $v : 2^N \to \mathbf{R}$, satisfying $v(\phi) = 0$, on the set 2^N of all subsets of N. A vector $x \in \mathbf{R}^N$ is said to be an *efficient payoff vector* for a game (N, v) if it meets the *efficiency* principle $\sum_{j \in N} x_j = v(N)$. The set of efficient payoff vectors for the game (N, v) is called the *pre-imputation set* for (N, v) and is denoted by $I^*(N, v)$.

A *value* concept for cooperative games is a function σ such that $\sigma(N, v) \in I^*(N, v)$ for any game (N, v). Any value can be interpreted as an allocation rule of the total profits among the players in a game. Here a value σ is called *standard for two-person games* if for all games $(\{i, j\}, v)$ with $i \neq j$

$$\sigma_i(\{i, j\}, v) = v(\{i\}) + \frac{1}{2}[v(\{i, j\}) - v(\{i\}) - v(\{j\})],$$

i.e., the "surplus" $v(\{i, j\}) - v(\{i\}) - v(\{j\})$ is equally divided among the two players according to the value. The class of all cooperative games with fixed player set N and a variable player set respectively are denoted by G^N and G. This paper is dedicated to the study of a special class of values which are introduced as follows.

In the first stage of the value for a game (N,v), each player i receives some kind of a yet unspecified *individual contribution* denoted by $IC_i(N,v)$. In the second stage, the amount of what is left of the total profits is equally divided among all the n players. That is, the so-called *egalitarian non-individual contribution* (ENIC-)value allocates an equal share of the *non-individual contribution* $NIC(N,v) := v(N) - \sum_{j\in N} IC_j(N,v)$ to each player and it is formally given by

$$\mathrm{ENIC}_i(N,v) := IC_i(N,v) + \frac{1}{n} NIC(N,v) \quad \text{for all } (N,v) \in G, \quad \text{all } i \in N.$$

Evidently, the ENIC-value meets indeed the efficiency principle and moreover, the ENIC-value is standard for two-person games if and only if the differences between the individual contribution and the individual worth in the game are the same for both players, i.e., $IC_i(\{i,j\},v) - v(\{i\}) = IC_j(\{i,j\},v) - v(\{j\})$ for all games $(\{i,j\},v)$ with $i \neq j$.

In addition to the study of the general ENIC-value, we pay attention to four special versions of the ENIC-value by specifying the notion of individual contribution. Firstly, if the individual contribution $IC_i(N,v)$ of any player i is chosen to be his individual worth $v(\{i\})$ in the game (N,v), then the ENIC-value becomes the *center* of gravity of the *imputation set*, called CIS-value. Here the *imputation set* $I(N,v)$ for and the CIS-value of a game (N,v) are formally given by

$$I(N,v) := \{x \in I^*(N,v) \mid x_i \geq v(\{i\}) \text{ for all } i \in N\},$$

$$\mathrm{CIS}_i(N,v) := v(\{i\}) + \frac{1}{n}\left[v(N) - \sum_{j\in N} v(\{j\})\right] \quad \text{for all } i \in N.$$

Secondly, if the individual contribution $IC_i(N,v)$ of any player i is chosen to be his marginal contribution from an $(n-1)$-person coalition to the grand coalition, called *separable contribution* and denoted by $SC_i(N,v) := v(N) - v(N \setminus \{i\})$, then the corresponding ENIC-value becomes the *egalitarian non-separable contribution* (ENSC-)value, which is formally given by

$$\mathrm{ENSC}_i(N,v) := SC_i(N,v) + \frac{1}{n}\left[v(N) - \sum_{j\in N} SC_j(N,v)\right] \quad \text{for all } i \in N.$$

Thirdly, if the individual contribution $IC_i(N,v)$ of any player i is chosen to be his *pairwise-averaged contribution* to the grand coalition, defined by

$$PAC_i(N,v) := \begin{cases} v(N) - \frac{1}{n-2}\sum_{j\in N\setminus\{i\}} v(N \setminus \{i,j\}) & \text{if } n \geq 3 \\ v(\{i\}) & \text{if } n = 1 \text{ or } n = 2, \end{cases}$$

then the corresponding ENIC-value becomes the *egalitarian non-pairwise-averaged contribution* (ENPAC-)value, which is formally given by

$$\mathrm{ENPAC}_i(N,v) := PAC_i(N,v) + \frac{1}{n}\left[v(N) - \sum_{j\in N} PAC_j(N,v)\right] \quad \text{for all } i \in N.$$

Fourthly, if the individual contribution $IC_i(N,v)$ of any player i is chosen to be the average of all his marginal contributions, called *Banzhaf contribution* and defined by

$$BC_i(N,v) := \frac{1}{2^{n-1}} \sum_{S \subseteq N \setminus \{i\}} [v(S \cup \{i\}) - v(S)] \quad \text{for all } n \geq 1,$$

then the corresponding ENIC-value becomes the *egalitarian non-Banzhaf contribution* (ENBC-) value, which is formally given by

$$\text{ENBC}_i(N,v) := BC_i(N,v) + \frac{1}{n}\left[v(N) - \sum_{j \in N} BC_j(N,v)\right] \quad \text{for all } i \in N.$$

Clearly, all four special versions of the ENIC-value are standard for two-person games and further, straightforward calculations yield that for three-person games (N,v)

$$\text{ENPAC}(N,v) = \text{CIS}(N,v) \text{ and } \text{ENBC}(N,v) = \frac{1}{2}\text{CIS}(N,v) + \frac{1}{2}\text{ENSC}(N,v).$$

Without going into technical details we also remark that

$$\sum_{j \in N} BC_j(N,v) = \frac{1}{2^{n-1}} \sum_{S \subseteq N} [2|S| - n]\, v(S).$$

Here $|S|$ denotes the cardinality of coalition S. The allocation rule based on the ENSC-value is well-known in the cost allocation literature and has been studied in many papers, e.g., Driessen [3],[4]; Funaki [9];Legros [14]. In Driessen and Funaki [6] it is established that the ENSC-value has some geometric relationships to both the Shapley value and the pre-nucleolus. To be more concrete, the Shapley value is located on the line segment between the CIS- and ENSC-values for some classes of games and further, the ENSC-value coincides with the pre-nucleolus for some other classes of games. The ENPAC-value was introduced in Driessen and Funaki [7]. The reasoning behind the notion of average contribution and some detailed properties of the ENPAC-value can be found in the same paper. The collinearity property among the ENPAC-, ENSC-, CIS-values and the Shapley value is investigated for a suitably defined class of TU-games in Dragan et al.[2].

Let's make some remarks referring to the notion of Banzhaf contribution. In the framework of voting situations, or its game theoretic setting by means of simple games (N,v) satisfying $v(S) \in \{0,1\}$ for all $S \subseteq N$ and $v(N) = 1$, Banzhaf [1] introduced a so-called power index η given by

$$\eta_i(N,v) := \sum_{S \subseteq N \setminus \{i\}} [v(S \cup \{i\}) - v(S)] \quad \text{for all } i \in N.$$

Two modified versions of this non-normalized Banzhaf index are presented in Dubey and Shapley [8]. One modified version is the normalized Banzhaf index which is given by

$$\beta_i(N,v) := \frac{\eta_i(N,v)}{\sum_{j \in N} \eta_j(N,v)} \quad \text{for all } i \in N.$$

The other is the absolute Banzhaf index which is given by

$$\beta_i(N,v) := \eta_i(N,v)/2^{n-1} \quad \text{for all } i \in N.$$

An axiomatization of the non-normalized Banzhaf index and the absolute Banzhaf index respectively can be found in Dubey and Shapley [8] and Owen [16]. Notice that the absolute Banzhaf index, extended to the class of all games, is our notion of Banzhaf contribution and does not meet in general the efficiency principle. Our approach on basis of the egalitarian division of the non-individual contribution $v(N) - \sum_{j\in N} BC_j(N,v)$ guarantees that the corresponding ENBC-value is indeed efficient and besides, it fits into our general model of ENIC-values. Further, straightforward calculations yield that the ENBC-value for three-person games equals the well-known Shapley value. Indeed the ENBC-value coincides with the so-called least square prenucleolus. Many properties of this value are investigated in Ruiz et al.[17].

We conclude this section with two elementary properties of values in general. A game (N,w) is called *strategically equivalent* to a game (N,v) if there exist a real number $\alpha > 0$ and a vector $\beta \in \mathbf{R}^N$ such that $w(S) = \alpha v(S) + \sum_{j\in S}\beta_j$ for all $S \subseteq N$, in short $w = \alpha v + \beta$ for some $\alpha > 0$ and $\beta \in \mathbf{R}^N$.

Axiom. A value σ on G is said to be *relatively invariant* under *strategic equivalence* (RISE) if

$$\sigma(N,w) = \alpha\sigma(N,v) + \beta \qquad \begin{array}{l}\text{for all games } (N,v) \in G \text{ and } (N,w) \in G\\ \text{satisfying } w = \alpha v + \beta \text{ for some } \alpha > 0 \text{ and } \beta \in \mathbf{R}^N.\end{array}$$

Axiom. A value σ on G is said to possess the *equal treatment property* (ETP) if

$$\sigma_i(N,v) = \sigma_j(N,v) \text{ for any substitutes } i,j \in N \text{ in a game } (N,v) \in G.$$

Here players i and j are called *substitutes* in the game (N,v) if

$$v(S \cup \{i\}) = v(S \cup \{j\}) \text{ for all } S \subseteq N \setminus \{i,j\}.$$

Both axioms are well-known and need no further explanation. Usually, RISE and ETP must be satisfied to conclude that the value is standard for two-person games.

Definition. The notion IC of individual contribution is said to be *relatively invariant* under *strategic equivalence* (RISE) if

$$IC(N,w) = \alpha IC(N,v) + \beta \qquad \begin{array}{l}\text{for all games } (N,v) \in G \text{ and } (N,w) \in G\\ \text{satisfying } w = \alpha v + \beta \text{ for some } \alpha > 0 \text{ and } \beta \in \mathbf{R}^N.\end{array}$$

Lemma 2.1 *(i) Suppose the notion IC of individual contribution satisfies RISE. Then the ENIC-value on G satisfies RISE.*

(ii) The notions of individual worth, separable-, pairwise-averaged- and Banzhaf-contribution respectively satisfy RISE.

(iii) The CIS-, ENSC-, ENPAC- and ENBC-values respectively satisfy RISE as well as ETP.

Proof. Let $(N,v) \in G$ and $(N,w) \in G$ be such that $w = \alpha v + \beta$ for some $\alpha > 0$ and $\beta \in \mathbf{R}^N$.

(i) Suppose $IC(N,w) = \alpha IC(N,v) + \beta$. Then it follows that

$$\begin{aligned} NIC(N,w) &= w(N) - \sum_{j\in N} IC_j(N,w) \\ &= \alpha v(N) + \sum_{j\in N} \beta_j - \sum_{j\in N}[\alpha IC_j(N,v) + \beta_j] \\ &= \alpha\Big[v(N) - \sum_{j\in N} IC_j(N,v)\Big] \\ &= \alpha NIC(N,v). \end{aligned}$$

So, for all $i \in N$,

$$\begin{aligned} \text{ENIC}_i(N,w) &= IC_i(N,w) + \frac{1}{n} NIC(N,w) \\ &= \alpha IC_i(N,v) + \beta_i + \frac{\alpha}{n} NIC(N,v) \\ &= \alpha\ \text{ENIC}_i(N,v) + \beta_i. \end{aligned}$$

Thus, $\text{ENIC}(N,w) = \alpha\ \text{ENIC}(N,v) + \beta$ whenever $w = \alpha v + \beta$.

(ii) It is left to the reader to verify that the notions of individual worth, separable- and Banzhaf-contribution satisfy RISE. Concerning the pairwise-averaged contribution in case $n \geq 3$, we obtain for all $i \in N$

$$\begin{aligned} PAC_i(N,w) &= w(N) - \frac{1}{n-2} \sum_{j\in N\setminus\{i\}} w(N\setminus\{i,j\}) \\ &= \alpha v(N) + \sum_{j\in N}\beta_j - \frac{1}{n-2}\sum_{j\in N\setminus\{i\}}\left[\alpha v(N\setminus\{i,j\}) + \sum_{l\in N\setminus\{i,j\}}\beta_l\right] \\ &= \alpha PAC_i(N,v) + \sum_{j\in N}\beta_j - \frac{1}{n-2}(n-2)\sum_{l\in N\setminus\{i\}}\beta_l \\ &= \alpha PAC_i(N,v) + \beta_i. \end{aligned}$$

(iii) In view of part (i)-(ii), all four values satisfy RISE. It is left to the reader to verify that the CIS-, ENSC- and ENBC-values satisfy ETP. Concerning the pairwise-averaged contribution in case $n \geq 3$, we obtain for any substitutes $i, j \in N$ in a game $(N,v) \in G$ that

$$\begin{aligned} PAC_i(N,v) &= v(N) - \frac{1}{n-2}v(N\setminus\{i,j\}) - \frac{1}{n-2}\sum_{k\in N\setminus\{i,j\}} v(N\setminus\{i,k\}) \\ &= v(N) - \frac{1}{n-2}v(N\setminus\{j,i\}) - \frac{1}{n-2}\sum_{k\in N\setminus\{j,i\}} v(N\setminus\{j,k\}) \\ &= PAC_j(N,v), \end{aligned}$$

where the second equality holds because of $v(N\setminus\{i,k\}) = v(N\setminus\{j,k\})$ for all $k \in N\setminus\{i,j\}$. From $PAC_i(N,v) = PAC_j(N,v)$ it follows immediately that $\text{ENPAC}_i(N,v) = \text{ENPAC}_j(N,v)$ for any substitutes i and j in (N,v). □

3 Axiomatizations of ENIC-values with respect to a fixed player set

Throughout this section, we fix the player set N and pay attention to an axiomatic characterization of the ENIC-value on the class G^N of games with player set N. First we put forward a modification of the classical equal treatment property for values in general.

Axiom. A value σ on G^N is said to possess the *equal treatment property* with respect to the notion of individual contribution (ETP_{ic}) if

$$\sigma_i(N,v) = \sigma_j(N,v) \quad \begin{array}{l}\text{for all } i,j \in N \text{ in a game } (N,v)\\ \text{satisfying } IC_i(N,v) = IC_j(N,v).\end{array}$$

The axiom states that players with the same individual contributions receive the same payoff according to the value involved. In other words, the axiom requires that the value is strongly dependent on the notion of individual contribution and as such, the ETP_{ic}-axiom is considered as a stronger axiom than the classical ETP-axiom. In particular, it turns out that ETP_{ic} implies ETP in the context of the CIS-, ENSC-, ENPAC- and ENBC-values. Obviously, the ENIC-value on G^N satisfies ETP_{ic}. The next theorem expresses that RISE and ETP_{ic} fully characterize the ENIC-value.

Theorem 3.1 *Suppose the notion IC of individual contribution satisfies RISE. Then the ENIC-value on G^N is the unique value on G^N which possesses RISE and ETP_{ic}.*

Proof. As noted before, the ENIC-value satisfies RISE and ETP_{ic}. Concerning the uniqueness part, let σ be a value on G^N that possesses RISE and ETP_{ic}. We prove that $\sigma(N,v) = \mathrm{ENIC}(N,v)$ for any game $(N,v) \in G^N$. Given the game (N,v), define the strategically equivalent game (N,w) by

$$w(S) := v(S) - \sum_{j\in S} IC_j(N,v) \quad \text{for all } S \subseteq N,$$

in short $w = v - IC(N,v)$. Due to RISE of the notion IC, we have $IC_i(N,w) = IC_i(N,v) - IC_i(N,v) = 0$ for all $i \in N$. Together with ETP_{ic} of the value σ, this yields $\sigma_i(N,w) = \sigma_j(N,w)$ for all $i,j \in N$. From this and the efficiency principle, we deduce that

$$\sigma_i(N,w) = \frac{w(N)}{n} = \frac{1}{n}\left[v(N) - \sum_{j\in N} IC_j(N,v)\right] = \mathrm{ENIC}_i(N,v) - IC_i(N,v)$$

for all $i \in N$. Now it follows from RISE of the value σ that

$$\sigma(N,v) = \sigma(N,w) + IC(N,v) = \mathrm{ENIC}(N,v).$$

□

The theorem above can easily be applied to the CIS-, ENSC-, ENPAC- and ENBC-values respectively. For that purpose we state four different axioms.

Axiom. Let $n \geq 3$. A value σ on G^N is said to possess the *equal treatment property* with respect to the notion of individual contribution concerning 1-, $(n-1)$-, sums of $(n-2)$-person coalitions and sums of arbitrary coalitions respectively if the following condition holds for all games (N, v):

ETP_{ic}^{1}: $\sigma_i(N,v) = \sigma_j(N,v)$ for all $i,j \in N$ with $v(\{i\}) = v(\{j\})$
ETP_{ic}^{n-1}: $\sigma_i(N,v) = \sigma_j(N,v)$ for all $i,j \in N$ with $v(N\setminus\{i\}) = v(N\setminus\{j\})$
ETP_{ic}^{n-2}: $\sigma_i(N,v) = \sigma_j(N,v)$
for all $i,j \in N$ with $\sum_{l\in N\setminus\{i\}} v(N\setminus\{i,l\}) = \sum_{l\in N\setminus\{j\}} v(N\setminus\{j,l\})$
ETP_{ic}^{*}: $\sigma_i(N,v) = \sigma_j(N,v)$
for all $i,j \in N$ with $\sum_{S\subseteq N\setminus\{i\}} v(S) = \sum_{S\subseteq N\setminus\{j\}} v(S)$.

The ETP_{ic}^{n-1}-, ETP_{ic}^{n-2}- and ETP_{ic}^{*}-axioms require that the value of any player is dependent on the worth of the $(n-1)$-person coalition, the sum of the worth of $(n-2)$-person coalitions or the sum of the worth of all coalitions, excluding the relevant player, respectively. Note that ETP_{ic}^{1} and ETP_{ic}^{n-2} are identical if $n = 3$. As a direct result of Theorem 3.1 and Lemma 2.1 (ii), we provide axiomatizations of the CIS-, ENSC-, ENPAC- and ENBC-values on G^N.

Corollary 3.2 *Let* $n \geq 3$.

(i) The CIS-value on G^N *is the unique value on* G^N *which possesses RISE and* ETP_{ic}^{1}.

(ii) The ENSC-value on G^N *is the unique value on* G^N *which possesses RISE and* ETP_{ic}^{n-1}.

(iii) The ENPAC-value on G^N *is the unique value on* G^N *which possesses RISE and* ETP_{ic}^{n-2}.

(iv) The ENBC-value on G^N *is the unique value on* G^N *which possesses RISE and* ETP_{ic}^{*}.

Proof. We merely prove part (iv). For all $i,j \in N$ in a game (N,v) we have the following equivalences:

$$
\begin{aligned}
IC_i(N,v) = IC_j(N,v) \;&\Leftrightarrow\; BC_i(N,v) = BC_j(N,v) \;\Leftrightarrow\\
\sum_{S\not\ni i}[v(S\cup\{i\}) - v(S)] &= \sum_{S\not\ni j}[v(S\cup\{j\}) - v(S)] \;\Leftrightarrow\\
\sum_{T\ni i} v(T) - \sum_{S\not\ni i} v(S) &= v\sum_{T\ni j} v(T) - \sum_{S\not\ni j} v(S) \;\Leftrightarrow\\
\sum_{T\subseteq N} v(T) - 2\sum_{T\not\ni i} v(T) &= \sum_{T\subseteq N} v(T) - 2\sum_{T\not\ni j} v(T) \;\Leftrightarrow\\
\sum_{T\not\ni i} v(T) &= \sum_{T\not\ni j} v(T).
\end{aligned}
$$

So, in the context of the Banzhaf contribution of any player, the general ETP_{ic}-axiom reduces to the ETP_{ic}^{*}-axiom and therefore, part (iv) follows immediately from Theorem 3.1 and Lemma 2.1 (ii). □

4 Axiomatizations of ENIC-values with respect to a variable player set

Throughout this section, we consider the ENIC-value on the class G of all games with a variable player set. In order to axiomatize the ENIC-value on G, we make use of the notion of a reduced game. A *reduced game* is deducible from a given game $(N, v), n \geq 2$, by removing one player on the understanding that the removed player i will be paid according to a proposed payoff vector $x \in \mathbf{R}^N$. The remaining players form the player set $N \setminus \{i\}$ of the reduced game; the characteristic function of which is composed of both the original characteristic function v and the proposed payoff vector x. At the moment we don't specify the characteristic function $v^x : 2^{N\setminus\{i\}} \to \mathbf{R}$ except for the requirement $v^x(N \setminus \{i\}) := v(N) - x_i$ in order to meet the efficiency principle of the value later on.

Axiom. A value σ on G is said to possess the *reduced game property* (RGP) with respect to a specified type of reduced games if

$$\sigma_j(N \setminus \{i\}, v^{\sigma(N,v)}) = \sigma_j(N, v) \qquad \begin{array}{l}\text{for all } (N,v) \in G \text{ with } n \geq 2,\\ \text{all } i \in N \text{ and all } j \in N \setminus \{i\}.\end{array}$$

The reduced game property for a value states that if all the players are supposed to be paid according to the value of the original game, then the players of the reduced game can achieve the same value payoff. In other words, there is no inconsistency in what the players of the reduced game can achieve, in either the original game or the reduced game. Thus, the reduced game property can be seen as a property of consistency. Notice that RGP is trivial involving two-person games $(\{i, j\}, v)$ with $i \neq j$ because the efficiency of a value implies $\sigma_j(\{j\}, v^{\sigma(N,v)}) = v^{\sigma(N,v)}(\{j\}) = v(N) - \sigma_i(N, v) = \sigma_j(N, v)$. According to the next theorem, the ENIC-value on G satisfies RGP if and only if the differences between the underlying individual contributions of any player in the original and reduced game are the same.

Theorem 4.1 *Let (N, v) be a game with $n \geq 3$, $i \in N$ and $(N\setminus\{i\}, v^x)$ some reduced game satisfying $v^x(N \setminus \{i\}) = v(N) - x_i$ for any $x \in \mathbf{R}^N$. Put $y := ENIC(N, v)$. The following two statements are equivalent.*

(i) $ENIC_j(N \setminus \{i\}, v^y) = ENIC_j(N, v)$ *for all* $j \in N \setminus \{i\}$.

(ii) There exists a constant $\alpha = \alpha(N, v, i) \in \mathbf{R}$ *such that*

$$IC_j(N \setminus \{i\}, v^y) = IC_j(N, v) + \alpha \quad \textit{for all } j \in N \setminus \{i\}.$$

Proof. For any $j \in N \setminus \{i\}$ we have the following equivalences:

$$\text{ENIC}_j(N \setminus \{i\}, v^y) = \text{ENIC}_j(N, v)$$

$$\Leftrightarrow \quad IC_j(N \setminus \{i\}, v^y) + \frac{1}{n-1}\left[v^y(N \setminus \{i\}) - \sum_{l \in N\setminus\{i\}} IC_l(N \setminus \{i\}, v^y)\right]$$

$$= IC_j(N, v) + \frac{1}{n} NIC(N, v)$$

$$\Leftrightarrow \quad IC_j(N\setminus\{i\},v^y) - IC_j(N,v)$$
$$= \frac{1}{n}NIC(N,v) - \frac{1}{n-1}\left[v(N) - y_i - \sum_{l\in N\setminus\{i\}} IC_l(N\setminus\{i\},v^y)\right].$$

The fact that $y_i = \mathrm{ENIC}_i(N,v) = IC_i(N,v) + \frac{1}{n}NIC(N,v)$ yields

$$\begin{aligned}
\frac{1}{n}NIC(N,v) - \frac{1}{n-1}[v(N) - y_i] &= \frac{1}{n-1}[NIC(N,v) - v(N) + IC_i(N,v)] \\
&= \frac{1}{n-1}\left[-\sum_{l\in N} IC_l(N,v) + IC_i(N,v)\right] \\
&= -\frac{1}{n-1}\sum_{l\in N\setminus\{i\}} IC_l(N,v).
\end{aligned}$$

¿From this we conclude that the following equivalences hold:

$$\mathrm{ENIC}_j(N\setminus\{i\},v^y) = \mathrm{ENIC}_j(N,v) \qquad \text{for all } j\in N\setminus\{i\}$$
$$\Leftrightarrow \quad IC_j(N\setminus\{i\},v^y) - IC_j(N,v) = \frac{1}{n-1}\sum_{l\in N\setminus\{i\}}\left[IC_l(N\setminus\{i\},v^y) - IC_l(N,v)\right]$$
$$\text{for all } j\in N\setminus\{i\}$$

$$\Leftrightarrow \quad IC_j(N\setminus\{i\},v^y) - IC_j(N,v) = \textit{constant} \qquad \text{for all } j\in N\setminus\{i\}.$$

□

In view of the characterization of the reduced game property for the ENIC-value mentioned in part (ii) of Theorem 4.1, we are interested in similar relations involving arbitrary vectors instead of a particular vector y.

Proposition 4.2 *Let (N,v) be a game with $n\geq 3$, $x\in\mathbf{R}^N$, $i\in N$ and $(N\setminus\{i\},v^x)$ some reduced game satisfying $v^x(N\setminus\{i\}) = v(N) - x_i$. Suppose that there exists a constant $\alpha = \alpha(N,v,i)\in\mathbf{R}$ such that*

$$IC_j(N\setminus\{i\},v^x) = IC_j(N,v) + \alpha \quad \textit{for all } j\in N\setminus\{i\}.$$

Then

(i) $NIC(N\setminus\{i\},v^x) = \frac{n-1}{n}NIC(N,v) + ENIC_i(N,v) - x_i - (n-1)\alpha$

(ii) $ENIC_j(N\setminus\{i\},v^x) = ENIC_j(N,v) + \frac{1}{n-1}[\,ENIC_i(N,v) - x_i]$ *for all* $j\in N\setminus\{i\}$

(iii) $ENIC_j(N\setminus\{i\},v^x) = ENIC_j(N,v)$ *for all* $j\in N\setminus\{i\}$ *whenever* $x = ENIC(N,v)$.

Proof. (i)

$$NIC(N\setminus\{i\},v^x) = v^x(N\setminus\{i\}) - \sum_{l\in N\setminus\{i\}} IC_l(N\setminus\{i\},v^x)$$

$$= v(N) - x_i - \sum_{l \in N \setminus \{i\}} IC_l(N,v) - (n-1)\alpha$$
$$= NIC(N,v) + IC_i(N,v) - x_i - (n-1)\alpha$$
$$= \frac{n-1}{n} NIC(N,v) + \text{ENIC}_i(N,v) - x_i - (n-1)\alpha.$$

(ii) We deduce from part (i) that for all $j \in N \setminus \{i\}$

$$\begin{aligned}
\text{ENIC}_j(N \setminus \{i\}, v^x) &= IC_j(N \setminus \{i\}, v^x) + \frac{1}{n-1} NIC(N \setminus \{i\}, v^x) \\
&= IC_j(N,v) + \alpha + \frac{1}{n} NIC(N,v) \\
&\quad + \frac{1}{n-1}[\,\text{ENIC}_i(N,v) - x_i] - \alpha \\
&= \text{ENIC}_j(N,v) + \frac{1}{n-1}[\,\text{ENIC}_i(N,v) - x_i].
\end{aligned}$$

This proves part (ii). Part (iii) is a direct consequence of part (ii). □

Our main goal is to axiomatize the ENIC-value on G. Due to Proposition 4.2 (iii), the ENIC-value on G satisfies RGP, provided that the notion of individual contribution possesses the so-called weak consistency property.

Definition. The notion IC of individual contribution is said to be *weakly consistent* (WCONS) with respect to a specified type of reduced games if for all $(N,v) \in G$ with $n \geq 3$, all $i \in N$, there exists a constant $\alpha = \alpha(N,v,i) \in \mathbf{R}$ such that

$$IC_j(N \setminus \{i\}, v^x) = IC_j(N,v) + \alpha \quad \text{for all } x \in \mathbf{R}^N, \text{ all } j \in N \setminus \{i\}.$$

Definition. The notion IC of individual contribution is said to possess the *equal treatment property* (ETP) if

$$IC_i(N,v) = IC_j(N,v) \;\; \text{for any substitutes } i,j \in N \text{ in a game } (N,v).$$

Obviously, the ENIC-value on G satisfies ETP if and only if the notion of individual contribution on G satisfies ETP. Now we are able to provide an axiomatization of the ENIC-value.

Theorem 4.3 *Suppose the notion IC of individual contribution satisfies RISE, ETP and WCONS with respect to a specified type of reduced games. Then the ENIC-value on G is the unique value on G which possesses RISE, ETP and RGP with respect to that reduced game.*

Proof. As noted before, the ENIC-value satisfies RISE, ETP and RGP. Concerning the uniqueness part, let σ be a value on G that possesses the following properties: RISE, ETP and RGP. We prove by induction on the number $|N|$ of players that $\sigma(N,v) = \text{ENIC}(N,v)$ for any game (N,v). The case $|N| = 1$ is trivial because of efficiency, while the case where $|N| = 2$ follows from the fact that any value satisfying RISE and ETP is standard for two-person games. Thus, let $(N,v) \in G$ with $|N| \geq 3$ and suppose that

$$\sigma(\overline{N}, w) = \text{ENIC}(\overline{N}, w) \quad \text{for all } (\overline{N}, w) \in G \text{ with } 1 \leq |\overline{N}| < |N|.$$

Put $x := \sigma(N,v)$. Now we obtain that for all $i \in N$, all $j \in N \setminus \{i\}$

$$\begin{aligned} \sigma_j(N,v) = & \quad \sigma_j(N \setminus \{i\}, v^x) = \text{ENIC}_j(N \setminus \{i\}, v^x) \\ = & \quad \text{ENIC}_j(N,v) + \tfrac{1}{n-1}[\text{ ENIC}_i(N,v) - x_i], \end{aligned}$$

where the equalities follow from RGP of σ, the induction hypothesis and Proposition 4.2 (ii) respectively. Hence,

$$\sigma_j(N,v) - \text{ ENIC}_j(N,v) = \frac{1}{n-1}[\text{ ENIC}_i(N,v) - \sigma_i(N,v)]$$

for all $i,j \in N, i \neq j$. By interchanging the roles of i and j, we also get

$$\sigma_i(N,v) - \text{ ENIC}_i(N,v) = \frac{1}{n-1}[\text{ ENIC}_j(N,v) - \sigma_j(N,v)]$$

for all $i,j \in N, i \neq j$. Combining both latter equalities yields

$$\sigma_i(N,v) - \text{ ENIC}_i(N,v) = \frac{1}{(n-1)^2}[\sigma_i(N,v) - \text{ ENIC}_i(N,v)]$$

for all $i \in N$. Together with $n \geq 3$, this implies $\sigma_i(N,v) = \text{ ENIC}_i(N,v)$ for all $i \in N$. So, we conclude $\sigma(N,v) = \text{ ENIC}(N,v)$ which completes the inductive proof of uniqueness. □

In the sequel we present specified types of reduced games in order to axiomatize the four special versions of the ENIC-value mentioned in Section 2. Let (N,v) be a game with $n \geq 2, x \in \mathbf{R}^N, i \in N$ and define the corresponding reduced game $(N \setminus \{i\}, v^x)$, satisfying $v^x(N \setminus \{i\}) = v(N) - x_i$, as follows:

Type 1:

$$v^x(S) := v(S) + \alpha \qquad \text{for all } S \subset N \setminus \{i\}, S \neq N \setminus \{i\}, \phi. \tag{1}$$

Type 2:

$$v^x(S) := v(S \cup \{i\}) - x_i + \alpha \qquad \text{for all } S \subset N \setminus \{i\}, S \neq N \setminus \{i\}, \phi. \tag{2}$$

where $\alpha = \alpha(N,v,i) \in \mathbf{R}$ is an arbitrary constant.

The players in the reduced game of (1) don't cooperate with the removed player, whereas the players in the reduced game of (2) do cooperate with the removed player i whose reward for his cooperation within any coalition is always equal to the payoff x_i. Besides, the worth of any non-trivial coalition may be increased or decreased by some constant. The notions of individual worth and separable contribution are both weakly consistent with respect to the reduced game of (1) and (2) respectively, e.g.,

$$\begin{aligned} SC_j(N \setminus \{i\}, v^x) &= v^x(N \setminus \{i\}) - v^x(N \setminus \{i,j\}) \\ &= v(N) - x_i - [v(N \setminus \{j\}) - x_i + \alpha] \\ &= v(N) - v(N \setminus \{j\}) - \alpha \\ &= SC_j(N,v) - \alpha \qquad \text{for all } j \in N \setminus \{i\} \text{ in case } n \geq 3. \end{aligned}$$

It is left to the reader to verify that both notions satisfy ETP. Consequently, Theorem 4.3 provides an axiomatization for the CIS-value as well as the ENSC-value.

Corollary 4.4 *(i) The CIS-value on G is the unique value on G which possesses RISE, ETP and RGP with respect to the reduced game of (1).*

(ii) The ENSC-value on G is the unique value on G which possesses RISE, ETP and RGP with respect to the reduced game of (2).

We emphasize that both axiomatizations arise from a general framework involving an axiomatization of a class of values. As a solitary case, the ENSC-value has already been axiomatized in Moulin [15] as well as Hart and Mas-Colell [12] and both used the same reduced game of (2) to axiomatize this value. Next we axiomatize the ENPAC-value.

Type 3: for all $S \subset N \setminus \{i\}, S \neq N \setminus \{i\}, \phi$.

$$v^x(S) := \begin{cases} \frac{|S|}{n-2}[v(S \cup \{i\}) - x_i] + \frac{n-2-|S|}{n-2}\left[\sum_{j \in N \setminus (S \cup \{i\})} v(S \cup \{j\})\right] - \alpha \\ \qquad\qquad \text{if } n \geq 4 \\ v(S) + \alpha \quad \text{if } n = 3 \end{cases} \tag{3}$$

where $\alpha = \alpha(N, v, i) \in \mathbf{R}$ is an arbitrary constant and $|S|$ denotes the cardinality of S.

In the reduced game of (3), the worth of a non-trivial coalition is some convex combination, depending on the coalition size, of the worth in the reduced game of (2) and some expression, involving the cooperation with single non-members different from the removed player i. As usual, the worth of any non-trivial coalition may be increased or decreased by some constant.

Corollary 4.5 *The ENPAC-value on G is the unique value on G which possesses RISE, ETP and RGP with respect to the reduced game of (3).*

Proof. Due to Theorem 4.3 and Lemma 2.1, it suffices to show that the notion of average contribution is weakly consistent with respect to the reduced game of (3). Let $(N, v) \in G$ with $n \geq 3$, $i \in N$ and $x \in \mathbf{R}^N$. If $n = 3$, say $N = \{i, j, k\}$, then

$$\begin{aligned} PAC_j(N, v) - PAC_j(N \setminus \{i\}, v^x) &= v(N) - v(\{i\}) - v(\{k\}) - v^x(\{j\}) \\ &= v(N) - v(\{i\}) - v(\{k\}) - v(\{j\}) - \alpha \\ &= v(N) - \sum_{l \in N} v(\{l\}) - \alpha \\ &= \textit{constant}. \end{aligned}$$

If $n \geq 4$, then we derive from (3) that for all $j \in N \setminus \{i\}$

$$\begin{aligned} & PAC_j(N \setminus \{i\}, v^x) \\ = \;& v^x(N \setminus \{i\}) - \frac{1}{n-3} \sum_{l \in N \setminus \{i,j\}} v^x(N \setminus \{i, j, l\}) \\ = \;& v(N) - x_i \\ & - \frac{1}{n-3} \sum_{l \in N \setminus \{i,j\}} \left[\frac{n-3}{n-2}[v(N \setminus \{j, l\}) - x_i] + \frac{v(N \setminus \{i, j\}) + v(N \setminus \{i, l\})}{n-2} + \alpha\right] \\ = \;& v(N) - x_i - \frac{1}{n-2} \sum_{l \in N \setminus \{i,j\}} v(N \setminus \{j, l\}) + x_i \end{aligned}$$

$$
\begin{aligned}
& -\frac{1}{n-3}[v(N \setminus \{i,j\}) + (n-2)\alpha + \frac{1}{n-2}\sum_{l \in N\setminus\{i,j\}} v(N \setminus \{i,l\})] \\
= \;& PAC_j(N,v) + \frac{1}{n-2}v(N \setminus \{i,j\}) \\
& -\frac{1}{n-3}[v(N \setminus \{i,j\}) + (n-2)\alpha - \frac{1}{n-2}v(N \setminus \{i,j\}) + v(N) - PAC_i(N,v)] \\
= \;& PAC_j(N,v) - \frac{1}{n-3}[(n-2)\alpha + v(N) - PAC_i(N,v)].
\end{aligned}
$$

So, $PAC_j(N\setminus\{i\}, v^x) - PAC_j(N,v) = -\frac{1}{n-3}[(n-2)\alpha + v(N) - PAC_i(N,v)] = \textit{constant}$ for all $j \in N \setminus \{i\}$. That is, the notion PAC is weakly consistent as was to be shown. □

In order to axiomatize the ENBC-value, we introduce the following type of a reduced game:

Type 4: for all $S \subset N \setminus \{i\}, S \neq N \setminus \{i\}, \phi$.

$$
v^x(S) := \begin{cases} \frac{1}{2}[v(S \cup \{i\}) - x_i] + \frac{1}{2}v(S) + \frac{|S| x_i}{2^{n-2}-n+1} + \alpha & \text{if } n \geq 4 \\ \frac{1}{2}[v(S \cup \{i\}) - x_i] + \frac{1}{2}v(S) + \alpha & \text{if } n = 3 \end{cases} \tag{4}
$$

where $\alpha = \alpha(N,v,i) \in \mathbf{R}$ is an arbitrary constant.

In the reduced game of (4), the worth of a non-trivial coalition is the average of the worth in the two reduced games of (1) and (2), plus, if $n \geq 4$, some payoff depending on the coalition size.

Corollary 4.6 *The ENBC-value on G is the unique value on G which possesses RISE, ETP and RGP with respect to the reduced game of (4).*

Proof. Because it turns out that the notion of Banzhaf contribution is *not* weakly consistent with respect to the two-person reduced game of (4), we are mainly interested in the adapted Banzhaf contribution of any player i in a game (N,v) defined by

$$
\overline{BC}_i(N,v) := \begin{cases} BC_i(N,v) & \text{if } n \neq 2 \\ \frac{1}{2}[v(\{i\}) - v(N \setminus \{i\})] & \text{if } n = 2. \end{cases}
$$

Clearly, EN$\overline{BC}$ (N,v) = ENBC (N,v) for any game (N,v) since both values are standard for two-person games. Notice that the notion $\overline{BC}$ of adapted Banzhaf contribution fails to possess RISE in the framework of two-person games, which is of no importance because the corresponding EN$\overline{BC}$-value is still standard for two-person games. Due to these remarks, Lemma 2.1 and Theorem 4.3, it suffices to show that the notion of adapted Banzhaf contribution is weakly consistent with respect to the reduced game of (4). Let $(N,v) \in G$ with $n \geq 3$, $i \in N$ and $x \in \mathbf{R}^N$. If $n = 3$, say $N = \{i,j,k\}$, then

$$
\begin{aligned}
\overline{BC}_j(N,v) &= \frac{1}{4}[v(\{j\}) + v(\{j,k\}) - v(\{k\}) + v(\{i,j\}) \\
&\qquad -v(\{i\}) + v(N) - v(\{i,k\})], \\
\overline{BC}_j(N \setminus \{i\}, v^x) &= \frac{1}{2}[v^x(\{j\}) - v^x(\{k\})] \\
&= \frac{1}{4}[v(\{i,j\}) + v(\{j\}) - v(\{i,k\}) - v(\{k\})]
\end{aligned}
$$

and so, $\overline{BC}_j(N, v) - \overline{BC}_j(N \setminus \{i\}, v^x) = \frac{1}{4}[v(N) + v(N \setminus \{i\}) - v(\{i\})] = constant.$

If $n \geq 4$, then we derive from (4) that for all $j \in N \setminus \{i\}$

$$\begin{aligned}
& \overline{BC}_j(N \setminus \{i\}, v^x) \\
= \; & BC_j(N \setminus \{i\}, v^x) \\
= \; & \frac{1}{2^{n-2}} \sum_{S \subseteq N \setminus \{i,j\}} [v^x(S \cup \{j\}) - v^x(S)] \\
= \; & \frac{1}{2^{n-2}} \sum_{S \subset N \setminus \{i,j\}, S \neq N \setminus \{i,j\}, \phi} \left[\frac{1}{2}[v(S \cup \{i,j\}) + v(S \cup \{j\}) - v(S \cup \{i\}) - v(S)] \right. \\
& \left. + \frac{x_i}{2^{n-2} - n + 1}\right] + \frac{1}{2^{n-2}}[v^x(N \setminus \{i\}) - v^x(N \setminus \{i,j\}) + v^x(\{j\})] \\
= \; & \frac{1}{2^{n-1}} \sum_{S \subseteq N \setminus \{i,j\}} [v(S \cup \{i,j\}) - v(S \cup \{i\}) + v(S \cup \{j\}) - v(S)] \\
& - \frac{1}{2^{n-1}}[v(N) - v(N \setminus \{j\}) + v(N \setminus \{i\}) - v(N \setminus \{i,j\}) + v(\{i,j\}) \\
& \qquad - v(\{i\}) + v(\{j\})] + \frac{(2^{n-2} - 2)}{2^{n-2}} \cdot \frac{x_i}{(2^{n-2} - n + 1)} \\
& + \frac{1}{2^{n-2}}\left[v(N) - x_i - \frac{1}{2}[v(N \setminus \{j\}) + v(N \setminus \{i,j\})] - \frac{(n-2)x_i}{2^{n-2} - n + 1} \right. \\
& \qquad \left. + \frac{1}{2}[v(\{i,j\}) + v(\{j\})] + \frac{x_i}{2^{n-2} - n + 1}\right] \\
= \; & \frac{1}{2^{n-1}} \sum_{S \subseteq N \setminus \{j\}} [v(S \cup \{j\}) - v(S)] + \frac{1}{2^{n-1}}[v(N) - v(N \setminus \{i\}) + v(\{i\})] \\
= \; & \overline{BC}_j(N, v) + \frac{1}{2^{n-1}}[v(N) - v(N \setminus \{i\}) + v(\{i\})],
\end{aligned}$$

so $\overline{BC}_j(N \setminus \{i\}, v^x) - \overline{BC}_j(N, v) =$ *constant* for all $j \in N \setminus \{i\}$. That is, the notion $\overline{BC}$ is weakly consistent as was to be shown. □

In Ruiz et al.[17], the least square prenucleolus which coincides with the ENBC-value is axiomatized by a reduced game property. Their reduced game becomes the average of the worth in the two reduced games of (1) and (2) without coalition size dependent term for our case of removing one player. Our reduced game is different and Theorem 4.3 could not be applied to their reduced game.

5 Concluding Remarks

Remark 5.1 *The axiomatic characterizations of the CIS-, ENSC-, ENPAC- and ENBC-values are essentially derived from Theorem 4.3 by verifying the weak consistency property for the relevant notion of individual contribution with respect to the reduced game in question. The weak consistency property for the notion of individual worth takes into account only the one-person coalitions in the reduced game and in fact, the worth of any k-person coalition,* $k \notin \{0, 1, n-1\}$, *in the reduced game may be chosen arbitrarily. Similarly, the worth of any* *k-person coalition, $k \notin \{0, n-2, n-1\}$, in the reduced game of (2) doesn't matter at all to check the weak consistency property for the notion of separable contribution, whereas it is necessary to require that*

$$v^x(N \setminus \{i,j\}) = v(N \setminus \{j\}) - x_i + \alpha \quad \textit{for all } j \in N \setminus \{i\}, \quad \textit{some constant } \alpha \in \mathbf{R}.$$

Because the notion of average contribution of an n-person game is mainly determined by the $(n-2)$-person coalitions, the worth of any k-person coalition, $k \notin \{0, n-3, n-1\}$, in the reduced game of (3) is irrelevant to check the weak consistency property for the notion of average contribution, whereas it is sufficient to require for any $(n-3)$-person coalition $N \setminus \{i,j,l\}$ that

$$v^x(N \setminus \{i,j,l\}) = \frac{n-3}{n-2}[v(N \setminus \{j,l\}) - x_i] + \frac{1}{n-2}[v(N \setminus \{i,j\}) + v(N \setminus \{i,l\})] + \alpha,$$

where $\alpha = \alpha(N,v,i) \in \mathbf{R}$ is an arbitrary constant. Concerning the reduced game of (4), the worth of any coalition is taken into account for the determination of the Banzhaf contribution and as such, there exists no such latitude for the definition of the reduced game as in the previous three cases.

Remark 5.2 *The well-known Shapley value (Shapley [20]) on the class G^N of games with fixed player set N can be fully characterized by the following three properties: ETP (or symmetry), the null player property and additivity. Here a value σ on G^N is said to possess the* additivity property *if $\sigma(N, v+w) = \sigma(N,v) + \sigma(N,w)$ for all games (N,v) and (N,w), where the sum game $(N, v+w)$ is defined by $(v+w)(S) := v(S) + w(S)$ for all $S \subseteq N$. It is easy to observe that the CIS-, ENSC-, ENPAC- and ENBC-values are additive because the corresponding notions of individual contributions are additive. It is well-known that the set of unanimity games $\{(N,u_T) \mid T \subseteq N, T \neq \phi\}$ forms a basis of G^N, where the unanimity game (N,u_T) is defined by $u_T(S) := 1$ whenever $S \supseteq T$ and $u_T(S) := 0$ otherwise. Members of T are substitutes in the game (N,u_T), whereas players outside T are called null players because of $u_T(S \cup \{i\}) = u_T(S)$ for all $S \subseteq N \setminus \{i\}$, all $i \in N \setminus T$. The null player property of a value requires that any null player in a game receives no payoff according to the value. Because the four distinct values above-mentioned, together with the Shapley value Φ, are additive on G^N and possess ETP by Lemma 2.1 (iii), they must differ somehow concerning the value payoff to any null player in the unanimity game (N,u_T).Straightforward, but tough calculations yield the following payoffs for players in the unanimity game (N,u_T) where $n \geq 3$. For convenience' sake, put $t := |T|$.*

	$i \in N \setminus T$	$i \in T$
$\Phi_i(N,u_T)$	0	$1/t$
$CIS_i(N,u_T)$	0 if $t=1$ $1/n$ if $t \geq 2$	1 if $t=1$ $1/n$ if $t \geq 2$
$ENSC_i(N,u_T)$	$-\frac{t-1}{n}$	$\frac{n-t+1}{n}$
$ENPAC_i(N,u_T)$	$-\frac{(t-1)(n-t-2)}{n(n-2)}$	$\frac{(n-t)^2+t-2}{n(n-2)}$
$ENBC_i(N,u_T)$	$\frac{2^{t-1}-t}{n \cdot 2^{t-1}}$	$\frac{2^{t-1}+n-t}{n \cdot 2^{t-1}}$

In case $|T| = 1$, the value payoff to any null player in the unanimity game (N, u_T) is zero for all five values. If $|T| \geq 2$, then the value payoff to a null player is strictly negative according to the ENSC- and ENPAC-values (the latter value as long as $|T| < n-2$). Now it follows that the difference between the value payoff to a player $j \in T$ and a null player $i \in N \setminus T$ in the unanimity game (N, u_T) with $2 \leq |T| \leq n-1$ is given by

$$\sigma_j(N, u_T) - \sigma_i(N, u_T) = \begin{cases} 1/t & \textit{if } \sigma = \Phi, \\ 0 & \textit{if } \sigma = \mathit{CIS}, \\ 1 & \textit{if } \sigma = \mathit{ENSC}, \\ \frac{n-t-1}{n-2} & \textit{if } \sigma = \mathit{ENPAC}, \\ 2^{1-t} & \textit{if } \sigma = \mathit{ENBC}. \end{cases}$$

In the context of consistency, Sobolev [21] proved that the Shapley value on the class G of all games with a variable player set is the unique value on G which possesses RISE, ETP and RGP with respect to the following type of a reduced game: for all $S \subseteq N \setminus \{i\}$

$$v^x(S) := \frac{|S|}{n-1}[v(S \cup \{i\}) - x_i] + \frac{n-1-|S|}{n-1} v(S).$$

Notice that, in the reduced game above, the worth of any coalition is some convex combination, depending on the coalition size, of the worth in the two reduced games of (1) and (2) applied to $\alpha = 0$. We emphasize the resemblance of the reduced game above to the reduced games of (3) and (4). Moreover, recall also that the maximum of the two reduced games of (1) and (2) is used for an axiomatic characterization of the pre-nucleolus and pre-kernel respectively as presented in Sobolev [22] and Peleg [18].

References

[1] Banzhaf, J.F.III (1965): Weighted Voting Doesn't Work: A Mathematical Analysis, *Rutgers Law Review* 19, pp 317-343.

[2] Dragan, I, Driessen, T.S.H. and Y. Funaki (1996): Collinearity between the Shapley Value and the Egalitarian Division Rules for Cooperative Games, *OR Spektrum* 18, pp 97-105.

[3] Driessen, T.S.H. (1985): Properties of 1-Convex n-Person Games, *OR Spektrum* 7, pp 19-26.

[4] Driessen, T.S.H. (1988): *Cooperative Games, Solutions and Applications*, Kluwer Academic Publishers, Dordrecht, the Netherlands.

[5] Driessen, T.S.H. (1991): A Survey of Consistency Properties in Cooperative Game Theory, *SIAM Review* 33, pp 43-59.

[6] Driessen, T.S.H. and Y. Funaki (1991): Coincidence of and Collinearity between Game Theoretic Solutions, *OR Spektrum* 13, pp 15-30.

[7] Driessen, T.S.H. and Y. Funaki (1996): The Egalitarian Non-Pairwise-Averaged Contribution (ENPAC-) Value for TU-Games, Memorandum, Department of Applied Mathematics, University of Twente, Enschede, The Netherlands.

[8] Dubey, P. and L.S. Shapley (1979): Mathematical Properties of the Banzhaf Power Index, *Math. Oper. Res.* 4, pp 99-131.

[9] Funaki, Y. (1986): Upper and Lower Bounds of the Kernel and Nucleolus, *Int. J. Game Theory* 15, pp 121-129.

[10] Funaki, Y. (1995): Dual Axiomatizations of Solutions of Cooperative Games, Discussion Paper No.13, Faculty of Economics, Toyo University, Tokyo, Japan.

[11] Funaki, Y. and T. Yamato (1995): The Core and Consistency Properties, Discussion Paper No.14, Faculty of Economics, Toyo University, Tokyo, Japan.

[12] Hart, S. and A. Mas-Colell (1985): The Potential: A New Approach to the Value in Multi-Person Allocation Problems, HIER D.P. -1157, Harvard University.

[13] Hart, S. and A. Mas-Colell (1989): Potential, Value and Consistency,*Econometrica* 57, pp 589-614.

[14] Legros, P. (1986): Allocating Joint Costs by Means of the Nucleolus, *Int. J. Game Theory* 15, pp 109-119.

[15] Moulin, H. (1985): The Separability Axiom and Equal-Sharing Methods, *Journal of Economic Theory* 36, pp 120-148.

[16] Owen, G. (1978): Characterization of the Banzhaf-Coleman Index, *SIAM J. Appl. Math.* 35, pp 315-327.

[17] Ruiz, L.M., Valenciano, F., and J.M.Zarzuelo (1996): The Least Square Prenucleolus and the Least Square Nucleolus. Two Values for TU Games Based on the Excess Vector, *Int. J. Game Theory* 25, pp 113-134.

[18] Peleg, B. (1986): On the Reduced Game Property and Its Converse, *Int. J. Game Theory* 15, pp 187-200; Correction, *Int. J. Game Theory* 16, (1987), pp 290.

[19] Peleg, B. (1989): An Axiomatization of the Core of Market Games, *Math. Oper. Res.* 14, pp 448-456.

[20] Shapley, L.S. (1953): A Value for n-Person Games, in *Contributions to the Theory of Games II*, H. Kuhn and A.W. Tucker, eds., *Ann. Math. Studies* 28, Princeton University Press, Princeton, NJ, pp 307-317.

[21] Sobolev, A.I. (1973): The Functional Equations that Give the Payoffs of the Players in an n-Person Game, in *Advances in Game Theory*, E. Vilkas, ed., Izdat. "Mintis" Vilnius, pp 151-153. (In Russian.)

[22] Sobolev, A.I. (1975): The Characterization of Optimality Principles in Cooperative Games by Functional Equations, in *Mathematical Methods in the Social Sciences* 6, N.N. Vorobev, ed., Vilnius, pp 94-151. (In Russian.)

[23] Young, H.P., N. Okada and T. Hashimoto (1982): Cost Allocation in Water Resources Development, *Water Resources Research* 18, pp 463-475.

T.S.H. Driessen
University of Twente
Department of Applied Mathematics
Enschede, The Netherlands

Y. Funaki
Faculty of Economics
Toyo University
Tokyo, Japan

AN IMPLEMENTATION OF THE CORE OF NTU-GAMES

Gustavo Bergantiños and Jos A. M. Potters

Abstract: *For each non transferable utility game a strategic game is introduced. All Nash equilibria of the strategic game are strict equilibria and the equilibrium payoffs are the same as the payoffs in core allocations of the NTU-game. Relations between the payoff map of the strategic game and the remainder map of Driessen and Tijs (1985) are derived.*

1 Introduction

In general the implementation of a cooperative solution concept is the task to design, for each cooperative game (N, V), a strategic game Γ_V with the property that the payoff allocations of (a certain class of) Nash equilibria coincide with the payoffs proposed by the solution concept. The meaning of such an exercise is that, in the strategic situation Γ_V, a fair and socially optimal payoff allocation of V can be obtained by decentralized strategic behavior. No difficult negotiations are necessary; as soon as the players accept the rules of the strategic game they can privately choose a strategy of which they think it gives them the highest payoff.

In this paper we design a strategic game or better a rule assigning a strategic game Γ_V to each NTU-game (N, V) such that the equilibrium payoffs of Γ_V are precisely the core allocations of the cooperative game (N, V). Moreover, it will turn out that all Nash equilibria in the game Γ_V are *strict* equilibria i.e., in a Nash equilibrium $x = (x_1, \ldots, x_n)$ the strategy x_i of each player i is the *unique best response to* x_{-i}.

In Borm and Tijs (1992) an other implementation of the core of NTU-games is given. In their 'claim games' each player chooses a coalition in which he will cooperate and a utility level he asks for his cooperation. A player obtains his claim if the other players in the coalition he took, choose the same coalition and the aggregate utility vector is feasible in this coalition. In Borm and Tijs (1992) it is shown that the payoffs of *strong* Nash equilibria correspond to the (strong) core allocations of the NTU-game. Strong Nash equilibrium means that no *coalition* can profitably deviate from the equilibrium strategy.

In the strategic game we will design the players only choose a utility level and the payoff function is chosen in such a way that only core allocations are Nash equilibria.

In Section 2 we introduce the strategic game associated with an NTU-game and formulate the Main Theorem. The proof of the Theorem is the contents of Section 3. In the final Section we compare the payoff function of the game we introduced with the remainder map of Bennett (1983) and Driessen and Tijs (1985).

2. A strategic game connected with an NTU-game.

Globally the concept of NTU-game used by different authors is the same but often some additional properties are assumed. In this paper an *NTU-game* will be a pair (N, V) wherein N is a finite player set and V is a rule which assigns to each nonempty coalition S a nonempty set $V(S) \subset \mathbf{R}^S$ with the following properties:

(a) There is a point $d \in \mathbf{R}^N$ (the disagreement outcome) such that
$$V(\{i\} := \{x_i \in \mathbf{R}^{\{i\}} \mid x_i \leq d_i\}.$$

(b) Each set $V(S)$ is a closed set in $\mathbf{R}^S$.

(c) The sets $V(S)_+ := \{x \in V(S) \mid x \geq d_{|S}\}$ are compact (bounded).

(d) If $x \in V(S)$ and $y \in \mathbf{R}^S$ is a point with $y \leq x$, then $y \in V(S)$ (comprehensiveness)

The properties (a) up to (d) are quite usual in the definition of an NTU-game. For our purposes we need a slight restriction of this concept.

(e) There is a point $x \in V(N)$ with $x > d$.

Property (c) and (e) imply that $u_i := \max\{x_i \mid x \in V(N)_+\}$ is well defined and $u_i > d_i$ for every player i. The vector $u = (u_i)_{i \in N}$ is called the *utopia vector* of the game (N, V).

Further we assume the following property:

(f) If $x \in V(S)_+$ and for some $y \in V(S)_+$ we have $y \geq x$, $y \neq x$, then there is a point $z \in V(S)_+$ with $z > x$ (strong comprehensiveness).

Property (f) states that a point $x \in V(S)_+$ is Pareto optimal ($y \in V(S)_+$, $y \geq x$ and $y \neq x$ does not exist) iff x is weakly Pareto optimal ($z \in V(S)_+$ with $z > x$ does not exist).

An NTU-game (N, V) is *normalized* if $d = 0$ and $u = e_N = (1, 1, \ldots, 1)$. Each NTU-game satisfying the properties (a) up to (f) can be normalized by a positive affine transformation of the utility functions of each of the players. In fact the transformations $x_i \rightarrow \frac{x_i - d_i}{u_i - d_i}$ will do the job. In the sequel we assume that the games (N, V) are normalized.

The set of (weakly) Pareto optimal points of $V(N)$, is denoted by $\partial V(N)$. The set of interior points of $V(N)$ is denoted by $V(N)^0 := V(N) \backslash \partial V(N)$. For a coalition S we define $\partial V(S)$ and $V(S)^0$ in the same way. A point $x \in V(N)$ is a *core allocation* if $x \notin V(S)^0$ for any coalition $S \subseteq N$.

Let (N, V) be a normalized NTU-game (satisfying the properties (a) up to (f)). We consider the following strategic game Γ_V:

The player set is N and strategy space is the segment $[0, 1]$ for each player $i \in N$. In a non-normalized situation the strategy space of player i is the segment $[d_i, u_i]$. Playing a strategy $x_i \in [0, 1]$ means 'claiming a normalized utility level x_i' (cf. the concept of aspiration level in Bennett and Wooders (1979) and Bennett (1983)). The payoff function $P: [0, 1]^N \rightarrow \mathbf{R}^N$ of the strategic game

Γ_V is defined in three steps. To do so we must introduce some new concepts. If $x \in [0,1]^N$ is a strategy N-tuple, player i is a *winner* (notation: $i \in W(x)$) if there is a coalition $S \ni i$ such that $x_{|S} \in V(S)$. To be a winner player i must be able to invoke the help of a coalition S to show that his 'claim' is reasonable. If he cannot show his claim is reasonable, he is a *loser* : $i \in L(x)$.
(I) If $L(x) \neq \emptyset$, the payoff of a winner is equal to his claim and the payoff of a loser is -1 (i.e. $d_i - u_i$ in the non-normalized case).
(II) If $L(x) = \emptyset$ and $x \in V(N)$, then $P(x) := x + t(u - x) \in V(N)$ where t is the largest number with the property $x + t(u - x) \in V(N)$.
(III) If $L(x) = \emptyset$ and $x \notin V(N)$, the definition of the payoff function is more complicated. The general idea is that players with the highest (normalized) claims see their claims reduced and the other players obtain their claim. So, let $M(x)$ be the set of players with $x_i = \max\{x_j \mid j \in N\}$.

(a) If $|M(x)| \geq 2$, we reduce the claims of the players in $M(x)$ equally i.e. we go along the line $\{x - t\,e_{M(x)} \mid t \in \mathbf{R}_+\}$ until we come in $V(N)$ or leave $\mathbf{R}^N_+$. This point is the payoff value in this case.

(b) If $|M(x)| = 1$, we need an—admittedly artificial—payoff function. First we define the excess $E_i(x)$ of the player $i \in M(x)$ in the point x as the minimum of the numbers $t \in \mathbf{R}_+$ with $x - te_i \in V(N)$ if this number exists and is at most x_i. Otherwise we define $E_i(x)$ to be x_i. Further, we need a function $f: (0,1] \to \mathbf{R}$ with the following properties:
(i) For all $t \in (0,1]$ we have $0 < f(t) < t$.
(ii) For all $t \in (0,1]$ there is a positive number $\delta(t) > 0$ such that the inequality
$s_2 - s_1 < f(s_2) - f(s_1) < 2(s_2 - s_1)$ holds for $t - \delta(t) \leq s_1 < s_2 \leq t$.
Notice that the conditions (i) and (ii) cannot be satisfied by a *continuous function* f but the figure gives a non-continuous example.
Now we define in case (III)-b the payoff function by $P_j(x) = x_j$ if $j \notin M(x)$ and $P_i(x) := x_i - f(E_i(x))$ if i is the player in $M(x)$. This finishes the definition of the payoff function P.

The idea behind the definition of the payoffs is the following:

If there are players with unreasonable claims they are punished severely and the other players get their claims (case (I)). If all claims are reasonable and the claims can be increased without leaving the set $V(N)$, the claims are increased proportional to $u - x$ until a Pareto optimal point is reached (case (II)). If there are no losers and the claims are too high, only the player(s) with the highest claim see their claims reduced with the same amount (case (III)).

See the diagram given before the reference

Notice that the payoffs of winners are always nonnegative and losers have a negative payoff. In case (III)-b this follows from $f(E_i(x)) < E_i(x) \leq x_i$.

The main theorem of this paper is the following

Theorem 1. *A point $x \in [0,1]^N$ is a core allocation of (N, V) if and only if the strategy n-tuple x is a Nash equilibrium of the strategic game Γ_V. Moreover, the game Γ_V has only strict Nash equilibria.*
The proof of this theorem will be the contents of section 3.

3. Proof of the main theorem.

The proof of the main theorem consists of four parts. In the first part we prove that a core allocation is a strict Nash equilibrium of Γ_V. Next we prove that a Nash equilibrium x has no losers. In the third part we prove that every Nash equilibrium x is efficient i.e., $x \in \partial V(N)$, the Pareto boundary of $V(N)$. In the final part we prove that $x \notin Core(V)$ implies that x is not a Nash equilibrium of Γ_V.

(A) *If $x \in Core(V)$, then x is a Nash equilibrium of Γ.*
Let $x \in Core(V)$. Then x can be understood as a strategy n-tuple and $L(x) = \emptyset$ (every player can invoke the coalition N to show that his claim is reasonable). We are in case (II) and $P(x) = x$. If some player i deviates from x_i to $\bar{x}_i = x_i + \delta$ with $\delta > 0$ his claim is no longer supported by coalition N and if it would be supported by a coalition $S \ni i$, we should have $x_{|S} \leq (\bar{x}_i | x_{|S\setminus\{i\}}) \in V(S)$ and therefore $x_{|S} \in V(S)^0$, because of property (f) and $x_{|S} \in V(S)_+$. Then $x \notin Core(V)$. So, player i becomes a loser with a negative payoff. His deviation gives a *strictly worse* payoff.
If some player i (with a positive claim) deviates from x_i to $\bar{x}_i = x_i - \delta$ with $\delta > 0$, then $L(\bar{x}_i | x_{-i}) = \emptyset$ and $\bar{x} \in V(N)^0$ by condition (f). If $P_i(\bar{x}) \geq P_i(x) = x_i$, then $P(\bar{x}) \geq x$ and no equality holds. If $P(\bar{x}) = x$, then $x_j = u_j$ for all $j \neq i$ and $\bar{x}$ is weakly Pareto optimal without being Pareto optimal. This means that $x \in V(N)^0$. This is impossible for a core element. The strategy n-tuple x is a Nash equilibrium. Notice that x_i is the unique best response to x_{-i} i.e. the Nash equilibrium is even *strict*.

(B) *If x is a Nash equilibrium of Γ, then $L(x) = \emptyset$.*
Suppose that $i \in L(x)$. If player i deviates from x_i to $\bar{x}_i = 0$, he is no longer a loser (as the coalition $\{i\}$ is supporting his claim). His negative payoff becomes nonnegative.

(C) *If x is a Nash equilibrium of Γ, then $x \in \partial V(N)$.*

(i) $x \in V(N)^0$. Take any player $i \in N$. As $x \in V(N)_+$ by (B), there is a real number $\bar{x}_i > x_i$ such that $\bar{x} := (\bar{x}_i | x_{-i})$ is a point of the Pareto boundary of $V(N)$. As x is a Nash equilibrium, $P_i(x) \geq P_i(\bar{x})$. This means that $P(x) = x + t(u - x) \geq \bar{x}$ and no equality holds as $u - x > 0$ by (f) and $t > 0$. Then $\bar{x}$ is not on the Pareto boundary (by condition (f)).

(ii) $x \notin V(N)$. If $|M(x)| \geq 2$, we take $i \in M(x)$ and diminish the claim of player i slightly ($x_i \to \bar{x}_i = x_i - \delta$). This is possible as $x_i > 0$ by condition (e). Then $i \notin M(\bar{x})$ and therefore $P_i(\bar{x}) = x_i - \delta$ and $P_i(x) = x_i - t$ with $t > 0$. If we take $\delta < t$ we have an improvement by deviation and x is not a Nash equilibrium. If $|M(x)| = 1$, a small deviation of player $i \in M(x)$ gives $P_i(\bar{x}) = x_i - \delta - f(E_i(x) - \delta) > x_i - f(E_i(x))$ when $\delta < \delta(t)$ for $t = E_i(x)$ (see property (ii) of the function f).

(D) *If $x \in \partial V(N) \setminus Core(V)$, then x is not a Nash equilibrium of Γ.*
If $x \in \partial V(N) \setminus Core(V)$, there is a coalition S with $x_{|S} \in V(S)^0$. Take any player i in S and take $\delta > 0$ such that $x_{|S} + \delta e_i \in V(S)$. Then $\bar{x} = x + \delta e_i \notin V(N)$. If $L(\bar{x}) \neq \emptyset$, player i (not a loser) obtains a payoff $x_i + \delta > P_i(x) = x_i$. Therefore we assume that $L(\bar{x}) = \emptyset$ (case (III)). If $i \in M(\bar{x})$ and $|M(\bar{x})| \geq 2$ we can take $\delta > 0$ slightly smaller such that $i \notin M(\bar{x})$. Hence we are left with the case that $M(\bar{x})$ consists of player i only (case (III)-b).

Notice that, in this case, $E_i(\bar{x}) = \delta$ and $P_i(\bar{x}) = x_i + \delta - f(\delta)$. As $f(t) < t$ for all $t \in (0,1]$, we also find a profitable deviation for player i in this case. ◁

4. Relation between the payoff function P and the Remainder Map R.

If (N, V) is an NTU-game, we define, for all $i \in N$,

$$U_i := \max_{S \ni i} \max \{x_i \mid x \in V(S)_+\}.$$

The vector U is a 'super' utopia point. If $x \in [d, U] := \prod_{i \in N} [d_i, U_i]$, if i is a player and S is a coalition containing player i we can define (cf. Driessen and Tijs (1985), E. Bennett (1983))

$$R_i(x_{-i} \mid S) := \sup \{y \in \mathbf{R} \mid (y, x_{|S \setminus \{i\}}) \in V(S)\} \quad \text{and} \quad R_i(x_{-i}) := \max_{S \ni i} R_i(x_{-i} \mid S).$$

The value of $R_i(x_{-i} \mid S)$ may be $-\infty$ but $R_i(x_{-i})$ is always bounded.
The map $x \in [d, U] \rightarrow R(x) = \{R_i(x_{-i})\}_{i \in N}$ is called the *remainder map*. Without any problem the payoff map P of the preceding sections can be extended to $[d, U]$. Note that $R(x) \in [d, U]$ if $x \in [d, U]$.
The relation between the remainder map R and the payoff function P of the game Γ_V of the preceding section will be the subject of the following proposition.
Proposition 2. *If (N, V) is an NTU-game and $x \in [d, U]$, the following statements are equivalent:*
(i) $P(x) = R(x)$.
(ii) $R(x) = x$ *and* $x \in V(N)$.
(iii) x *is a core element of* (N, V).
Proof: First we prove that, if $x \in V(N)^0$, then $R(x) > P(x)$. If $x \in V(N)^0$, then the payoff $P(x) = x + t(u - x) \in \partial V(N)$. Take $i \in N$ arbitrarily and let y_i be the largest number such that $(y_i \mid x_{-i}) \in V(N)$. Note that $(y_i \mid x_{-i}) \in \partial V(N)$ and $x_i < y_i \leq R_i(x)$. If $P_i(x) \geq R_i(x)$ then $P(x) \geq (y_i \mid x_{-i})$ and there is equality as $(y_i \mid x_{-i}) \in \partial V(N)$. This means that $x_{-i} = u_{-i}$. Since $x \in V(N)^0$, there is a point $z \in V(N)$ with $z > x$ and in particular, $z_j > x_j = u_j$ for $j \neq i$. This is in contradiction with the definition of u_j.

(i) → *(ii)*. If $P(x) = R(x)$, then $L(x) = \emptyset$ and $R(x) \geq x$. Hence, $P(x) \geq x$ and therefore, $x \in V(N)$. We have seen that $x \notin V(N)^0$ and therefore, $x \in \partial V(N)$ and $x = P(x) = R(x)$.

(ii) → *(iii)*. If $R(x) = x$, there is no coalition S with $x_{|S} \in V(S)^0$. Further, $x \in V(N)$ and we have $x \in Core(V)$.

(iii) → *(i)*. If $x \in Core(V)$, we have $x_{|S} \notin V(S)^0$ and $x \in V(N)$ and therefore $x = R(x)$. Further, we have $x \in \partial V(N)$ and $P(x) = x$. So we find $P(x) = R(x) (= x)$. ◁

Remark: (cf. I.M. Bomze (1985)) *The remainder map has always a fixed point in $[d, U]$ but these fixed points may lie outside $\partial V(N)$.*
Proof: Take a point $x \in [d, U]$ with $R(x) \leq x$. The super utopia point U is a good candidate. As long as $R(x) \neq x$ we repeate the following algorithmic step.
Choose $i \in N$ with $R_i(x_{-i}) < x_i$ and replace x by $x - t\, e_i$ wherein t is the minimal change which brings x in some $V(S)$ with $S \ni i$.
Notice that this action keeps the inequality $R(x) \leq x$ intact and the set of players with $R_j(x_{-j}) = x_j$ increases with at least one player, player i.
Repeating this process finitely many times gives a fixed point of R. ◁

Addendum after revision.

This paper circulated as a discussion paper during the last three years. One of the points of criticism we heard was that not each play of the strategic game (and not just Nash equilibrium behavior) gives a feasible point in the NTU-game. In this addendum we will try to meet this requirement. In fact, we will show that we can change the payoff function in such a way that any play of the game leads to a point in $V(N)$. In all cases the new rules for the payoff functions will be of the form:

'Do what we did in the paper' and 'add a harmless action to bring the point in $V(N)$'.

The 'harmless' action we mean is: 'Diminish the payoffs of **all winners** proportionally till we enter $V(N)$. This means: if $P(x)$ is the payoff in the paper and $P(x) \notin V(N)$, the new payoff $\bar{P}(x)$ gives to the losers again the payoff -1 and to winners the payoff $\rho P_i(x)$ where $\rho \in (0,1)$ is chosen such that $\bar{P}(x) \in \partial V(N)$. This is possible because of the properties (d) and (e).

Of course we have to show that our proofs are still valid. So, we have to check the points where the claims or one-person deviations from the claims lead to points outside $V(N)$.

In part (A) of the proof this only happens when a player deviates unilaterally from the core allocation and *increases* his claim. Then he becomes a loser (as we showed) and his payoff will become negative. In the core allocation his payoff was nonnegative.

In the parts (B) and (C)-(i) there does not occur claims outside $V(N)$. So, in these parts nothing has to be changed.

In the parts (C)-(ii) and (D) of the proof we must be more careful.

(C)-(ii) $x \notin V(N)$.

If $|M(x)| \geq 3$, we have to compare the payoff of a player $i \in M(x)$ under the strategy profile x and under the strategy profile $x - \delta e_i$ for small $\delta > 0$. If player i obtains a payoff zero under the rules in the paper, he also receives nothing under the new rule. Then it is better to deviate to $x_i - \delta > 0$. Then he receives a positive payoff. Therefore we assume that under the old rule (as well as under the new rule) the payoff to player i is $x_i - t > 0$ and $x - t\, e_{M(x)} \in \partial V(N)$. If player i deviates to $x_i - \delta$ with $\delta < t$ and he would obtain the same payoff $x_i - t$ or less, the factor $\rho = (x_i - t)/(x_i - \delta) < 1$ and all players outside $M(x)$ suffer from his action. Therefore the players in $M(x) \backslash \{i\}$ must obtain a really higher payoff than $x_j - t = x_i - t$. This means $x - \delta e_i - t\, e_{M(x) \backslash \{i\}} \in V(N)$. This is impossible because of $\delta < t$ and (f).

If $M(x) = \{i, j\}$, we must compare $x - t\, e_{\{i,j\}} \in \partial V(N)$ with $\rho \big(x - \delta e_i - f(E_j(x - \delta e_i)) e_j\big) \in \partial V(N)$. As the number ρ is between zero and one, all players $k \notin M(x)$ suffer from i's action and, if player i himself would not have advantage of his deviation, player j must have advantage because of (f). So,

$$\rho \left(x_j - f\left(E_j\left(x - \delta e_i\right)\right)\right) > x_j - t = x_i - t \geq \rho\left(x_i - \delta\right) = \rho\left(x_j - \delta\right).$$

Then $\delta > f(E_j(x - \delta e_i))$. So we are left to prove that we can choose $\delta > 0$ with $\delta \leq f(E_j(x - \delta e_i)$.

Let $f(E_j(x)) =: 2\eta > 0$. As f is continuous on a segment $[E_j(x) - \alpha, E_j(x)]$ for some positive number α, there is a number $\beta > 0$ such that $f(t) > \eta$ for $t \in (E_j(x) - \beta, E_j(x)]$. We are done if we can choose $\delta < \eta$ with $E_j(x - \delta e_i) > E_j(x) - \beta$. Suppose $E_j(x - \delta e_i) \leq E_j(x) - \beta$ for all $\delta \in (0, \eta)$. Then

$x - \delta e_i - (E_j(x) - \beta) e_j \in V(N)$ for all $\delta \in (0, \eta)$. As $V(N)$ is closed, also $x - (E_j(x) - \beta) e_j \in V(N)$. This is in contradiction with the definition of $E_j(x)$.

If $M(x) = \{i\}$, we have to compare

$$\rho(x - f(E_i(x)) e_i) \in \partial V(N) \text{ and } \rho'(x - \delta e_i - f(E_i(x - \delta e_i)) e_i) \in \partial V(N).$$

In the paper we proved that $\delta + f(E_i(x - \delta e_i) < f(E_i(x))$. If player i would not have advantage of his deviation, we must have $\rho' < \rho$ and all players $k \neq i$ would suffer from player i's action. This contradicts property (f).
This finishes the proof of (C)-(ii).

(D) Suppose $x \in \partial V(N) \backslash Core(V)$. As in the paper there is a player i who can deviate from x to $\tilde{x} := x + \delta e_i$ and still be a winner. If player i does not have the highest claim under x, we assume that $x_i + \delta$ is still not the highest claim. Then $\tilde{x} \notin V(N)$. In case $L(\tilde{x}) \neq \emptyset$, the losers under $\tilde{x}$ see their payoffs diminished and the winners under $\tilde{x}$ get less by the factor $\rho < 1$. Therefore player i must have advantage of his action.
Therefore, we may assume that $L(\tilde{x}) = \emptyset$ and $i \notin M(\tilde{x})$ or $M(\tilde{x}) = \{i\}$.
In both cases no player $k \neq i$ has advantage of player i's action and at least one player really suffers from his action. Then player i must have advantage of his deviation.
This finishes the proof of the main theorem under the alternative payoff function $\bar{P}$.

An other problem raised is about the discontinuity of the payoff functions. Indeed the payoff functions we gave are discontinuous functions of the claims. However, the strategic games we gave in the paper, are based on the idea of punishment or penalties. In our opinion penalties are typically not continuous. Surpassing a borderline leads to punishment, often not continuous in the 'size of the transgression'.

References.

Bennett E (1983) The Aspiration Approach to Prediction of Coalition Formation and Payoff Distribution in Sidepayment Games, *International Journal of Game Theory* **12** 1–28.

Bennett E, Wooders M (1979) Income Distribution and Firm Formation, *J. Comparative Economics* **3**, 304–317.

Bomze IM (1985) A Note on Aspirations in Non Transferable Utility Games, *International Journal of Game Theory* **17**, 193–200.

Borm P, Tijs SH (1992) Strategic Claim Games Corresponding to an NTU-Game, *Games and Economic Behavior* **4**, 58–71.

Driessen TES, Tijs SH (1985) The τ-Value, the Core and Semiconvex Games, *International Journal of Game Theory* **14**, 229–247.

Gustavo Bergantiños
University of Vigo
Spain

Jos A. M. Potters
Department of Mathematics
University of Nijmegen

6525 ED Nijmegen, The Netherlands.

PURE-STRATEGY NASH EQUILIBRIUM POINTS IN NON-ANONYMOUS GAMES

M. Ali Khan, Kali P. Rath and Yeneng Sun

Abstract: *We present an example of a nonatomic game without pure Nash equilibria. In the example, the set of players is modelled on the Lebesgue unit interval with an equicontinuous family of payoff functions, and an identical action set given by* $[-1, 1]$. *This example is sharper than that recently presented by Rath-Sun-Yamashige in that the relationship between societal responses and individual payoffs is linear. We also present a theorem on the existence of pure strategy Nash equilibria in nonatomic games in which the set of players is modelled on a nonatomic Loeb measure space.*

1 Introduction

Consider a game with a "large" number of players, each with an identical action set, and with each player's payoff function depending on individual actions as well as on the distribution of the actions of all of the others. A Nash equilibrium of such a game is a function from the set of players, more precisely players' names, to the set of actions such that the action corresponding to each player is optimal given the distribution of actions induced by the function. If the distribution is seen as formalizing "societal responses," then individual equilibrium actions induce precisely those societal responses under which they are optimal. Since there is no recourse to randomization by any individual player, this is a Nash equilibrium in pure strategies, and since the set of players' names is made explicit, this is a non-anonymous game. In such a game, relative to the games considered by Nash (1950, 1951), the finite set of N players has been replaced by a "large" set, and a player's dependence on all of the individual responses replaced by the dependence on an aggregate measure of those responses.

It is now known that if the set of players' names is taken to be an abstract nonatomic probability space $(T, \mathcal{T}, \lambda)$, the identical action set A to be a compact metric and countably infinite space, and the association of payoff functions to a players' names to be measurable in a suitable sense, there does exist a Nash equilibrium in pure strategies.[1] The measurability hypothesis in the specification of the game is necessitated by the requirement that the Nash equilibrium function induce a probability measure on the common action set. The precise result is not difficult to state. One works with the set of probability measures $\mathcal{M}(A)$ on the action set A, and the space of real-valued continuous functions $\mathcal{U}_A^m$ on $A \times \mathcal{M}(A)$, continuity understood to be in terms of the product topology generated by the original topology on A and the induced relative weak* topology on $\mathcal{M}(A)$. Since A is compact, $\mathcal{M}(A)$ is also compact, $\mathcal{U}_A^m$ can be endowed with its sup norm topology, and thence with the Borel σ-algebra $\mathcal{B}(\mathcal{U}_A^m)$ generated by this topology.[2] Under this measurable structure on the space of payoff functions, a "large" non-anonymous game $\mathcal{G}$ can be defined simply as a measurable function (random variable) from T to $\mathcal{U}_A^m$. A precise statement of a result due to Schmeidler (1973) and Khan-Sun (1996a) can now be given.

Theorem 1 *For any large non-anonymous game, which is to say any measurable mapping $\mathcal{G} : T \longrightarrow \mathcal{U}_A^m$, there exists a measurable mapping $f : T \longrightarrow A$ such that for λ-almost $t \in T$,*

$$u_t(f(t), \lambda \cdot f^{-1}) \geq u_t(a, \lambda \cdot f^{-1}) \quad \textit{for all } a \in A,$$

where λf^{-1} is the distribution induced on the set A by f, and $u_t \equiv \mathcal{G}(t)$.

An answer to the question as to whether the countable restriction on the action set A can be dispensed with is negative. Such an answer can be furnished in the context of a game in which the set of players' names are given by the Lebesgue unit interval and the action sets by the interval $[-1, 1]$. This game is due to Rath-Sun-Yamashige (1995), and here again, the basic outline of the specification and the argumentation is not difficult to state. Let $\mathcal{G}_1$ be a mapping from $[0, 1]$ to $\mathcal{U}_{[-1,1]}$ such that for any player $t \in [0, 1]$,

$$\mathcal{G}(t)(a, \nu) = u_t(a, \nu) = h(a, \nu) - |t - |a||,$$

where $h(\cdot, \cdot) : [-1, 1] \times \mathcal{M}([-1, 1]) \longrightarrow \mathbb{R}$ is a jointly continuous function. We leave it to the reader to check that $\mathcal{G}_1$ is a continuous function from T into $\mathcal{U}_A^m$, and that the family $\{\mathcal{G}_1(t) : t \in T\}$ is equicontinuous. Thus $\mathcal{G}_1$ is $(\mathcal{T}, \mathcal{B}(\mathcal{U}_A^m))$-measurable, and therefore clearly a large non-anonymous game. Under the further specification that $h(a, \rho^\star) = 0$ irrespective of the value of $a \in A$, $\rho^\star$ the uniform distribution on $[-1, 1]$, one can conclude that $\rho^\star$ cannot be induced in equilibrium. If it was, the best response correspondence

[1] Throughout this paper we shall use actions and strategies as synonyms.

[2] This is a consequence of Prohorov's theorem; see Parthasarathy (1967) or Billingsley (1968). The reader can refer to these texts, and to Rudin (1974), for additional details and terminology.

would be given by $t \longrightarrow \{t, -t\}$, and there is no way to choose a measurable selection from this correspondence[3] that induces $\rho^\star$; see Figure 1.[4] The point, however, is that a further specification of $h(\cdot,\cdot)$ forces the conclusion that $\mathcal{G}_1$ does not have *any* Nash equilibrium! The argument revolves around the Prohorov metric[5] $d(\cdot,\cdot)$ generating the weak* topology. Any distribution ρ different from $\rho^\star$ generates a best response correspondence that induces another distribution ρ' such that $d(\rho', \rho^\star) < d(\rho, \rho^\star)$. Thus, any distribution different from the uniform distribution gravitates during the "course of play" to the uniform distribution, which, as has been shown already, can never be an equilibrium. There is an absence of closure in a particular class of measurable functions, and it is this that is responsible for the lack of existence of Nash equilibrium.[6]

The individual payoff functions in the game $\mathcal{G}_1$ depend non-linearly on societal responses parametrized by distributions on the action set. The notion of players in a game-theoretic situation maximizing their expected utilities goes back at least to Nash (1950, 1951), and it is natural to consider a concept of equilibrium in which an individual simply integrates out the societal variable in his payoff function by using his beliefs of the distribution of societal plays. A Nash equilibrium would then be a situation when these beliefs both coincide among players and are generated and sustained by their individual actions. For a formal development of this question, consider the space $\mathcal{U}_A^2$ of real-valued continuous functions defined on $A \times A$, and endowed with the product topology and the associated Borel σ-algebra $\mathcal{B}(\mathcal{U}_A^2)$. For any measurable mapping $\mathcal{G}^e : T \longrightarrow \mathcal{U}_A^e$, the superscript e denoting the maximization of expected payoffs, one can ask whether there exists a measurable mapping $f : T \longrightarrow A$ such that for λ-almost $t \in T$,

$$\int_A u_t(f(t), a_\sigma) d(\lambda \cdot f^{-1})(a_\sigma) \geq \int_A u_t(a, a_\sigma) d(\lambda \cdot f^{-1})(a_\sigma) \quad \text{for all } a \in A,$$

where, as before, λf^{-1} is the distribution induced on the set A by f, and $u_t \equiv \mathcal{G}^e(t)$.

This question can also be posed from the perspective of Glicksberg's 1952 theorem on the existence of mixed-strategy Nash equilibria in finite N-person games. Glicksberg (1952) considered finite player games based on action sets which are compact Hausdorff spaces, and

---

[3]This correspondence is a canonical example in general equilibrium theory; see Hart-Kohlberg (1974), Hart-Hildenbrand-Kohlberg (1974) and Artstein (1983); also Claim 4 below.

[4]It is a good exercise for the reader to prove this fact for herself; a proof is nevertheless furnished at the end of Section 2.

[5]Recall from Billingsley (1968; p. 237-238) that this metric on the space of probability measures is defined as $d(\rho, \nu) = \inf\{\epsilon > 0 : \rho(E) \leq \nu(B_\epsilon(E)) + \epsilon \text{ and } \nu(E) \leq \rho(B_\epsilon(E)) + \epsilon\}$, for all Lebesgue measurable sets E in $[-1, 1]$, and where, for any $\epsilon > 0$, $B_\epsilon(E) = \{x \in [-1, 1] : |x - y| < \epsilon,\ y \in [-1, 1]\}$.

[6]See Rath-Sun-Yamashige (1995), and also Khan-Rath-Sun (1995), for detailed computations, and a complete argument.

on payoff functions which are generated by continuous functions defined on the Cartesian product of these action sets; and showed the existence of mixed strategy Nash equilibria as a consequence of what subsequently came to be called the Fan-Glicksberg fixed point theorem.[7] in. Specifically, in the context of a two-player game with an identical compact Hausdorff action set A, his theorem can be stated as follows.

Theorem 2 *For any pair of continuous functions $u_t : A \times A \longrightarrow \mathbb{R}$, $t = 1, 2$, there exist a pair of probability measures $\nu_t \in \mathcal{M}(A)$, $t = 1, 2$, such that*

$$\int_A \int_A u_1(a_1, a_2) d\nu_1 d\nu_2 \geq \int_A \int_A u_1(a_1, a_2) d\mu d\nu_2 \quad \textit{for all } \mu \in \mathcal{M}(A),$$

$$\int_A \int_A u_1(a_1, a_2) d\nu_1 d\nu_2 \geq \int_A \int_A u_1(a_1, a_2) d\nu_1 d\mu \quad \textit{for all } \mu \in \mathcal{M}(A).$$

Each player chooses a mixed strategy on the basis of his beliefs regarding the other player's actions, and equilibrium outcomes are those in which these beliefs are sustained. The matching pennies game[8] offers a simple example of a game in which there is no equilibrium in pure strategies, which is to say a situation in which ν_1 and ν_2 are Dirac point measures.

The question that we pose extrapolates this situation to a setting in which, for each individual player, the "other" is constituted not by one or a finite set of players, but a multiplicity, and rather than keeping track of all of the measures representing individual plays, each player takes cognisance of only the "societal" measure. The players constituting this multiplicity, now individually strategically negligible, are collectively significant for individual payoffs in precisely the same way that a single opponent was in Glicksberg's setting. This is to say that the payoff of a particular player t is given by

$$\int_A \int_A u_t(a, a_\sigma) d\nu_\sigma(a_\sigma) d\nu(a),$$

where ν represents the mixed strategy of the individual player and ν_σ her beliefs regarding society's plays. The question then is whether there exists an equilibrium in the sense that each individual's actions induce the equilibrium societal beliefs which led him to take those actions, or alternatively, does there exist a set of beliefs that is macroscopically sustainable by microscopic individual actions? However, the question is not precisely specified until we are clear on how to connect individual actions and the individual beliefs representing the distribution of societal plays. One possibility is to assume that each player plays a pure

[7] See Fan (1952) and Glicksberg (1952). This fixed point theorem has been an essential tool for existence proofs in the theory of "large" games.

[8] For this and other examples, the reader can se Fudenberg-Tirole (1991).

strategy, and the distribution induced by the function listing these plays constitute society's plays. The other possibility is to allow mixed strategies for each individual player, and let some suitable integral of these strategies constitute society's plays. It is the first possibility that furnishes the question posed above.[9]

Irrespective of the perspective from which this question is posed, what is essential to the problem is that the distribution of society's plays enters the individual payoffs in a linear fashion. Put differently, each player is taking the expectation of the payoff function with respect to a probability measure, and it is this linearity property embodied in this integral aspect that connects the work of Glicksberg and Schmeidler and furnishes the question that we investigate. In any case, the objective of this paper is to present a negative answer to it. The simplicity of the game that that is presented here is perhaps surprising, and relative to the work of Rath-Sun-Yamashige (1995), it frees us from computations involving the Prohorov metric. Our example is also sharper in the sense that it covers situations with structure additional to theirs.

The paper is organized as follows. In Section 2, we present the nonatomic game that serves as the counterexample to the question that we pose. We characterize the best response correspondence, show how it depends on a summary statistic h, and present a complete argument for the non-existence of a Nash equilibrium. In Section 3, we present an existence theorem that can be obtained nevertheless provided we restrict ourselves to a special class of measure spaces, nonatomic Loeb measure spaces introduced in Loeb (1975).[10] Section 4 concludes the paper.

2 The Counterexample

Consider a game $\mathcal{G}_1^e$ in which the set of players is the unit interval $[0,1]$ endowed with Lebesgue measure λ, and the action set A is the interval $[-1,1]$. For the specification of the payoff functions, consider a function $z : [0,1] \times [-1,1] \longrightarrow \mathbb{R}$ such that for all $t \in [0,1]$,

$$\begin{aligned} z(t,a) &= a && \text{if } 0 \leq a \leq t \\ &= t && \text{if } t < a \leq 1 \\ &= -z(t,-a) && \text{if } a < 0 \end{aligned}$$

[9] However, we shall return to the second possibility at the end of Section 3 below.

[10] See Anderson (1991, 1992), Rashid (1987) and Loeb-Rashid (1987) for further explication, and for applications in mathematical economics.

This function of two variables can also be seen an uncountable family of functions on $[-1,1]$, indexed by points in $[0,1]$. Two representatives of this family, indexed by r and $(r+\epsilon)$, are shown in Figure 2. Note that the specification implies that $z(0,a)=0$ for all a in $[-1,1]$. We shall now use this family of functions to define the payoff functions for each player. For any player t, let $u_t : [-1,1] \times [-1,1] \longrightarrow \mathbb{R}$ be given by

$$u_t(a, a_\sigma) = -|t - |a|| + (t-a)z(t, a_\sigma).$$

It is easy to check that each u is a jointly continuous function in its three arguments (t, a, a_σ), and that the family of utility functions indexed by the name t is an equicontinuous family. This allows us to conclude that the mapping $\mathcal{G}_1^e$ from the unit interval to the space of all continuous functions on $[-1,1] \times [-1,1]$ is itself continuous, and therefore measurable when the latter is equipped with its Borel σ-algebra. The specification of the large non-anonymous game $\mathcal{G}_1^e$ is complete.

For the reader's convenience, we give a formal definition of Nash equilibrium for the game $\mathcal{G}_1^e$.

Definition 1 *An equilibrium of $\mathcal{G}_1^e$ is a measurable function f from $[0,1]$ to $[-1,1]$ such that for λ-almost all $t \in [0,1]$,*

$$\int_{-1}^{1} u_t(f(t), a_\sigma) d(\lambda \cdot f^{-1})(a_\sigma) \geq \int_{-1}^{1} u_t(a, a_\sigma) d(\lambda \cdot f^{-1})(a_\sigma) \quad \textit{for all } a \in [-1,1].$$

We shall show that there is no equilibrium in the sense of Definition 1 for the game $\mathcal{G}_1^e$.

We shall need the functions $U_t : A \times \mathcal{M}(A) \longrightarrow \mathbb{R}$ given by $U_t(a,\nu) = \int_A u_t(a, a_\sigma) d\nu(a_\sigma)$, and $h : T \times \mathcal{M}(A) \longrightarrow \mathbb{R}$ given by $h(t,\nu) = \int_{-1}^{1} z(t,a) d\nu(a)$. The function U_t is standard in the literature, and is used to lift the individual payoff functions from the space $A \times A$ to the space $A \times \mathcal{M}(A)$. For any given distribution ν on the action set A, the function $h(t,\nu)$ represents the relevant societal statistic, and along with its derivative, it plays a crucial role in the analysis to follow. Geometrically, $h(t,\nu)$ is the ν-weighted area under $z(t,\cdot)$; Figure 2 illustrates the case when the values of t are given by ϵ and $(r+\epsilon)$. The basic non-existence argument is tailored around the fact that in equilibrium $\nu^\star$, $h(t,\nu^\star)$ must be zero for almost all players, leading to the fact that $\nu^\star$ must be the uniform measure on $[-1,1]$, an impossibility for the same reason as in the Rath-Sun-Yamashige example; namely the absence of a measurable selection from the correspondence pictured in Figure 1.

The crucial and non-routine part of the argument concerns the claim that equilibrium value of $h(t,\cdot) = 0$ for λ-almost all players in T. Before we turn to this, we chart out the implication of the value of $h(t,\nu)$ for the determination of the best response correspondence.

Proposition 1 *For any $\nu \in \mathcal{M}([-1,1])$, and for any player $t \in [0,1]$,*

$$argmax_{a \in A} U_t(a,\nu) = \begin{cases} \{t, -t\} & \text{if } h(t,\nu) = 0 \\ t & \text{if } h(t,\nu) < 0 \\ -t & \text{if } h(t,\nu) > 0 \end{cases}$$

In words, for all non-zero values of $h(t,\nu)$, there is a unique best action for each player identical to his name in magnitude but opposite in sign to that of $h(t,\nu)$. If the latter is zero, the best response is a doubleton set shown in Figure 1. Figure 3 illustrates this case by depicting the values of the payoff functions over the entire action set. The routine computations underlying these assertions are relegated to the Appendix.

Next we consider for any distribution $\nu \in \mathcal{M}([-1,1])$, the value of the difference of the tails of the distribution. This is to consider the function $d : [0,1] \times \mathcal{M}([-1,1]) \longrightarrow \mathbb{R}$ where

$$d(t,\nu) = \nu([t,1]) - \nu([-1,-t]). \tag{1}$$

It is curious analytical property of the functions $z(t,\cdot)$ that inspite of the kink at t, the function $h(t,\nu)$ is differentiable at t, and that its differential is given by $d(t,\nu)$. Given the best response correspondence, the argument involves elementary analysis and makes no additional reference to game-theoretic ideas. The intuition is clear. For any $r \in [0,1]$, and any positive ϵ less than r, the difference $h(r+\epsilon,\nu) - h(r,\nu)$ is given by the sum of the shaded areas in Figure 2. Note that in this figure, the areas are ν-weighted, but the measure ν is not specified. Now the interval $[-(r+\epsilon), -r]$ has at most full ν-measure, in which case the interval $[r, r+\epsilon]$ has zero ν-measure, and by computing areas of relevant rectangles and triangles, we obtain

$$h(r+\epsilon,\nu) - h(r,\nu) \leq \epsilon\nu([r,1]) + (1/2)\epsilon^2 - \epsilon\nu([-1,-r]) = \epsilon d(r,\nu) - (1/2)\epsilon^2.$$

On dividing throughout by ϵ and by taking limits, we have a claim for the value of a right-sided derivative. A similar argument furnishes the following

$$h(r-\epsilon,\nu) - h(r,\nu) \leq -\epsilon\nu([r,1]) + (1/2)\epsilon^2 + \epsilon\nu([-1,-r]) = -\epsilon d(r,\nu) + (1/2)\epsilon^2,$$

and we have a complementary claim about the left-sided derivative. Putting the two claims together, we obtain the value for the derivative at all internal points in the interval. The complete argument formal is relegated to the Appendix, and we present a formal statement of the claim.

Lemma 1 *Let f be a measurable selection from the best reply correspondence $\{t, -t\}$ and ν the induced measure $\lambda \cdot f^{-1}$. Then for any $r \in (0,1)$, $h(r,\nu)$ is differentiable with its* *derivative equal to* $d(r,\nu)$. *Furthermore,* $h'_+(0,\nu) = d(0,\nu)$, *and* $h'_-(1,\nu) = d(1,\nu)$, *where* $h'_+(0,\nu)$ *and* $h'_-(1,\nu)$ *respectively denote the right and left derivatives of* h *at* 0 *and* 1.

We shall now develop the non-existence argument in a series of claims. Suppose $f : T \longrightarrow A$ is an equilibrium of the game $\mathcal{G}_1^e$, and that $\nu = \lambda \cdot f^{-1} \in \mathcal{M}(A)$ is the induced distribution on $[-1,1]$.

Claim 1 $h(t,\nu) = 0$ *for all* $t \in [0,1]$.

Suppose to the contrary that there exists $x \in [0,1]$ such that $h(x,\nu) \neq 0$. Since $h(0,\nu) = 0$, certainly $x > 0$. Let

$$S_1 = \{t \in [0,x] : h(t,\nu) = 0\} \text{ and } S_2 = \{t \in [x,1] : h(t,\nu) = 0\}$$

Since $h(0,\nu) = 0$, S_1 is nonempty. Let $r = \sup S_1$. By the continuity of $h(\cdot,\nu)$, $h(r,\nu) = 0$ and therefore $r < x$. If S_2 is empty, let $s = 1$; otherwise, let $s = \inf S_2$. Clearly, $r < x \leq s$ and on the interval (r, s), $h(\cdot,\nu)$ is nonzero and does not change sign.

First consider the possibility that $h(x,\nu) > 0$. Since $h(r,\nu) = 0$ and $h(t,\nu) > 0$ for all $t \in (r, s)$, $h'_+(r,\nu) \geq 0$. If $s < 1$, then $h(s,\rho) = 0$ and $h(t,\nu) > 0$ for all $t \in (r, s)$ implies that $h'_-(s,\nu) \leq 0$. If $s = 1$, then $d(1,\nu) = 0$, and therefore $h'_-(1,\nu) = 0$. Thus, $h'_-(s,\nu) \leq 0$. Since $h(t,\nu) > 0$ for all $t \in (r, s)$, the best response correspondence assures us that the action of any player $t \in (r,s)$ is $-t$, and hence $\nu((r,s)) = 0$, and $\nu((-s,-r)) = s - r$. An appeal to the fact that

$$d(r,\nu) = d(s,\nu) + \nu([r,s)) - \nu((-s,-r]), \quad 0 \leq r \leq s \leq 1, \tag{2}$$

allows us to assert that $h'_+(r,\nu) = d(r,\nu) = d(s,\nu) + \nu([r,s)) - \nu((-s,-r]) = d(s,\nu) - (s-r)$ $= h'_-(s,\nu) - (s-r) < 0$. But this contradicts the fact that $h'_+(r,\nu) \geq 0$.

All that remains is the possibility that $h(x,\nu) < 0$. In this case, we simply mimic the above argument to assert that $h'_+(r,\nu) \leq 0$ and $h'_-(s,\nu) \geq 0$. Since $h(t,\nu) < 0$ for all $t \in (r, s)$, from the best response correspondence the action of any player $t \in (r,s)$ is t, and hence $\nu((r,s)) = s - r$, and $\nu((-s,-r)) = 0$. On using this and (20, we obtain $h'_+(r,\nu) = d(r,\nu)$ $= d(s,\nu) + \nu([r,s)) - \nu((-s,-r]) = d(s,\nu) + (s-r) = h'_-(s,\nu) + (s-r) > 0$. But this contradicts the fact that $h'_+(r,\nu) \leq 0$ and completes the proof.

Claim 2 $d(t,\nu) = 0$ *for all* $t \in [0,1]$.

By the differentiability property of $h(\cdot,\nu)$, $d(t,\nu) = 0$ for all $t \in (0,1)$. Since ν is the image under f of the Lebesgue measure λ, certainly $d(1,\nu) = 0$. These two facts imply[11] that $\nu((0,1]) = \nu([-1,0))$, and hence that $d(0,\nu) = 0$.

Claim 3 *ν is the uniform distribution on* $[-1,1]$.

¿From the best response correspondence $\nu([0,t]) + \nu([-t,0]) = t$. Claim 2 implies that $\nu([0,t]) = \nu([-t,0])$ for any t, and hence $\nu([0,t]) = \nu([-t,0]) = t/2$ for all $t \in [0,1]$.

We now show the impossibility of inducing the uniform measure by a measurable selection from the best response correspondence, as mentioned in the introduction; the elementary proof is taken from Khan-Rath-Sun (1996).

Claim 4 *For any Lebesgue measurable subset F of $[0,1]$, let $f : T \longrightarrow A$ be such that $f(t) = t$ if $t \in F$ and $f(t) = -t$ if $t \notin F$. Then the induced measure $\nu = \lambda \cdot f^{-1}$ is not the uniform distribution on $[-1,1]$.*

Let f be such a measurable selection. Then $\nu(F) = \lambda \cdot (f^{-1}(F)) = \lambda(F)$. Since $\nu = (1/2)\lambda$, $\lambda(F) = 0$, and hence $\nu([-1,0]) = \lambda \cdot f^{-1}([-1,0]) = \lambda(\{t \notin F\}) = 1$, a contradiction.

This completes the non-existence argument.

3 An Existence Theorem

In the light of this counterexample, a natural question arises as to the possibility of a positive result. In this section, we show that this is indeed the case if we model the set of players names by a measure space with additional properties. We present an existence theorem based on nonatomic measure spaces introduced in Loeb (1975), and now commonly referred to as hyperfinite Loeb measure spaces. The importance of these standard measure spaces for mathematical economics is fully discussed in Anderson (1991); also see Rashid (1987).

Let $(T, \mathcal{T}, \lambda)$ denote a hyperfinite internal probability space and $(T, L(\mathcal{T}), L(\lambda))$ its standardization – the Loeb space. We shall assume that this Loeb space is atomless. Loeb spaces are constructed as a simple consequence of Caratheodory's extension theorem and the $\aleph_1$-saturation property of the nonstandard models. However, in any application, one can ignore the construction of hyperfinite sets and Loeb measures in much the same way that a user of Lebesgue measure spaces can afford to ignore the Dedekind set-theoretic construction

[11] This follows from an elementary property of measures; see Rudin (1974; p. 17).

of real numbers and the particular construction of Lebesgue measure. One simply appeals to those special properties of Loeb spaces not shared by general measure spaces. It also bears emphasis any result established for an abstract measure space applies *a fortiori* to Loeb spaces; $L(\mathcal{T})$ is a σ-algebra in the standard sense of being closed under complementation and countable unions, and $L(\lambda)$ is a measure in the standard sense of being countably additive. Atomless Loeb measure spaces also fulfill other important methodological criteria for the modelling of game-theoretic and other economic phenomena; these concern measurability, homogeneity,[12] and asymptotic implementability.[13] Here we shall be solely concerned with Loeb spaces being a vehicle for the formalization of strategic negligibility.

Let A be a compact metric space and all measurability notions understood with respect to the measurable spaces $(T, L(\mathcal{T}))$, $(\mathcal{U}_A^e, \mathcal{B}(\mathcal{U}_A^e))$, and $(A, \mathcal{B}(A))$. We can now state

Theorem 3 *For any measurable mapping* $\mathcal{G}_\ell^e : T \longrightarrow \mathcal{U}_A^e$, *there exists a measurable mapping* $f : T \longrightarrow A$ *such that for* $L(\lambda)$*-almost* $t \in T$,

$$\int_A u_t(f(t), a_\sigma) d(L(\lambda) \cdot f^{-1})(a_\sigma) \geq \int_A u_t(a, a_\sigma) d(L(\lambda) \cdot f^{-1})(a_\sigma) \quad \textit{for all } a \in A,$$

where $L(\lambda) f^{-1}$ *is the distribution induced on the set* A *by* f, *and* $u_t \equiv \mathcal{G}_\ell^e(t)$.

Given the generality of the action sets, previous work based on abstract measure spaces has only been able to furnish approximate existence results even for the idealized limit setting; see Khan (1986) and Pascoa (1993). Theorem 3 is a simple corollary of Theorem 1 in Khan-Sun (1995b). However, the reader can supply a direct proof based on the Fan-Glicksberg fixed point theorem by setting up a mapping from $\mathcal{M}(A)$ to itself, by utilizing the convexity and upper semicontinuity results from Sun (1996), and by the topological and measure-theoretic supplementation from Berge (1959) and Castaing-Valadier (1977). Yet another alternative argument can be developed on the basis of the results on Gel'fand integration developed in Sun (1993).

In both the counterexample and the existence theorem, players choose points from the action set A, and the equilibrium distribution is induced from this function f summarizing the collection of pure strategies. As discussed in Khan-Rath-Sun (1996), the induced distribution of any random variable $f(\cdot)$ is the Gel'fand integral of the measure-valued function $\delta_{f(\cdot)}$, where δ_a, is the Dirac point measure at a in A. We can then consider an equilibrium concept

[12] See von Neumann (1932) for the observation that Lebesgue measures do not satisfy the homogeneity property, and Khan-Sun (1996b) for the game-theoretic implications.

[13] These criteria have been discussed in Anderson (1991) and, more specifically in the case of non-cooperative game theory, in Khan-Sun (1996b).

in which each player randomizes and plays mixed strategies, which is to say that his action is in $\mathcal{M}(A)$ rather than in A, and societal responses are furnished by the Gel'fand integral of the function summarizing these choices. The fact that there exists a mixed-strategy Nash equilibrium in this sense is straightforward; there are no existence difficulties when the action set is convex and the payoff function is quasi-concave on it.[14] What makes Theorem 3 interesting is that concerns a setting where neither of these hypotheses hold.

4 Concluding Question

We conclude this paper by asking what is it about Lebesgue measure that makes existence of equilibrium problematic; or to put the matter another way, what is it about an atomless Loeb measure that overcomes these obstacles? We hope to return to this question in future work.

5 Appendix

We begin with the proof of the characterization of the best response correspondence.

Proof of Proposition 1: If $h(t,\nu) = 0$, $U_t(a,\nu) < 0 = U_t(t,\nu) = U_t(-t,\nu)$ for all $a \in [-1,1]$, $a \neq t, -t$.

If $h(t,\nu) < 0$, $U_t(a,\nu) - U_t(t,\nu) = -\mid t - \mid a \mid\mid + (t-a)h(t,\nu)$. If $t = 1$, the above expression is negative for all $a \neq t$. Suppose $t < 1$. Now, for all $a < t$, the right hand side of the above equality is negative; and for all $a > t$, it reduces to $(t-a)[1 + h(t,\nu)]$ which is also negative by virtue of

$$|h(t,\nu)| = |\int_{-1}^{1} z(t,a)d\nu(a)| \leq \int_{-1}^{1} |z(t,a)|d\nu(a) \leq t. \tag{3}$$

If $h(t,\nu) > 0$,

$$\begin{aligned} U_t(a,\nu) - U_t(-t,\nu) &= -\mid t - \mid a \mid\mid + (t-a)h(t,\nu) - (2t)h(t,\nu) \\ &= -\mid t - \mid a \mid\mid - (t+a)h(t,\nu) \end{aligned}$$

If $t = 1$, then the above expression is negative for all $a \neq -t$. Suppose $t < 1$. Now, for all $a \geq -t$, the last expression is negative; and for all $a < -t$, it reduces to $(t+a)[1 - h(t,\nu)]$ which, given (3), is also negative.

This completes the proof. ■

[14] More specifically, convex-valued correspondences do not require their domain to be a Loeb space for the upper semi-continuity result to hold, as can be seen by modifying the relevant argument in Sun (1993b).

Next, we turn to the differentiability property of $h(\cdot, \nu)$.

Proof of Lemma 1: Let r and t belong to [0, 1]. From the definitions of the functions $z(t, \cdot)$ and $z(r, \cdot)$

$$\begin{aligned} h(t,\nu) &= td(t,\nu) + \int_0^t z(t,\cdot)d\nu + \int_{-t}^0 z(t,\cdot)d\nu \\ h(r,\nu) &= rd(r,\nu) + \int_0^r z(r,\cdot)d\nu + \int_{-r}^0 z(r,\cdot)d\nu \end{aligned}$$

Suppose that $r < t$. Since $z(t, \cdot)$ and $z(r, \cdot)$ are identical on $[0, r]$ and on $[-r, 0]$, one obtains

$$\begin{aligned} & h(t,\nu) - h(r,\nu) \\ &= td(t,\nu) - rd(r,\nu) + \int_r^t z(t,\cdot)d\nu + \int_{-t}^{-r} z(t,\cdot)d\nu \\ &= (t-r)d(r,\nu) - t\nu([r,t)) + t\nu((-t,-r]) + \int_r^t z(t,\cdot)d\nu + \int_{-t}^{-r} z(t,\cdot)d\nu \end{aligned}$$

where the latter equality has been obtained by substituting, from (2) above, for $d(t,\nu) = d(r,\nu) - \nu([r,t)) + \nu((-t,-r])$. Since $|z(t,\cdot)| \leq t$, $\int_r^t z(t,\cdot)d\nu - t\nu([r,t))$ is nonpositive and $t\nu((-t,-r]) + \int_{-t}^{-r} z(t,\cdot)d\nu$ is nonnegative. Transposition of some terms yields

$$(t-r)d(r,\nu) - t\nu([r,t)) + \int_r^t z(t,\cdot)d\nu \leq h(t,\nu) - h(r,\nu) \leq (t-r)d(r,\nu) + t\nu((-t,-r]) + \int_{-t}^{-r} z(t,\cdot)d\nu.$$

Note that on $[r,\ t]$, $z(t,\cdot) \geq r$ and on $[-t,\ -r]$, $z(t,\cdot) \leq -r$. Thus, $\int_r^t z(t,\cdot)d\nu \geq r\nu([r,t))$ and $\int_{-t}^{-r} z(t,\cdot)d\nu \leq -r\nu((-t,-r])$. Therefore,

$$(t-r)d(r,\nu) - (t-r)\nu([r,t)) \leq h(t,\nu) - h(r,\nu) \leq (t-r)d(r,\nu) + (t-r)\nu((-t,-r]).$$

By the assumption on ν, $\nu([r,t)) \leq t - r$ and $\nu((-t,-r]) \leq t - r$, and hence,

$$(t-r)d(r,\nu) - (t-r)^2 \leq h(t,\nu) - h(r,\nu) \leq (t-r)d(r,\nu) + (t-r)^2. \tag{4}$$

On dividing (4) throughout by $(t - r) > 0$, and on letting t tend to r, we obtain $h'_+(r,\nu) = d(r,\nu)$ for all $r \in [0,\ 1)$.

Next, we consider the case $t < r$. By interchanging t and r above, we obtain

$$\begin{aligned} & h(r,\nu) - h(t,\nu) \\ &= (r-t)d(t,\nu) - r\nu([t,r)) + r\nu((-r,-t]) + \int_t^r z(r,\cdot)d\nu + \int_{-r}^{-t} z(r,\cdot)d\nu. \end{aligned}$$

Since $t < r$, $d(t,\nu) = d(r,\nu) + \nu([t,r)) - \nu((-r,-t])$, and the relevant substitution yields,

$$\begin{aligned} & h(t,\nu) - h(r,\nu) \\ &= (t-r)d(r,\nu) + t\nu([t,r)) - t\nu((-r,-t]) - \int_t^r z(r,\cdot)d\nu - \int_{-r}^{-t} z(r,\cdot)d\nu \end{aligned}$$

On the interval $[t, r]$, $z(r,\cdot) \geq t$ and on the interval $[-r, -t]$, $z(r,\cdot) \leq -t$. Thus, $\int_t^r z(r,\cdot)d\nu \geq t\nu([t,r))$, and $\int_{-r}^{-t} z(r,\cdot)d\nu \leq -t\nu((-r,-t])$. Therefore,

$$(t-r)d(r,\nu)+t\nu([t,r))-\int_t^r z(r,\cdot)d\nu \leq h(t,\nu)-h(r,\nu) \leq (t-r)d(r,\nu)-t\nu((-r,-t])-\int_{-r}^{-t} z(r,\cdot)d\nu.$$

Since $|z(r,\cdot)| \leq r$, $t\nu([t,r)) - \int_t^r z(r,\cdot)d\nu \geq (t-r)\nu([t,r))$ and $-t\nu((-r,-t]) - \int_{-r}^{-t} z(r,\cdot)d\nu \leq -(t-r)\nu((-r,-t])$. By invoking the facts that both $\nu([t,r))$ and $\nu((-r,-t])$ are less than or equal to $r-t$, we obtain $(t-r)\nu([t,r)) \geq -(t-r)^2$ and $\nu((-r,-t]) \leq (t-r)^2$. Therefore,

$$(t-r)d(r,\nu) - (t-r)^2 \leq h(t,\nu) - h(r,\nu) \leq (t-r)d(r,\nu) + (t-r)^2. \tag{5}$$

On dividing (5) throughout by $(t-r) < 0$, and on letting t tend to r, we obtain $h'_-(r,\nu) = d(r,\nu)$ for all $r \in (0,1]$.

This completes the proof. ■

ACKNOWLEDGEMENTS: This research was conceived while the second and third authors were visiting the Department of Economics at Johns Hopkins during parts of the years 1994-1996. A preliminary version was presented at the *International Conference on Game Theory and Economic Applications* held in Bangalore, January 2-6, 1996; stimulating conversations with Professors K. Chatterjee, T. Parthasarathy, D. Ramachandran and R. Sundaram are gratefully acknowledged.

References

1. Anderson, R. M. (1991). "Non-standard Methods in Economics," in W. Hildenbrand and H. Sonnenschein (eds.) *Handbook of Mathematical Economics.* Amsterdam: North Holland Publishing Company.

2. Anderson, R. M. (1992), "The Core in Perfectly Competitive Economies," in R. J. Aumann and S. Hart (eds.) *Handbook of Game Theory,* Volume 2. Amsterdam: North-Holland Publishing Company.

3. ARTSTEIN, Z.(1983). "Distributions of random sets and random selections." *Israel Journal of Mathematics* **46**, 313-324.

4. Berge, C. (1959) *Topological Spaces.* London: Oliver & Boyd.

5. Billingsley, P. (1968). *Convergence of Probability Measures.* New York: John Wiley.

6. Castaing, C., and Valadier, m. (1977). *Convex Analysis and Measurable Multifunctions,* Lecture Notes in Mathematics no. **580**, Berlin and New York: Springer-Verlag, 1977.

7. Fan, K. (1952). "Fixed Points and Minimax Theorems in Locally Convex Linear Spaces." *Proc. Nat. Acad. Sci. U.S.A* **38**, 121-126.

8. Fudenberg, D., and J. Tirole (1991). *Game Theory.* Cambridge: MIT Press.

9. Glicksberg, I. (1952). "A Further Generalization of Kakutani's Fixed Point Theorem with Application to Nash Equilibrium Points." *Proc. Amer. Math. Soc.* **38**, 170-172.

10. Hart, S. and E. Kohlberg (1974), Equally Distributed Correspondences." *Jour. Math. Econ.* **1**, 167-174.

11. Hart, S., W. Hildenbrand and E. Kohlberg (1974), "On Equilibrium Allocations as Distributions on the Commodity Space", *Jour. Math. Econ.* **1**, 159-166.

12. Khan, M. Ali (1986). "Equilibrium Points of Nonatomic Games over a Banach Space." *Trans. Amer. Math. Soc.* **293**, 737-749.

13. Khan, M. Ali, Rath, K. P., and Sun, Y. N. (1994) "On Games with a Continuum of Players and Infinitely Many Pure Strategies." *Johns Hopkins Working Paper* No. 322. *Jour. Ec. Theory* forthcoming.

14. Khan, M. Ali, Rath, K. P., and Sun, Y. N. (1995) "On Private Information Games without Pure Strategy Equilibria." *Johns Hopkins Working Paper* No. 352.

15. Khan, M. Ali and Sun, Y. N. (1995a). "Pure Strategies in Games with Private Information." *J. Math. Econ.* **24**, 633-653.

16. Khan, M. Ali and Sun, Y. N. (1995b) "Non-Cooperative Games on Hyperfinite Loeb Spaces." *Johns Hopkins Working Paper* No. 359.

17. Loeb, P. A. (1975). "Conversion from Nonstandard to Standard Measure Spaces and Applications in Probability Theory." *Trans. Amer. Math. Soc.* **211**, 113-122.

18. Loeb, P. A., and Rashid, S. (1987). "Non-standard Analysis," in J. Eatwell at al. (eds.) *The New Palgrave.* London: The MacMillan Publishing Co.

19. Nash, J. F. (1950). "Equilibrium Points in N-person Games." *Proc. Natl. Acad. Sci. U.S.A.* **36**, 48-49.

20. Nash, J. F. (1951), "Noncooperative Games." *Ann. Math.* **54**, 286-295.

21. Parthasarathy, K. R. (1967). *Probability Measures on Metric Spaces.* New York: Academic Press.

22. Pascoa, M. R. (1993). "Approximate Equilibrium in Pure Strategies for Non-atomic games." *J. Math. Econ.* **22**, 223-241.

23. Rashid, S. (1987). *Economies with Many Agents.* Baltimore: The Johns Hopkins University Press.

24. Rath, K. (1992). "A Direct Proof of the Existence of Pure Strategy Equilibria in Games with a Continuum of Players." *Ec. Theory* **2**, 427-433.

25. Rath, K., Sun, Y., and Yamashige, S. (1995). "The Nonexistence of Symmetric Equilibria in Anonymous Games with Compact Action Spaces." *J. Math. Econ.* **24**, 331-346.

26. Rudin, W. (1974). *Real and Complex Analysis.* New York: McGraw Hill.

27. Sun, Y. N. (1993a) "Distributional Properties of Correspondences on Loeb Spaces." *J. Func. Anal.* **139**, 68-93.

28. Sun, Y. N. (1993b) "Integration of Correspondences on Loeb Spaces." *Trans. Amer. Math. Soc.* forthcoming.

29. Schmeidler, D. (1973) "Equilibrium Points of Nonatomic Games." *J. Stat. Phys.* **7**, 295-300.

30. Von Neumann, J. (1932). "Einige Sätze über Messbare Abbildungen." *Ann. Math.* **33**, 574-586.

M. Ali Khan
Department of Economics
The Johns Hopkins University
Baltimore, MD 21218,USA

Kali P. Rath
Department of Economics
University of Notre Dame
Notre Dame, IN 46556, France

Yeneng Sun
Department of Mathematics
National University of Singapore
Singapore 119260

AND

Cowles Foundation
Yale University
New Haven, 06520, USA

EQUILIBRIA IN REPEATED GAMES OF INCOMPLETE INFORMATION THE DETERMINISTIC SYMMETRIC CASE

Abraham Neyman and Sylvain Sorin

Abstract: *Every two person game of incomplete information in which the information to both player is identical and deterministic has an equilibrium.*

1 Introduction

This note collates two results: the reduction of a class of incomplete information two person zero sum games to games with absorbing states (Kohlberg and Zamir, 1974; Kohlberg 1974) and the existence of equilibrium payoffs for two person non zero sum games with absorbing states (Vrieze and Thuisjman, 1989) to obtain the existence of equilibrium payoffs for a class of two person non zero sum incomplete information games.

We consider a situation where there is a finite set K of states and for each k in K a bi-matrix game G^k defined by $I \times J$ real valued payoff matrices A^k, B^k and $I \times J$ "signalling matrices" H^k with value in some space H.

To each initial distribution p on K is associated a game $\Gamma(p)$ played as follows: The state k is chosen once and for all according to p but is not transmitted to the players. The game is played an infinite number of stages where at stage n, player I (resp. player II) chooses $i_n \in I$ (resp. $j_n \in J$). The vector payoff at stage n is thus $x_n = (A^k_{i_n j_n}, B^k_{i_n j_n})$ (for player I and II respectively), but is not announced. Rather the players are told the "public signal" $h_n = H^k_{i_n j_n}$. We want that the signal contains all the information of the players at that stage and that perfect recall holds, hence the signal contains the moves: $i \neq i'$ or $j \neq j'$ implies $H^k_{ij} \neq H^{k'}_{i'j'}$.

2 The Result

Any pair of strategies σ of player 1 and τ of player 2, together with the initial probability p, induces a probability distribution on the set of histories $k, i_1, j_1, \ldots, i_n, j_n, \ldots$ and therefore it also defines a probability distribution on the stream of payoffs $x_1, \ldots, x_n, \ldots$ where $x_t = (a^k_{i_t j_t}, b^k_{i_t j_t})$. Let $x_t(\sigma, \tau) = E_{p,\sigma,\tau}(a^k_{i_t j_t}, b^k_{i_t = j_t})$, and set $\bar{x}_n(\sigma, \tau) = (1/n) \sum_{t=1}^{n} x_t(\sigma, \tau)$.

The set of equilibrium payoffs in $\Gamma(p)$, E_0, is defined as $\cap_{\varepsilon > 0} E_\varepsilon$, where E_ε is the set of all payoff vectors $d = (d^1, d^2) \in I\!R^2$ for which there exist strategies σ of player 1 and τ of player 2 and a positive integer N such that for any pair of strategies σ' of player 1 and τ' of player 2, and $n \geq N$,

$$\bar{x}^1_n(\sigma, \tau) + \varepsilon \geq d^1 \geq \bar{x}^1_n(\sigma', \tau) - \varepsilon$$

and

$$\bar{x}^2_n(\sigma, \tau) + \varepsilon \geq d^2 \geq \bar{x}^1_n(\sigma, \tau') - \varepsilon$$

(see Mertens, Sorin and Zamir (1994), p. 403).

A pair of strategies in the infinitely repeated game, σ for player 1 and τ for player 2, is called an *ε-uniform equilibrium*, if there is N such that, for all $n, m \geq N$ and every strategy pair, σ' of player 1 and τ' of player 2,

$$\bar{x}_n^1(\sigma, \tau) \geq \bar{x}_m^1(\sigma', \tau) - \varepsilon$$

and

$$\bar{x}_n^2(\sigma, \tau) \geq \bar{x}_m^1(\sigma, \tau') - \varepsilon.$$

Note that E_0 is not empty if and only if, for every $\varepsilon > 0$, there exists an ε-uniform equilibrium.

Theorem *For any two person game with symmetric and deterministic information, E_0 is non empty.*

Proof

As in Kohlberg and Zamir (1974), the proof is by induction on the number of "active" states, i.e. on the size m of the support of p, and we assume the result true for $\Gamma(p)$ for all $m < \#K$.

For every couple of moves (i, j), the signals in $H(ij) = \{H_{ij}^k, k \in K\}$ induce a partition of K. For any signal h, let $K(h)$ denote the set of k's compatible with it and $p(h)$ be the corresponding conditional probability. Note that if $K(h) \neq K$, the induction hypothesis implies that the game $\Gamma(p(h))$ has equilibrium payoffs, hence we can choose one and we will denote it by $(a(h), b(h))$.

Define now a game $\Gamma'(p)$ as a repeated game with absorbing states (and standard signalling) with payoffs:

$$G'(p)_{ij} = \begin{cases} \sum_k p^k (A_{ij}^k, B_{ij}^k) & \text{if } K(h) = K, \\ \left(\sum_{h \in H(ij)} \sum_{k \in K(h)} p^k (a(h), b(h))\right)^* & \text{otherwise} \end{cases}$$

where, as usual, a * denotes an absorbing payoff.

By the theorem of Vrieze and Thuijsman (1989) (see also Mertens, Sorin and Zamir (1994), p. 406-408), $\Gamma'(p)$ has at least one equilibrium payoff, say (a, b).

We claim that it is also an equilibrium payoff of $\Gamma(p)$: in fact it is clear that playing the "equilibrium" strategies in $\Gamma'(p)$ as long an absorbing state is not reached and playing then, if the signal h is observed, the "equilibrium" strategies inducing the equilibrium payoff $(a(h), b(h))$ in $\Gamma(p(h))$ will define alltogether "equilibrium" strategies. ∎

3 Comments and Open Problems

This research is part of a general program which purpose is to characterize the information structures for which equilibria do exist.

For two person games with lack of information on one side, the existence has been recently proved by Simon, Spiez and Torunczyk (1995).

Recall that in the framework of lack of information on both sides, already in the zero sum case the value may not exist.

This paper provides a positive answer for a class of two person non zero sum games with symmetric incomplete information. A proof for the n person case would follow in the same way from the proof of existence of equilibria for n person games with absorbing states.

Games with absorbing states have also been used by Forges (1982) to generalize Kohlberg and Zamir's result and to prove the existence of a value in the case of zero-sum games with symmetric but random signals; however the method does not extend directly to get equilibria in the non zero sum case.

References

1 Forges F. (1982) Infinitely repeated games of incomplete information: symmetric case with random signals, *International Journal of Game Theory*, 11,3-213.

2 Kohlberg E. (1974) Repeated games with absorbing states, *The Annals of Statistics*, 2, 724-738.

3 Kohlberg E. and S. Zamir (1974), Repeated games of incomplete information: the symmetric case, *The Annals of Statistics*, 2, 1040-1041.

4 Mertens J.-F., S. Sorin and S. Zamir (1994), Repeated games, Core D.P. 94, 9421, 9422.

5 Simon R.S., S. Spiez and H. Torunczyk (1995) The existence of an equilibrium in games of incomplete information on one side and a theorem of Borsuk-Ulam type, *Israel Journal of Mathematics*, 92, 1-21.

6 Vrieze O.J. and F. Thuijsman (1989) On equilibria in repeated games with absorbing states, *International Journal of Game Theory*, 187, 293-310.

Abraham Neyman
Institute of Mathematics
The Hebrew University of Jerusalem
Givat Ram, Jerusalem 91904
Israel

AND

Institute for Decision Sciences
State University of New York
Stony Brook, New York 11794
USA

Sylvain Sorin
Laboratoire d'Econométrie
Ecole Polytechnique
1 rue Descartes
75005 Paris, France

AND

MODAL'X, U.F.R. SEGMI
Université Paris X Nanterre
200 Avenue de la République
92001 Nanterre Cedex, France

ON STABLE SETS OF EQUILIBRIA

A.J. Vermeulen, J.A.M. Potters and M.J.M. Jansen

Abstract: *A new kind of perturbations of normal form games is introduced and the stability concept related to these perturbations is investigated. The CQ-sets obtained in this way satisfy the properties of the Kohlberg-Mertens program except Invariance. In order to overcome this problem our solution concept is modified in such a way that all properties formulated by Kohlberg and Mertens are satisfied.*

1 Introduction

The paper of Kohlberg and Mertens (1986) marked a turning point in the research of equilibria for extensive form games. In at least four respects the paper introduced new fields of attention.

(1) The search for stability was no longer directed to individual equilibria but to sets of equilibria. In the paper the authors argue convincingly that the restriction to undominated equilibria and the requirement that iterated deletion of dominated strategies 'preserves stability', inevitably leads to set-valued solution concepts.

(2) The authors propagate the idea that the normal form (and even the reduced normal form) of an extensive form game contains all (or at least enough) information to make rational decisions. In this paper we take this idea seriously and we only investigate normal form games.

(3) Furthermore, a list of properties that a solution concept for normal form games should satisfy, is formulated. This list of properties, namely Existence, Connectedness, Admissibility, Backward Induction, Independence of inadmissible strategies and Invariance, will be called the Kohlberg–Mertens program in this paper. Later, in subsequent papers of Mertens (1987),(1989), it was adapted and extended with other properties.

(4) In the paper three stability concepts are introduced following the same pattern of definition. First a class of perturbations of a normal form game is given together with a metric that measures the size of the perturbation. Then a nonempty closed subset S of strategy profiles is called *stable with respect to the given class of perturbations* if, for every open neighborhood V of S, there exists a number $\delta > 0$ such that the equilibrium set of every perturbed game intersects V as long as the size of the perturbation is smaller than δ. The authors call *minimal* sets with the forementioned stability property hyperstable, fully stable or stable, respectively.

In subsequent papers of Mertens (1989),(1991) the definition of stability takes a completely different direction. Stability is now related to the local combinatorial-topological behavior of the graph of the Nash equilibria over the set of small perturbations of the normal form game. For this (these) stability concept(s) Mertens proves the properties of the Kohlberg–Mertens program and several properties more. The construction of stable sets contains many nasty details and uses a sophisticated mathematical machinery. Moreover, the definition moves rather far away from game theoretical intuitions. Therefore, we assume, there is room for a new definition of stability that is closer to these intuitions. Finally, it is important to note that in Mertens (1989), (1991) 'minimality' is no longer required in full strength. Stable sets must be "small enough to exclude unreasonable equilibria and large enough to be invariant". The problems with the invariance property are, in our opinion, under-estimated in the original Kohlberg–Mertens paper.

Next, Hillas (1990) introduced two other types of perturbations defining quasi-stability and – what we will call – H-stability. In this paper we introduce a new kind of perturbation (inspired by both types of Hillas' perturbations) and therewith a new type of stability, *continuous quasi-stability*. For this stability concept we prove all the properties of the (original) Kohlberg-Mertens program but Invariance. It turns out that Invariance causes difficulties but we give a general method to 'make certain solution concepts invariant'. In this paper we shall not give the details of this method but refer to an other paper of the authors.

Now we will further discuss four subjects that provide a background for subsequent sections:

(α) We give a general definition of (pre-)stability and show that hyperstability, full stability, stability in the sense of Kohlberg and Mertens, quasi-stability and stability in the sense of Hillas fit into this general frame work. We will also introduce new names as we now have a proliferation of stability concepts.

(β) We introduce the concept of CQ-stability in an informal way.

(γ) We give the properties of the Kohlberg-Mertens program in the context of normal form games.

(δ) We discuss problems connected with Invariance, in particular in relation to the minimality condition required for stability concepts.

(α) A (pre)stability concept can be defined as soon as we know against what kind of perturbations a set of strategy profiles should be stable and what perturbations are considered to be small. Generally speaking, a perturbation of a normal form game changes some of the features of the game and a perturbation is small if the change is small. In Kohlberg and Mertens perturbations of the payoff functions are considered as well as perturbations of the strategy spaces. In Hillas we find perturbations of the strategy spaces and perturbations of the best reply correspondence. The perturbations we will use to introduce our stability concept, change the set of strategies that can be used as best reply by a player. It can also be seen as a perturbation of the best reply correspondence.
Suppose that we are investigating a class of perturbations. Let us call them Z-perturbations. A nonempty closed subset S of strategy profiles of a normal game Γ is called ***Z-pre-stable*** or a ***Z-set*** if, for each neighborhood V of S, there is a number $\delta > 0$ such that, the Z-perturbed 'game' has an 'equilibrium' in V, whenever the size of the perturbation is less than δ. If we have a perturbation of the best reply correspondence, the perturbation of a game is no longer a game and the term 'equilibrium' looses its meaning. In that case we replace it by 'fixed point of the perturbed best reply correspondence'. A nonempty closed subset S of strategy profiles of a game Γ is called ***Z-stable*** if S is a ***minimal*** Z-(pre-stable) set.
In Kohlberg and Mertens (1986) one finds three classes of perturbations:

Perturbation of the payoff functions: E-perturbations

A normal form game Γ' is an E-perturbation of the game Γ if Γ' has the same player set and the same strategy sets as Γ but different payoff functions. If u and u' are the payoff functions of Γ and Γ' respectively, we introduce a distance function (metric) by

$$d(\Gamma', \Gamma) := \max_{x \in \Delta} ||u'(x) - u(x)||.$$

Accordingly we can define E-sets and E-stable sets. If a set is E-stable and consists of *one* equilibrium, it is an essential equilibrium in the sense of Yia Yiang-He and Wu Wen-Tsun (1962). The definition is closely related to the definition of 'hyperstability' in Kohlberg and Mertens (1986). The only difference is that we do not try to make the concept 'invariant' by including a phrase like '... for all perturbations of games equivalent with Γ ... (i.e. with the same reduced normal form)'.

Perturbation of the strategy spaces: F-perturbations

An F-perturbation of a normal form game Γ is based on a restriction of the strategy spaces of the players. Each player i can only use the strategies in a polytope P_i in the interior of his strategy space Δ_i in Γ. The distance between the perturbed game and the original game Γ is measured by the Hausdorff distance of $\prod_i P_i$ and Δ. F-sets and F-stable sets are defined following the general idea. The F-stable sets are exactly the 'fully stable sets' of Kohlberg and Mertens.

A compromise: KM-perturbations

A KM-perturbation of a game Γ is a game Γ' wherein each player i must play each pure strategy at least with some given positive weight. The new strategy spaces P_i are simplices with facets parallel to the facets of Δ_i. If we take the extreme points of P_i as pure strategies in the perturbed game, a KM-perturbation can also be understood as a perturbation of the payoff function (E-perturbation). The same distance function as before can be used to measure KM-perturbations. The KM-stable sets are the 'stable sets' in Kohlberg and Mertens.

In Hillas (1990) we find two types of perturbations:

Another perturbation based on restriction of the strategy spaces: Q-perturbations

Also Q-perturbations are based on restrictions of the strategy spaces. The set of admissible strategy spaces is a special family of polytopes in the strategy spaces of the players. In this sense the set of Q-perturbations is a (proper) subset of the set of F-perturbations and contains the family of KM-perturbations. The quasi-stable sets in the sense of Hillas are the Q-stable sets in our general frame work.

Perturbation of the best reply correspondence: H-perturbations

A normal form game Γ induces a compact- and convex-valued USC correspondence $BR: \Delta \to \to \Delta$, the best reply correspondence. Nash equilibria are exactly the fixed points of this correspondence. An H-perturbation is a compact- and convex-valued USC correspondence from Δ to Δ. In this case the perturbed game is no longer a game. The fixed points of the perturbation (existing according to Kakutani's fixed point theorem) take the place of equilibria in the definition of pre-stability. On the set of H-perturbations we take the distance function d_∞ defined by $d_\infty(\varphi, BR) := \max_{x \in \Delta} d_H(\varphi(x), BR(x))$, the uniform Hausdorff distance. The H-stable sets are the stable sets in the sense of Hillas (1990) if we skip the phrase '... for all games Γ' equivalent with Γ ...' also from his definition,

(β) The perturbations we will introduce are continuous versions of Q-perturbations. For every player i and every proper subset T of pure strategies of player i, we define a continuous function $\varepsilon_{i,T}: \Delta \to [0, 1]$. The set $\Delta_i[\varepsilon](x)$ is the set of mixed strategies that put at least total weight $\varepsilon_{i\,T}(x)$ on the strategies in T. If $\Delta[\varepsilon](x) := \prod_i \Delta_i[\varepsilon](x)$ is not empty for all strategy profiles $x \in \Delta$, we call the correspondence $x \twoheadrightarrow \Delta[\varepsilon](x)$ a *CQ-perturbation* of (the strategy spaces of) the game Γ. Accordingly, the best reply correspondence $BR[\varepsilon]$ assigns to a strategy profile $x \in \Delta$ the best replies *in* $\Delta[\varepsilon](x)$ to the profile x. As turns out, the correspondence $BR[\varepsilon]$ is an H-perturbation of the best reply correspondence of the game Γ and in this sense the collection of CQ-perturbations is a subset of the set of H-perturbations. The main subject of this paper will be the properties of (minimal) CQ-sets defined as (minimal) nonempty and closed subsets of Δ with the property that, for each neighborhood V of S, there exists a number $\delta > 0$ such that the fixed point set of $BR[\varepsilon]$ intersects V if $\|\varepsilon\| := \max_x \max_{i,T} \varepsilon_{i,T}(x) < \delta$.

(γ) For CQ-sets we will prove the properties of the Kohlberg-Mertens program:
(1) Existence: every finite normal form game has at least one minimal CQ-set.
(2) Connectedness: every minimal CQ-set is connected.
(3) Admissibility: every minimal CQ-set consists of *perfect* equilibria only. This implies that in a minimal CQ-set only undominated pure strategies are used.
(4) Independence of inadmissible strategies: a minimal CQ-set for a game Γ is a CQ-set (but not necessarily a minimal CQ-set) in the game Γ' obtained by deleting a pure strategy that is not an admissible best reply against S.
(5) Backward induction: every CQ-set of a game contains at least one *proper* equilibrium of the game.

The last property in the (original) Kohlberg-Mertens program is Invariance, saying loosely that games with the same reduced normal form have the 'same' (minimal) CQ-sets. In Kohlberg and Mertens (1986) (the definition of hyperstable sets) and Hillas (1990) problems with Invariance are 'avoided' by taking 'invariance' as a part of the definition of stability. In the next section of the introduction we will see that this offers no relief.

(δ) In the seminal paper of Kohlberg and Mertens (1986) it is argued that solutions of extensive form games and normal form games should only depend on the *reduced normal form* of the game. Hillas (1990) is even more explicit and requires that extensive form games with the same reduced normal form should have the 'same' stable sets. Unfortunately, it is by no ways clear what is meant by 'the same stable sets'. Equivalent games will normally have different sets of strategy profiles and there is no *canonical way* to identify strategy profiles of both games. Therefore, it is hard to see whether 'stable sets are the same'. Even in a completely trivial one-person game one can see what the difficulties are.
Let Γ_1 be a one-person game with three pure strategies Δ, ∇ and $\odot$ and equal payoffs 1 for each of these strategies. The game Γ_2 has two pure strategies ♠ and ♢ and also payoff 1 for both pure strategies. It is clear that the reduced normal form Γ_0 of both games has one strategy (let us call it Ω) with payoff 1. Now, if a stability concept satisfies Existence, the game Γ_0 has one stable set: $\{\Omega\}$. If Γ_1 or Γ_2 'have the same stable sets', one has to discriminate (completely arbitrarily) between the strategies Δ, ∇ and $\odot$ in Γ_1 an ♠ and ♢ in Γ_2. The only reasonable solution can be that every one-point set or the whole set of mixed strategies is stable (also the mixed strategies, as Kohlberg and Mertens do not want to distinguish between pure and mixed strategies). In the first case Γ_i, $(i = 1, 2)$ has *more* stable sets than Γ_0, in the second case it has *larger* stable sets.

Another problem with invariance is that identification of **all** equivalent strategies leads to strategic games with polytopes as strategy spaces. This gives again a lot of trouble. In this paper we have chosen for a partial identification of equivalent strategies, that is, we only identify *pure* strategies with a mixture of the other strategies, if they are equivalent. Such an identification defines a projection from the set of strategy profiles of the original game to the set of strategy profiles of a reduced game. Now we can introduce at least three invariance properties, namely

(a) If S is stable in a reduced game, the inverse image under the projection is pre-stable,

(b) If S is stable in the original game, the projection is stable in any reduced game,

(c) If S is stable in a reduced game and x is a point in the inverse image of S, there is a stable set in the inverse image of S that contains x.

In this paper we use a method (cf. Vermeulen e.a. (1995)) to make solutions invariant in the sense of property (a), (b) and (c), if the solution satisfies a property that we called the "projection-stability property". The application of this procedure implies that "minimality" can no longer be maintained as a property for a stable set. This procedure does not disturb the other properties of the Kohlberg–Mertens program if these properties hold true before the application of the procedure.

An outline of the paper is the following. After the preliminaries we introduce CQ-perturbations in section 3 and prove some elementary properties of this type of perturbations. In the following section we introduce the concepts of CQ-set and CQ-stable set. In section 5 we prove the properties of the Kohlberg–Mertens program except Invariance. In the same section we briefly describe the method to make a solution concept invariant in the sense of property 1, 2 and 3 and change the solution from CQ-stable sets into CQ^*-stable sets.

Notation For $n \in \mathbb{N} := \{1, 2, \ldots\}$, $\mathbb{R}^n$ is the vector space of n-tuples of real numbers and $N := \{1, 2, \ldots, n\}$. If T is a finite set, $\Delta(T)$ is the set of probability distributions on T. If $x \in \mathbb{R}^n$ and $\epsilon > 0$, $\|x\|_\infty := \max_{i \in N} |x_i|$ and $B_\epsilon(x) := \{y \in \mathbb{R}^n | \; \|x - y\|_\infty < \epsilon\}$. For $A \subset \mathbb{R}^n$ we denote by conv(A) the convex hull of A and by cl(A) the closure of A.

2. Preliminaries

A *finite n-person game* (in normal form) is a pair $\Gamma = \langle M, u \rangle$, where $M := \prod_{i \in N} M_i$ is a product of finite sets and $u = (u_1, \ldots, u_n)$ is an n-tuple of functions $u_i : M \rightarrow \mathbb{R}$. Here M_i is the set of (*pure*) *strategies* of player i and u_i is his *payoff function*. To simplify notation, M_i will be seen as a subset of $\Delta(M_i)$. For a strategy profile $x = (x_1, x_2, \ldots, x_n) \in \Delta := \prod_{i \in N} \Delta(M_i)$ we define, as usual, the (expected) payoff function of player i by

$$u_i(x) := \sum_{(k_1, k_2, \ldots, k_n) \in M} \prod_{j \in N} x_{jk_j} u_i(k_1, k_2, \ldots, k_n).$$

We also write Δ_i (Δ_M) instead of $\Delta(M_i)$ (Δ), while $\Delta_{-i} := \prod_{j \neq i} \Delta_i$ is the set of strategy profiles for the opponents of player i. Furthermore, $(x_{-i} | \; y_i) \in \Delta$ is the strategy profile where player i uses $y_i \in \Delta_i$ and his opponents use the strategies in $x_{-i} \in \Delta_{-i}$. For player i and an element $x_{-i} \in \Delta_{-i}$

$$BR_i(x_{-i}) := \{y_i \in \Delta_i | \; u_i(x_{-i} | \; y_i) \geq u_i(x_{-i} | \; z_i) \quad \text{for all } z_i \in \Delta_i\}$$

is the set of *best replies* to x_{-i}. The correspondence $BR: \Delta \twoheadrightarrow \Delta$ with $BR(x) := \prod_{i \in N} BR_i(x_{-i})$ is called the *best reply correspondence* of Γ. A strategy profile $x \in \Delta$ is called a (*Nash*) *equilibrium* of Γ if $x \in BR(x)$. The set of all equilibria of the game Γ is denoted by $E(\Gamma)$.

Next we describe the refinements of the equilibrium concept as given by Selten (1975) and Myerson (1978). Let $\eta > 0$ and let $x \in \Delta$ be a completely mixed vector (i.e. all coordinates are positive). Then x is called *η-perfect* if the inequality $u_i(x_{-i} | \; k) < u_i(x_{-i} | \; l)$ implies that $x_{ik} \leq \eta$. If this inequality implies that $x_{ik} \leq \eta \cdot x_{il}$, then x is called *η-proper*. A profile $x \in \Delta$ is called *perfect* (*proper*) if there exist a sequence $(\eta_t)_{t \in \mathbb{N}}$ of positive real numbers converging to zero and a sequence $(x^t)_{t\in\mathbb{N}}$ in Δ converging to x, such that x^t is η_t-perfect (η_t-proper) for all t. The set of all perfect (proper) strategy profiles of Γ is denoted by $PE(\Gamma)$ ($PR(\Gamma)$). It is well known that these sets are non-empty, that every perfect strategy profile is an equilibrium and that every proper strategy profile is also perfect.

Finally we describe a stability concept related to, but different from, the one introduced by Hillas (1990). The Hausdorff distance of two compact subsets S and T of Δ is defined as

$$d_H(S,T) := \inf\{\eta > 0 |\ S \subset B_\eta(T), T \subset B_\eta(S)\},$$

where $B_\eta(S) := \bigcup_{x\in S} B_\eta(x)$. Note that d_H is a metric on the class of all compact subsets of Δ. Together with this metric this class is a compact metric space. For two compact and convex valued upper semicontinuous correspondences $\varphi, \psi: \Delta \twoheadrightarrow \Delta$

$$d_\infty(\varphi,\psi) := \max\{d_H(\varphi(x),\psi(x)) |\ x \in \Delta\}$$

and $\text{fix}(\varphi) := \{x \in \Delta |\ x \in \varphi(x)\}$ is the set of fixed points of φ. Note that d_∞ is a metric on the class of all compact and convex valued upper semicontinuous correspondences $\varphi: \Delta \twoheadrightarrow \Delta$. For a game Γ, a closed set $S \subset E(\Gamma)$ is an H-set if for any open set V containing S there exists a $\delta > 0$ such that $\text{fix}(\varphi) \cap V \neq \phi$ if $d_\infty(BR,\varphi) < \delta$. An H-set not properly containing another H-set is called an *H-stable set.*

3. Perturbations

In this section we introduce the perturbations that are central in this paper. Usually, the best response correspondence gives the optimal reactions to a proposed strategy profile x. In the perturbations in this paper the set wherein the best reply to x must be chosen is a polytope continuously depending on x. Accordingly, if a strategy profile x is given, there is also a polytope $\Delta[\varepsilon](x)$ given and the best replies to x inside this polytope will be the value of the best response correspondence $BR[\varepsilon]$ in x.

Definition 1. For each player $i \in N$ and each proper subset T of M_i, let $\varepsilon_{i,T}: \Delta \to [0,1]$ be a continuous function. The finite family $\varepsilon := \{\varepsilon_{i,T}\}_{i\in N, T\subset M_i}$ is called a *perturbation* if for all $i \in N$ and all $x \in \Delta$ the polytope

$$\Delta_i[\varepsilon](x) := \{y_i \in \Delta_i |\ y_i(T) \geq \varepsilon_{i,T}(x) \text{ for all } T \subset M_i\}$$

is non-empty, where $y_i(T) := \sum_{k\in T} y_{ik}$. The collection of all perturbations is endowed with the norm $\|\varepsilon\| := \max_{x\in\Delta} \max_{i\in N, T\subset M_i} \varepsilon_{i,T}(x)$.

Now for a perturbation ε we consider the correspondence $\Delta[\varepsilon]$ that assigns the polytope $\Delta[\varepsilon](x) := \prod_{i\in N} \Delta_i[\varepsilon](x)$ to an element $x \in \Delta$. Sometimes we also call the correspondence $\Delta[\varepsilon]$ a perturbation. Note that $\Delta[\varepsilon](x)$ is the perturbed strategy space corresponding to the Q-perturbation $\varepsilon(x) = (\varepsilon_{i,T}(x))_{i\in N, T\subset M_i}$ (cf. Hillas (1990)).

Theorem 1. *For each perturbation ε, $\Delta[\varepsilon]$ is a continuous correspondence.*

Proof. Since only the right hand side of the inequalities defining the polytope $\Delta[\varepsilon](x)$ depends on x, there exists a matrix A such that $\Delta[\varepsilon](x) = \{y \in \Delta |\ yA \geq \varepsilon(x)\}$ for all perturbations ε and strategy profiles x. Here $\varepsilon(x) = (\varepsilon_{i,T}(x))_{i\in N, T\subset M_i}$. By application of theorem 13 of the Appendix for the special case $c = 0$, one can find a constant D_A which only depends on the matrix A such that for all $x', x'' \in \Delta$

$$d_H(\Delta[\varepsilon](x'), \Delta[\varepsilon](x'')) \leq D_A \|\varepsilon(x') - \varepsilon(x'')\|_\infty.$$

So there is also a number η such that $d_H(\Delta[\varepsilon](x'), \Delta[\varepsilon](x'')) \leq \delta$ if $\|x' - x''\|_\infty < \eta$. $\triangleleft$

The correspondences obtained in this way will be used to introduce 'perturbed games' as follows. For a perturbation ε, $\Gamma(\varepsilon)$ is the 'game', where player i given a strategy profile $x \in \Delta$ has to respond to x_{-i} by choosing a best reply in the set $\Delta_i[\varepsilon](x)$. Since this is a non-empty compact set and $y_i \mapsto u_i(x_{-i}| \, y_i)$ is continuous, there is at least one best reply. This leads to

Definition 2. For a perturbation ε the best-reply correspondence of the 'perturbed game' $\Gamma(\varepsilon)$ is given by $BR[\varepsilon] = \prod_{i \in N} BR_i[\varepsilon]$, where for $i \in N$ and $x \in \Delta$

$$BR_i[\varepsilon](x) = \{y_i \in \Delta_i[\varepsilon](x) | \; u_i(x_{-i}| \, y_i) \geq u_i(x_{-i}| \, z_i) \text{ for all } z_i \in \Delta_i[\varepsilon](x)\}.$$

Note that the linearity of $y_i \mapsto u_i(x_{-i}| \, y_i)$ together with the fact that $\Delta_i[\varepsilon](x)$ is compact and convex implies the compactness and convexity of $BR[\varepsilon](x)$. Application of the maximum theorem to the continuous function $f\colon \Delta \times \Delta_i \to \mathbb{R}$ defined by $f(x, y_i) = u(x_{-i}| \, y_i)$ and the continuous correspondence $\Delta_i[\varepsilon]$ leads to the upper semicontinuity of the correspondence $BR_i[\varepsilon]$. Hence

Theorem 2. *For each perturbation ε, $BR[\varepsilon]$ is an upper semicontinuous correspondence.*

For later purposes we need

Theorem 3. *The mapping BR is Lipschitz continuous: there is a constant D such that $d_\infty(BR[\delta], BR[\varepsilon]) \leq D\|\delta - \varepsilon\|$, for all perturbations δ and ε.*

Proof. Note that for some matrix A_i, $\Delta_i[\varepsilon](x) = \{y_i \in \Delta_i | \; y_i A_i \geq \varepsilon_i(x)\}$, where $\varepsilon_i(x) = (\varepsilon_{i,T}(x))_{T \subset M_i}$. By theorem 13 in the Appendix, there are constants D_{A_i} such that for all perturbations δ and ε

$$\begin{aligned} d_\infty(BR[\delta], BR[\varepsilon]) &\leq \max_x \max_i d_H(BR_i[\delta](x), BR_i[\varepsilon](x)) \\ &\leq \max_x \max_i D_{A_i} \|\delta_i(x) - \varepsilon_i(x)\|_\infty \leq \max_i D_{A_i} \|\delta - \varepsilon\|. \end{aligned}$$

$\triangleleft$

4. On continuously quasi-stable sets

In view of theorem 2 and Kakutani's fixed point theorem, $\text{fix}(BR[\varepsilon])$ is non-empty and closed for each perturbation ε. So it makes sense to define a CQ-set as a set of strategy profiles such that for all small perturbations the perturbed games have an equilibrium close to this set.

Definition 3. A closed, non-empty subset S of Δ is a *continuously quasi-stable set* – CQ-set for short – if for each open neighborhood V of S there is a number $\eta > 0$ such that $\text{fix}(BR[\varepsilon]) \cap V \neq \phi$ for every perturbation ε with $\|\varepsilon\| < \eta$.

Comments If all the functions $\varepsilon_{i,T}$ are constant and positive, we obtain the perturbation system Hillas (1990) introduced in order to define quasi-stable sets. Hence, every CQ-set contains a quasi-stable set. Furthermore, theorem 2 implies that each best reply correspondence $BR[\varepsilon]$ is in fact a perturbation of the best reply correspondence of Γ in the sense of Hillas. Hence, in view of theorem 3, every H-stable set is a CQ-set. Another consequence is the existence of CQ-sets. A direct proof of this fact can be based on the following lemma which is also important for later purposes.

Lemma 1. *The correspondence $\varepsilon \mapsto \text{fix}(BR[\varepsilon])$ is upper semicontinuous.*

Proof. Take a sequence $(\varepsilon^t)_{t\in\mathbb{N}}$ of perturbations converging to ε and a sequence $(x^t)_{t\in\mathbb{N}}$ converging to x such that x^t in fix($BR[\varepsilon^t]$) for all t. We will show that x is an element of fix($BR[\varepsilon]$).
Let $\eta > 0$. Because $\|\varepsilon^t - \varepsilon\| < \eta$ for large t, theorem 3 implies that $d_\infty(BR[\varepsilon^t], BR[\varepsilon]) < \eta$ for large t. Hence, by the definition of d_∞ and the fact that x^t is an element of $BR[\varepsilon^t](x^t)$, $x^t \in B_\eta(BR[\varepsilon](x^t))$ for large t. Furthermore, by theorem 2, $BR[\varepsilon](x^t) \subset B_\eta(BR[\varepsilon](x))$ for large t. Combination of these statements leads to $x^t \in B_{2\eta}(BR[\varepsilon](x))$ for large t. Since $(x^t)_{t\in\mathbb{N}}$ converges to x, this implies that $x \in \text{cl}\Big(B_{2\eta}(BR[\varepsilon](x))\Big)$. Now η is arbitrary and $BR[\varepsilon](x)$ is a closed set, so x is an element of $BR[\varepsilon](x)$. ◁

Since $\Gamma(0) = \Gamma$ and $E(\Gamma) =$ fix($BR[0]$), lemma 1 implies

Theorem 4. *The set $E(\Gamma)$ of equilibria is a CQ-set.*

In order to formulate two characterizations of CQ-sets, we introduce the following notions.

A perturbation $\varepsilon = \{\varepsilon_{i,T}\}$ is called *completely mixed* if $\varepsilon_{i,T}(x) > 0$ for all $x \in \Delta, i \in N$ and $T \subset M_i$.
If $(\varepsilon^t)_{t\in\mathbb{N}}$ is a sequence of perturbations converging to zero (i.e. $\lim_{t\to\infty} \|\varepsilon^t\| = 0$), we denote with $\limsup_t$ fix($BR[\varepsilon^t]$) the set of those $x \in \Delta$ that are a *limit point* of a sequence $(x^t)_{t\in\mathbb{N}}$ with x^t is a fixed point of $BR[\varepsilon^t]$ for every $t \in \mathbb{N}$.

Lemma 2. *For a non-empty and closed set S, the following three statements are equivalent:*
(1) S is a CQ-set
(2) for any open set V containing S there is a number $\eta > 0$ such that fix($BR[\varepsilon]$) $\cap V \neq \phi$ *for every completely mixed perturbation ε with $\|\varepsilon\| < \eta$*
(3) for any sequence $(\varepsilon^t)_{t\in\mathbb{N}}$ of completely mixed perturbations converging to zero,

$$S \cap \limsup_t \ \text{fix}(BR[\varepsilon^t]) \neq \phi.$$

Proof. We will only prove the implications (2) → (3) and (3) → (1).
(a) Assume that (2) is true. Take a sequence $(\varepsilon^t)_{t\in\mathbb{N}}$ of completely mixed perturbations converging to zero, and an open set V in Δ containing S. By (2) there is a $K \in \mathbb{N}$ such that fix($BR[\varepsilon^t]$) $\cap V$ is non-empty, whenever $t \geq K$. Hence, the intersection of the set of limit points of $(\text{fix}(BR[\varepsilon^t]))_{t\in\mathbb{N}}$ and the closure of V is non-empty. Since this holds for any neighborhood V of S and the set of limit points is closed, it is clear that the intersection of S and the set of limit points of $(\text{fix}(BR[\varepsilon^t]))_{t\in\mathbb{N}}$ is non-empty. This establishes (3).
(b) Suppose that the condition in (3) is satisfied and that S is not a CQ-set. Then there exists a $\delta > 0$ such that for any $t \in \mathbb{N}$ there is a perturbation ε^t with $\|\varepsilon^t\| < \frac{1}{t}$ and fix($BR[\varepsilon^t]$) $\cap B_\delta(S) = \phi$.
Now for a given t, we consider for $s \in \mathbb{N}$ the completely mixed perturbation $\varepsilon^{t,s}$ defined by $\varepsilon^{t,s}_{i,T}(x) := (1 - \frac{1}{s})\varepsilon^t_{i,T}(x) + \frac{1}{s}$, where $i \in N$, T a proper subset of M_i and $x \in \Delta$. By lemma 1, fix($BR[\varepsilon^{t,s}]$) $\subset B_{\frac{1}{2}\delta}(\text{fix}(BR[\varepsilon^t]))$, for large s. So we can choose, for each $t \in \mathbb{N}$, a number $s(t) \geq t$ such that fix($BR[\varepsilon^{t,s(t)}]$) $\cap B_{\frac{1}{2}\delta}(S) = \phi$. By construction $\varepsilon^{t,s(t)} \to 0$ as $t \to \infty$, while $\limsup_t$ fix($BR[\varepsilon^{t,s(t)}]$) does not intersect with S. This contradicts (3). ◁

5. The Kohlberg-Mertens properties

In this section we will investigate, for CQ-sets, the properties of the Kohlberg-Mertens program as mentioned in the introduction. Some proofs of these properties depend on technical details. Those details that are used more frequently will be elaborated in separate lemmas.

5.1. ON EXISTENCE AND CONNECTEDNESS

In this section we deal with the existence and connectedness of minimal CQ-sets.

Theorem 5. (Existence) *Each finite n-person game has a minimal CQ-set.*

Proof. By theorem 4, the collection of CQ-sets of Γ is non-empty. For a sequence $S_1 \supset S_2 \supset S_3 \supset \cdots$ of CQ-sets we consider the non-empty and closed set $S := \bigcap_j S_j$. In order to show that S is a CQ-set, let V be an open neighborhood of S.
First we will show that $S_j \subset V$ for large j. If no S_j is contained in V, we can choose for each index j a point $x^j \in S_j \setminus V$. Since Δ is compact, we may suppose that the sequence $x^1, x^2, \ldots$ converges, say to $x \in \Delta$. As $x^j \in S_{j_0}$ for all indices $j \geq j_0$, we have $x \in S_j$ for all j. Then however $x \in S$, which is impossible because $x \notin V$.
So there is an index j with $S_j \subset V$. Because S_j is a CQ-set, there is a number $\eta > 0$ such that $\text{fix}(BR[\varepsilon]) \cap V \neq \phi$ whenever $||\varepsilon|| < \eta$. Hence S is a CQ-set.
Now Zorn's lemma guarantees the existence of minimal elements (w.r.t. inclusion) in the collection of CQ-sets. ◁

Theorem 6. (Connectedness) *Every minimal CQ-set is connected.*

Proof. Suppose that a minimal CQ-set S is not connected. Then there are two disjoint, non-empty closed sets S_1 and S_2 with $S = S_1 \cup S_2$. Let $\eta > 0$. Now S_1 and S_2 are not CQ-sets, so there are open neighborhoods V_1 of S_1 and V_2 of S_2 and perturbations ε^1 and ε^2 with $||\varepsilon^1|| < \eta$ and $||\varepsilon^2|| < \eta$ such that $BR[\varepsilon^1]$ and $BR[\varepsilon^2]$ have no fixed points in V_1 and V_2, respectively. We may assume without loss of generality that also $\text{cl}(V_1) \cap \text{cl}(V_2)$ is empty. By Urysohn's lemma there is a continuous function $\alpha: \Delta \to [0,1]$ with value one on $\text{cl}(V_1)$ and value zero on $\text{cl}(V_2)$. Then the family $\varepsilon := \{\varepsilon_{i,T}\}_{i \in N, T \subset M_i}$ with $\varepsilon_{i,T} := \alpha \cdot \varepsilon^1_{i,T} + (1-\alpha) \cdot \varepsilon^2_{i,T}$ is a perturbation. Since ε equals ε^i on V_i for $i = 1, 2$, it is clear that $BR[\varepsilon]$ has no fixed points in V_i. Therefore $BR[\varepsilon]$ has no fixed points on $V_1 \cup V_2$. Furthermore, it is straightforward to show that $||\varepsilon|| < \eta$. However, $V_1 \cup V_2$ is an open neighborhood of S. Hence, we get a contradiction with the fact that there is an $\eta > 0$ such that for each perturbation ε with supnorm smaller than η there is a fixed point of $BR[\varepsilon]$ contained in $V_1 \cup V_2$. ◁

5.2. ON ADMISSIBILITY AND BACKWARD INDUCTION

In this section we show that the intersection of a CQ-set with the set of perfect equilibria is also a CQ-set and that each CQ-set contains a proper equilibrium. To prove the admissibility of a CQ-set we need

Lemma 3. *For a completely mixed perturbation ε all the elements of* $\text{fix}(BR[\varepsilon])$ *are $||\varepsilon||$-perfect.*

Proof. Take a completely mixed perturbation ε and an element x of $\text{fix}(BR[\varepsilon])$. Now suppose that m is not a (pure) best reply of player i in Δ_i to x_{-i} and k is a best reply. Since x is completely mixed (because ε is completely mixed), the proof is complete if we can show that $x_{im} \leq ||\varepsilon||$.
First we will prove that there is a $T \subset M_i \setminus \{k\}$ containing m such that $x_i(T) = \varepsilon_{i,T}(x)$. So suppose that $x_i(T) > \varepsilon_{i,T}(x)$ for any subset T of M_i with $m \in T$ and $k \notin T$. Then we can

find a $\lambda > 0$ such that $x_i + \lambda(k - m)$ is an element of $\Delta_i[\varepsilon](x)$. The calculation

$$u_i(x_{-i}|\ x_i + \lambda(k - m)) = u_i(x_{-i}|\ x_i) + \lambda u_i(x_{-i}|\ k - m) > u_i(x_{-i}|\ x_i)$$

then shows that x_i is not a best reply for player i in $\Delta_i[\varepsilon](x)$ to x_{-i}. However, this is in contradiction with the fact that x is a fixed point of $BR[\varepsilon]$. Hence, there exists a subset T of M_i with $m \in T$ and $k \notin T$ such that $x_i(T) = \varepsilon_{i,T}(x)$. Since m is an element of T, this equality implies that $x_{im} \leq \varepsilon_{i,T}(x) \leq \|\varepsilon\|$. ◁

Now we can prove

Theorem 7. (Admissibility) *For every CQ-set S of Γ, $PE(\Gamma) \cap S$ is also a CQ-set of Γ.*

Proof. Suppose S is a CQ-set. Then statement (3) of lemma 2 holds for S. We prove that statement (3) also holds for the closed set $PE(\Gamma) \cap S$. Take a sequence $(\varepsilon^t)_{t\in\mathbf{N}}$ of completely mixed perturbations converging to zero. By statement (3) there is a limit point of $(\text{fix}(BR[\varepsilon^t]))_{t\in\mathbf{N}}$ that is contained in S. By lemma 3 we know that this limit point is perfect. So $S \cap PE(\Gamma) \neq \phi$ and also the intersection of the set of limit points of $(\text{fix}(BR[\varepsilon^t]))_{t\in\mathbf{N}}$ with $S \cap PE(\Gamma)$ is non-empty. Hence, statement (3) holds for $S \cap PE(\Gamma)$. In view of lemma 2 we may conclude that $S \cap PE(\Gamma)$ is a CQ-set. ◁

As the next item of the Kohlberg-Mertens program we consider backward induction.

Let η be a real number in $(0, 1)$. We consider a perturbation consisting of constant functions only. To be precise, let ε be the perturbation with $\varepsilon_{i,T}(x) := v_i(T)$, where

$$v_i(T) := \frac{1-\eta}{1-\eta^{|M_i|}} \sum_{k=1}^{|T|} \eta^{|M_i|-k} \qquad (i \in N, T \subset M_i, T \neq M_i).$$

In order to describe the polytope $\Delta_i[\varepsilon](x)$, we introduce the vector

$$p(\eta) := \frac{1-\eta}{1-\eta^{|M_i|}}(1, \ldots, \eta^{|M_i|-1})$$

and for a permutation σ of M_i the vector $p^\sigma(\eta)$ with $p(\eta)_{\sigma(k)}$ as k-th coordinate. We call these vectors *marginal vectors.*

Lemma 4. *For all $i \in N$ and $x \in \Delta$*

$$\Delta_i[\varepsilon](x) = \text{conv}\{p^\sigma(\eta)|\ \sigma \text{ is a permutation of } M_i\}.$$

Proof. If we define $v_i(\phi) := 0$ and $v_i(M_i) := 1$, then $\langle M_i, v_i\rangle$ can be seen as a TU-game. Since $v_i(S) + v_i(T) \leq v_i(S \cap T) + v_i(S \cup T)$ for all subsets S and T of M_i, it is in fact a convex game. Obviously, $\Delta_i[\varepsilon](x)$ is the core $C(v_i)$ of this game. According to Shapley (1971) $y \in C(v_i)$ is an extreme point if and only if there exists an increasing family of coalitions $\phi = T_0 \subset T_1 \subset \cdots \subset T_{|M_i|} = M_i$ with $|T_k - T_{k-1}| = 1$, for all $k = 1, 2, \ldots, |M_i|$ such that $y_l = v_i(T_k) - v_i(T_{k-1})$ where $T_k \setminus T_{k-1} = \{l\}$. Hence $y \in C(v_i) = \Delta_i[\varepsilon](x)$ is an extreme point if and only if $y = p^\sigma(\eta)$ where σ is a permutation of M_i. This completes the proof. ◁

With the help of this lemma we can prove

Lemma 5. *Let $\eta \in (0, 1)$. If ε is the perturbation as constructed before, then any fixed point of $BR[\varepsilon]$ is η-proper.*

Proof. Let x be a fixed point of $BR[\varepsilon]$. Suppose that $u_i(x_{-i}|\ k) < u_i(x_{-i}|\ l)$ for some $i \in N$, $k,l \in M_i$. Since x is completely mixed (because ε is), it is sufficient to prove that $x_{ik} \leq \eta \cdot x_{il}$.
Since x_i is an element of $\Delta_i[\varepsilon](x)$, it is, by lemma 4, a convex combination of marginal vectors. Obviously these marginal vectors are best replies to x_{-i} in $\Delta_i[\varepsilon](x)$, because x_i is such a best reply.
Now suppose that the marginal vector $p^\sigma(\eta)$ is a best reply in $\Delta_i[\varepsilon](x)$ to x_{-i} and that $p_k^\sigma(\eta) > \eta \cdot p_l^\sigma(\eta)$. Then $p_k^\sigma(\eta) > p_l^\sigma(\eta)$. Consider the permutation τ of M_i defined by

$$\tau(h) = \begin{cases} \sigma(h) & \text{if } h \neq k,l \\ \sigma(l) & \text{if } h = k \\ \sigma(k) & \text{if } h = l. \end{cases}$$

Then the inequalities $u_i(x_{-i}|\ k) < u_i(x_{-i}|\ l)$ and $p_k^\sigma(\eta) \geq p_l^\sigma(\eta)$ lead to

$$\begin{aligned} u_i\big(x_{-i}|\ p^\tau(\eta)\big) &= \sum_{h\neq k,l} u_i(x_{-i}|\ h)\, p_h^\tau(\eta) + u_i(x_{-i}|\ l)\, p_l^\tau(\eta) + u_i(x_{-i}\ \ k)\, p_k^\tau(\eta) \\ &= \sum_{h\neq k,l} u_i(x_{-i}|\ h)\, p_h^\sigma(\eta) + u_i(x_{-i}|\ l)\, p_k^\sigma(\eta) + u_i(x_{-i}\ \ k)\, p_l^\sigma(\eta) \\ &> \sum_{h\neq k,l} u_i(x_{-i}|\ h)\, p_h^\sigma(\eta) + u_i(x_{-i}|\ l)\, p_l^\sigma(\eta) + u_i(x_{-i}\ \ k)\, p_k^\sigma(\eta) \\ &= u_i\big(x_{-i}|\ p^\sigma(\eta)\big). \end{aligned}$$

This contradicts the fact that $p^\sigma(\eta)$ is a best reply for player i to x_{-i}. Hence, $x_{ik} \leq \eta \cdot x_{ih}$ must hold, since it holds for each marginal vector in the convex decomposition of x_i. ◁

This leads to

Theorem 8. (Backward Induction) *Every CQ-set contains a proper equilibrium.*

Proof. Let S be a CQ-set. Take a sequence $(\eta_t)_{t\in\mathbb{N}}$ of positive real numbers converging to zero. For every η_t we can construct the completely mixed perturbation ε^t according to the procedure above. By the previous lemma we know that any fixed point of $BR[\varepsilon^t]$ is η_t-proper. Since η_t converges to zero, every limit point of the sequence $(\text{fix}(BR[\varepsilon^t]))_{t\in\mathbb{N}}$ is a proper equilibrium. Furthermore, by checking the construction of ε^t it can be seen that $(\varepsilon^t)_{t\in\mathbb{N}}$ also converges to zero. Hence, by lemma 2, S contains a proper equilibrium. ◁

5.3 ON THE DELETION OF A STRATEGY

This section deals with the deletion of a pure strategy that is not an admissible best reply against any element of a given CQ-set. A pure strategy m of player j is an *admissible* best reply against an element $x_{-j} \in \Delta_{-j}$ – denoted by $m \in PB_j^a(x_{-j})$ – if there is a sequence $(x_{-j}^t)_{t\in\mathbb{N}}$ of completely mixed strategy profiles in Δ_{-j} converging to x_{-j} such that $m \in BR_j(x_{-j}^t)$, for all t. For a subset S of Δ, $PB_j^a(S) := \bigcup_{x\in S} PB_j^a(x_{-j})$ is the set of admissible pure best replies against S.

When dealing with the deletion of an inadmissible pure strategy of one of the players, we need a way to extend a perturbation of a game with one (pure) strategy deleted to a perturbation of the original game.

Definition 4. Let Γ be the finite n-person game $\langle M, u\rangle$. Let m be a pure strategy of player j, who has at least two pure strategies. The game Γ' induced by the deletion of m is by definition $\langle M', u' \rangle$, where u'_i is the restriction of u_i to $\prod_i M'_i$ and

$$M'_i = \begin{cases} M_i & \text{if } i \neq j \\ M_j \setminus \{m\} & \text{if } i = j. \end{cases}$$

Next we consider the linear map $\pi_j \colon \Delta_j \to \Delta'_j$ defined by

$$\pi_{jk}(x_j) = x_{jk} + \frac{1}{|M_j| - 1} x_{jm}.$$

Then $\pi(x) := (\pi_1(x_1), \ldots, \pi_n(x_n))$ defines a continuous mapping $\pi \colon \Delta \to \Delta'$, where π_i denotes the identity on Δ_i for $i \neq j$. We will use this projection mapping to extend a perturbation ε' of the game Γ' to a perturbation ε of the game Γ: if

$$\varepsilon_{i,T}(x) := \begin{cases} \varepsilon'_{i,T}(\pi(x)) & \text{if } i \neq j \text{ or } i = j \text{ and } m \notin T \neq M'_j \\ 0 & \text{otherwise} \end{cases} \qquad (x \in \Delta),$$

then it is evident that ε indeed is a perturbation and that $\|\varepsilon\| \leq \|\varepsilon'\|$. Finally, we need the following notation: for a strategy profile $y \in \Delta'$ we introduce the *lift* $\tilde{y}$ of y as the strategy profile in Δ obtained by adding to y_j zero as the m-th coordinate.

Lemma 6. *Let $z \in \Delta$. Then for π and ε as defined before, the following holds*
(1) if $z \in \Delta[\varepsilon](z)$, then $\pi(z) \in \Delta[\varepsilon'](\pi(z))$
(2) if $y \in \Delta[\varepsilon'](\pi(z))$, then $\tilde{y}$ is an element of $\Delta[\varepsilon](z)$.

Proof. (a) Let $z \in \Delta[\varepsilon](z)$. For any proper subset T of M'_i,

$$\pi(z)_i(T) = \begin{cases} z_i(T) & \text{if } i \neq j \\ z_i(T) + z_{im} \dfrac{|T|}{|M_i| - 1} x_{im} & \text{if } i = j \end{cases} \geq z_i(T) \geq \varepsilon_{i,T}(z) = \varepsilon'_{i,T}(\pi(z)).$$

(b) If $i \neq j$ and T is a proper subset of M_i, or $i = j$ and T is a proper subset of M_j with $m \notin T$ and $T \neq M'_j$, then

$$\tilde{y}_i(T) = y_i(T) \geq \varepsilon'_{i,T}(\pi(z)) = \varepsilon_{i,T}(z).$$

Otherwise, $\tilde{y}_j(T) \geq 0 = \varepsilon_{j,T}(z)$. ◁

Theorem 9. (Independence of inadmissible strategies) *Let S be a CQ-set of the game Γ. Let Γ' denote the game induced by the deletion of a strategy m of player j that is inadmissible against S. Then $\pi(S)$ is a CQ-set of the game Γ'.*

Proof. First of all we note that $\pi(S)$ is closed and non-empty. So if we can prove (2) of lemma 2 for $\pi(S)$, we have the desired result. Take an open set V' containing $\pi(S)$. Then $\pi^{-1}(V')$ is an open neighborhood of S. Since m is not an element of $PB^a_j(S)$, there is an open neighborhood $V \subset \pi^{-1}(V')$ of S such that $m \notin PB^a_j(V)$. Since S is a CQ-set, there is a number $\eta > 0$ such that the intersection of V and $\text{fix}(BR[\varepsilon])$ is non-empty, whenever $\|\varepsilon\| < \eta$. Now take a completely mixed perturbation ε' of Γ' with $\|\varepsilon'\| < \eta$. Then $\|\varepsilon\| \leq \|\varepsilon'\| < \eta$, where ε is the extension of ε'. So there is a fixed point z of $BR[\varepsilon]$ contained in V. Obviously, $\pi(z) \in V'$. The proof is complete if we can show that $\pi(z)$ is a fixed point of $BR[\varepsilon']$.
(a) First we will prove that $z_{jm} = 0$. So suppose that $z_{jm} > 0$. Since $z \in V$ and $m \notin PB^a_j(V)$, $m \notin PB_j(z_{-j})$ since z_{-j} is completely mixed. Take $k \in PB_j(z_{-j})$. Then $y_j := z_j + z_{jm}(k - m)$ is an element of $\Delta[\varepsilon]_j(z)$ with $u_j(z_{-j} | \, y_j) > u_j(z_{-j} | \, z_j)$. This however contradicts the fact that $z_j \in BR_j[\varepsilon](z)$. Hence, $z_{jm} = 0$.

(b) Next, part (1) of lemma 6 implies that $\pi(z) \in \Delta[\varepsilon'](\pi(z))$. Finally, if $y \in \Delta[\varepsilon'](\pi(z))$, then $\tilde{y} \in \Delta[\varepsilon](z)$ by part (2) of lemma 6. So, with the help of (a),

$$u'_i(\pi(z)) = u_i(z) \geq u_i(z_{-i}|\ \tilde{y}_i) = u'_i(\pi(z)_{-i}|\ y_i).$$

Hence, $\pi(z) \in BR[\varepsilon'](\pi(z))$. ◁

In the special case, where S is a minimal CQ-set, theorem 7 implies that $x_{jm} = 0$ for all $x \in S$. Therefore, in that case, $\pi(S)$ is the set of strategy profiles obtained by deleting the m-th coordinate of x_j for each $x \in S$.

5.4 ON INVARIANCE

In this section we will investigate the invariance of the mapping σ that assigns to a game Γ the collection of all (not necessarily minimal) CQ-sets of Γ. Such a mapping is called a *solution*. Kohlberg and Mertens (1986) called a solution invariant if, roughly speaking, it only depends on the reduced game. In order to give a precise definition, let $\Gamma = \langle M, u\rangle$ be an n-person game.

Definition 5. For each player $i \in N$, let $P_i = \{p_i^k \in \Delta_i|\ k \in K_i\}$ be a *finite* set of strategies. We call the set of vectors $P := \bigcup P_i$ an *extension set* for Γ. For such a set P we introduce the *P-extended game* $\Gamma_P = \langle L, v\rangle$, where $L_i := M_i \cup K_i$ (supposing that the sets M_i and K_i do not have elements in common). In order to define the payoff functions of this game, we consider the linear map $\pi_i: \Delta(L_i) \rightarrow \Delta(M_i)$ with

$$\pi_i(l) := \begin{cases} l & \text{if } l \in M_i \\ p_i^l & \text{if } l \in K_i. \end{cases}$$

The payoff function $v_i: \Delta_L \rightarrow \mathbb{R}$ is then defined by $v_i := u_i \circ \pi$, where $\pi := (\pi_1 \ldots, \pi_n)$. The projection $\pi: \Delta_L \rightarrow \Delta_M$ defined in this way is sometimes denoted by π_P to avoid a possible misinterpretation.

Definition 6. A solution τ is *invariant* if for any game Γ and any extension set P for Γ
(1) $\tau(\Gamma) = \{\pi(S)|\ S \in \tau(\Gamma_P)\}$
(2) $\pi^{-1}(T) = \bigcup\{S \in \tau(\Gamma_P)|\ \pi(S) = T\}$ for all $T \in \tau(\Gamma)$.

Unfortunately the authors do not know whether σ is an invariant solution. Therefore they developed a procedure (cf. Vermeulen, Jansen and Potters (1995)) to modify projection-stable solutions in such a way that the resulting solution is invariant. A solution τ is called *projection-stable* if for any game Γ

(1) $\tau(\Gamma) \neq \phi$ (non-emptiness property)
(2) if a sequence $S_1, S_2, \ldots$ of elements of $\tau(\Gamma)$ d_H-converges to S, then S is also an element of $\tau(\Gamma)$ (closedness property)
(3) every closed set T containing an element of $\tau(\Gamma)$ is also an element of $\tau(\Gamma)$ (monotonicity property)
(4) the projection $\pi(T)$ of an element T of $\tau(\Gamma_P)$ is an element of $\tau(\Gamma)$ for all extension sets P for the game Γ (projection property).
(5) every minimal element of $\tau(\Gamma)$ is a connected subset of $PE(\Gamma)$ (perfect-valuedness property).

Furthermore they have shown that this modified solution possesses all the Kohlberg-Mertens properties if the original solution does.

In order to show that σ is p-stable, we introduce a method to extend perturbations of the game Γ. The extension of such a perturbation ε is the perturbation $\tilde{\varepsilon}$ of Γ_P defined by

$$\tilde{\varepsilon}_{i,T}(x) := \begin{cases} \varepsilon_{i,T\cap M_i}(\pi(x)) & \text{if } \phi \neq T \cap M_i \neq M_i \\ 0 & \text{otherwise} \end{cases} \qquad (x \in \Delta_L).$$

In the following definition we describe how a strategy of a player in the game Γ can be extended to a strategy in the extended game.

Definition 7. The *lift* of a vector $x_i \in \Delta(M_i)$ is the vector $\tilde{x}_i$ in $\Delta(L_i)$ obtained by adding, at the appropriate places, zero coordinates to the vector x_i.

Lemma 7. *Let $z \in \Delta_L$. The perturbation $\tilde{\varepsilon}$ has the following properties:*
(1) if $z \in \Delta[\tilde{\varepsilon}](z)$, then $\pi(z) \in \Delta[\varepsilon](\pi(z))$
(2) if $y \in \Delta[\varepsilon](\pi(z))$, then the lift $\tilde{y}$ of y is an element of $\Delta[\tilde{\varepsilon}](z)$.

Proof. (a) For $i \in N$ and a non-empty proper subset T of M_i,

$$\pi_i(z_i)(T) \geq z_i(T) \geq \tilde{\varepsilon}_{i,T}(z) = \varepsilon_{i,T\cap M_i}(\pi(z)) = \varepsilon_{i,T}(\pi(z)).$$

Hence, $\pi(z) \in \Delta[\varepsilon](\pi(z))$.
(b) For $i \in N$ and a non-empty proper subset T of L_i with $T \cap M_i \neq \phi$,

$$\tilde{y}_i(T) = y_i(T \cap M_i) \geq \varepsilon_{i,T\cap M_i}(\pi(z)) = \tilde{\varepsilon}_{i,T}(z).$$

If however $T \cap M_i = \phi$, then $\tilde{y}_i(T) = 0 = \tilde{\varepsilon}_{i,T}(z)$. Hence, $\tilde{y} \in \Delta[\tilde{\varepsilon}](z)$. ◁

Theorem 11. *The solution σ that assigns to a game the collection of all CQ-sets is p-stable.*

Proof. In view of theorem 4, 6 and 7 and the fact that the properties (2) and (3) are obvious for σ, the proof is complete if we can show that the solution σ has the projection property.
Let T be a CQ-set of the game Γ_P. In order to prove that $\pi(T)$ is a CQ-set of the game Γ, we first note that $\pi(T)$ is a closed, non-empty subset of Δ_M. Let V be an open set containing $\pi(T)$. Since T is a CQ-set of Γ_P and $\pi^{-1}(V)$ is an open neighborhood of T, there is an $\eta > 0$ such that

$$\text{fix}(BR[\xi]) \cap \pi^{-1}(V) \neq \phi$$

for every perturbation ξ of the game Γ_P with $\|\xi\| < \eta$. If ε is a perturbation of the game Γ with $\|\varepsilon\| < \eta$, then $\|\tilde{\varepsilon}\| < \eta$. So $\text{fix}(BR[\tilde{\varepsilon}]) \cap \pi^{-1}(V) \neq \phi$ and we can take a $z \in \text{fix}(BR[\tilde{\varepsilon}]) \cap \pi^{-1}(V)$. Since $\pi(z) \in V$, the proof is complete if we can show that $\pi(z)$ is a fixed point of $BR[\varepsilon]$.
By part (1) of lemma 7, $\pi(z) \in \Delta[\varepsilon](\pi(z))$. Part (2) of lemma 7 implies that the lift $\tilde{y}$ of a $y \in \Delta[\varepsilon](\pi(z))$ is an element of $\Delta[\tilde{\varepsilon}](z)$. Hence

$$u_i(\pi(z)_{-i}|\ \pi_i(z_i)) = u_i(\pi(z)) = v_i(z) \geq v_i(z_{-i}|\ \tilde{y}_i) = u_i(\pi(z)_{-i}|\ \pi_i(\tilde{y}_i)) = u_i(\pi(z)_{-i}|\ y_i).$$

So $\pi(z) \in BR[\varepsilon](\pi(z))$. ◁

The modified solution of σ is the solution σ^* that assigns to a game Γ the collection of so-called CQ^*-sets, where a set $S \in \sigma(\Gamma)$ is a CQ^**-set* if
(1) $\pi^{-1}(S)$ is a CQ-set for every extension set P
(2) S is a connected subset of $PE(\Gamma)$.

The result of Vermeulen, Jansen and Potters leads to the final objective of this paper.

Theorem 12. *CQ^*-sets have the following properties*
(1) *Every game has CQ^*-sets.*
(2) *Every CQ^*-set is connected.*
(3) *Every CQ^*-set of a game Γ is contained in $PE(\Gamma)$.*
(4) *Every CQ^*-set of a game Γ contains a proper equilibrium.*
(5) *Let S be a CQ^*-set of a game Γ and let Γ' be induced by the deletion of an inadmissible strategy m of player j against S. Then $S' := \{x \in \Delta' \mid \tilde{x} \in S\}$ contains a CQ^*-set of the game Γ'.*
(6) *The solution that assigns to a game the collection of all CQ^*-sets is invariant.*

APPENDIX

Let, for a $p \times q$-matrix A, Y_A be the set of those $b \in \mathbb{R}^p$ for which $\varphi(b) := \{x \in \mathbb{R}^q \mid Ax \leq b\}$ is non-empty. For a $b \in Y_A$ we define

$$X_A(b) := \{c \in \mathbb{R}^q \mid \text{the problem 'maximize } \langle c, x\rangle \text{ under the condition } Ax \leq b\text{' is solvable}\}.$$

In view of the duality theorem for linear programming, the set $X_A(b)$ consists of the vectors $c \in \mathbb{R}^q$ such that the linear equation $yA = c$ has a nonnegative solution. So, $X_A(b)$ does not depend on b and we write X_A instead. Next, for $c \in X_A$, we consider the correspondence $\psi_c: Y_A \twoheadrightarrow \mathbb{R}^q$ with

$$\psi_c(b) := \{x \in \mathbb{R}^q \mid \langle c, x\rangle \text{ is maximal on } \varphi(b)\} \qquad (b \in Y_A).$$

Now theorem 10.5 of Schrijver (1986) implies

Theorem 13. *There is a constant D_A only dependent on A such that*

$$d_H(\psi_c(b), \psi_c(b')) \leq D_A \|b - b'\|_\infty$$

for all $b, b' \in Y_A$ and $c \in X_A$.

References

1 Hillas, J. (1990) *On the definition of the strategic stability of equilibria*, Econometrica **58**,
1365 – 1391.

2 Kohlberg, E. and J.F. Mertens (1986) *On strategic stability of equilibria*, Econometrica **54**, 1003 – 1037.

3 Mertens, J.F. (1987) *Ordinality in noncooperative games*, Core Discussion Paper 8728, CORE Louvain de la Neuve, Belgium.

4 Mertens, J.F. (1989) *Stable equilibria – a reformulation. Part I: Deinitions and basic properties*, Math. Oper. Res. **14**, 575 – 625.

5 Mertens, J.F. (1991) *Stable equilibria – a reformulation. Part II: Discussion of the definition and futher results*, Math. Oper. Res. **16**, 694 – 753.

6 Myerson, R.B. (1978) *Refinements of the Nash equilibrium concept*, Internat. J. Game Theory **7**, 73 – 80.

7 Nash, J.F. (1950) *Equilibrium points in n-person games*, Proc. Nat. Acad. Sci. U.S.A. **36**,
48 – 49.

8 Selten, R. (1975) *Reexamination of the perfectness concept for equilibrium points in extensive games*, Internat. J. Game Theory **4**, 25 – 55.

9 Schrijver, A. (1986) *Theory of linear and integer programming*, John Wiley & Sons, New York.

10 Shapley, L.S. (1971) *Cores of convex games*, Int. J. Game Theory **1**, 11 – 26.

11 Vermeulen, A.J., Jansen, M.J.M. and J.A.M. Potters (1995) *Making solutions invariant*, Report 9519, Department of Mathematics, University of Nijmegen, The Netherlands.

12 Wilson, R. (1992). *Computing simply stable equilibria*, Econometrica **60**, 1039 – 1070.

D. Vermeulen
University of Maastricht
Faculty of Economics
PO Box 616, 6200 MD Maastricht
The Netherlands

J. A. M. Potters
Department of Mathematics
University of Nijmegen
Toernooiveld, 6525 ED Nijmegen
The Netherlands

M. J. M Jansen
University of Maastricht
PO Box 616, 6200 MD Maastricht
The Netherlands

A CHAIN CONDITION FOR Q_0-MATRICES

A. K. Biswas and G. S. R. Murthy

Abstract: *The class Q_0 of real square matrices is one of the fundamental classes encountered in the study of Linear Complementarity Problem. There are no nice methods to verify whether a given general matrix is in Q_0 or not. In this note, we present a simple and elegant proposition which provides a necessary condition on Q_0-matrices. We demonstrate the usefulness of this result in studying a number of examples and in answering Stone's conjecture that principal minors of fully semimonotone Q_0-matrices are nonnegative in the affirmative for 5×5 matrices.*

INTRODUCTION. The linear complementarity problem (LCP) with data $A \in R^{n\times n}$ and $q \in R^n$, denoted by (q, A), is to find a vector z in $F(q, A) = \{z \in R^n_+ : Az + q \geq 0\}$ such that $z^t(Az + q) = 0$. The set of solutions to (q, A), $\{z \in R^n{}_+ : Az + q \geq 0\}$, denoted by $S(q, A)$. A number of matrix classes have been identified in connection with the study of LCP (see [4,10]). The class Q_0, introduced by Parsons [12], is one of the fundamental classes of LCP and consists of all real square matrices A for which the following holds: for every q with $F(q, A) \neq \emptyset$, $S(q, A) \neq \emptyset$. There is no efficient method to check whether a given general matrix is in Q_0 or not. This is so even in the case of 3×3 matrices. A finite but inefficient method is described in Murty [8]. In this note we present a simple and elegant proposition (Prop.3) which provides a necessary codition on Q_0-matrices. We cite some examples of non-Q_0 matrices from the literature through which we demonstrate the elegance of our proposition. The proposition has also been extremely useful in answering Stone's conjecture in the affirmative for 5×5 matrices (see below).

The class of fully semimonotone matrices was introduced by Cottle and Stone [5]. Throughout this note we follow the notations of Cottle, Pang and Stone [4]. A matrix A is said to be semimonotone (E_0) if (q, A) has a unique solution for every $q > 0$. The matrix A is said to be fully semimonotone (E_0^f) if every principal pivotal transform (PPT) of A is in E_0 (see [4] for definition of PPT). Stone conjectured that if A is in $E_0^f \cap Q_0$, then $A \in P_0$, that is, the principal minors of A are nonnegative. Murthy and Parthasarathy [7] showed that this conjecture is true when the order of the matrix is 4 or less, and in a number of special cases of general order. The current artilce is actually an outcome of the authors' attempt to establish Stone's conjecture for 5×5 matrices. Using proof techniques similar to those used in [7] we have been able to show that $R^{5\times 5} \cap E_0^f \cap Q_0 \subseteq P_0$. Since the proof is long, we will only outline the proof technique. Interested readers may refer to [3] for the complete proof. Our main interest here is to record this result.

Definition 1 (Chain). Suppose $A \in R^{n\times n}$. Say that A has a chain if there exist distinct indices $i_1, i_2, \ldots, i_k$ such that

(i) $a_{i_1 i_2}$ is the only positive entry in $A_{i_1 .}$,

(ii) for each $j \in \{2, 3, \ldots, k-1\}$ $a_{i_j i_{j+1}}$ is the only positive (or the only negative) entry in row $A_{i_j .}$,

(iii) $A_{i_k .} \geq 0$.

We shall denote such a chain by $i_1,\ i_2,\ \ldots,\ i_k$.

Example 2. Consider the matrix

$$A = \begin{bmatrix} 0 & -1 & -4 & 2 & -1 & 0 & 0 \\ 0 & 1 & -3 & 2 & 1 & 0 & 1 \\ 0 & 4 & 2 & 1 & 0 & 1 & 2 \\ 1 & 4 & 2 & 0 & 1 & -3 & 0 \\ -4 & 6 & -2 & 1 & -3 & 0 & 1 \\ -3 & 2 & 0 & -1 & -2 & -1 & 0 \\ -3 & 5 & 4 & -2 & 1 & -2 & 2 \end{bmatrix}.$$

Note that A has a chain, viz., (1,4,6,2,3).

Proposition 3 (The Chain Condition). Suppose $A \in \boldsymbol{R}^{n\times n} \cap \boldsymbol{Q}_0$. Then A cannot have any chain.

Proof. Suppose A has a chain $i_1,\ i_2,\ \ldots,\ i_k$. Define $q \in \boldsymbol{R}^n$ with $q_{i_1} = -1$, $q_{i_k} = 1$ and $q_{i_j} = -\lambda a_{i_j i_{j+1}}$ for $j \in \{2, 3, \ldots, k-1\}$. It is easy to check that for all large (positive) values of λ, (q, A) has a feasible solution but no complementary solution. This contradicts that $A \in \boldsymbol{Q}_0$. Thus A cannot have any chain. □

Example 4. Consider the matrix A given in Example 2. Since A has a chain, A does not belong to $\boldsymbol{Q}_0$.

Example 5. The class $\boldsymbol{U}$ consists of matrices A for which (q, A) has a unique solution for every q which is in the interior of the union of complementary cones. Stone [13] showed that $\boldsymbol{U} \cap \boldsymbol{Q}_0 \subseteq \boldsymbol{P}_0$ and constructed the following example (a $\boldsymbol{U}$-matrix) to show that $\boldsymbol{U}$ is not a subclass of $\boldsymbol{P}_0$.

$$A = \begin{bmatrix} 0 & 0 & -1 & 0 \\ 0 & 0 & 0 & 1 \\ 1 & 0 & 0 & 0 \\ 0 & 1 & 0 & 0 \end{bmatrix}$$

Since $\det A < 0$, by Stone's result $A \notin \boldsymbol{Q}_0$. This fact can be directly observed as A has a chain (2,4).

We say that a positive entry a_{ij} of a matrix $A \in \boldsymbol{R}^{n\times n}$ leads to a chain if there is a chain in matrix B with the first index equal to i, where the matrix B is obtained by replacing all the positive entries other than a_{ij} in $A_{i\cdot}$ by zero.

Proposition 6. Suppose $A \in \boldsymbol{R}^{n\times n}$. Assume that for some i each positive entry of $A_{i\cdot}$ leads to a chain. Then A does not belong to $\boldsymbol{Q}_0$.

Example 7. The following example was constructed by Murthy, Parthasarathy and Ravindran [8] while trying to examine the above mentioned Stone's conjecture.

$$A = \begin{bmatrix} 2 & -1 & 1 & 2 \\ -2 & 1 & -1 & 1 \\ -1 & 2 & 1 & -1 \\ 2 & -1 & -2 & 2 \end{bmatrix}.$$

Matrix A is a $\boldsymbol{E}_0^f$-matrix with negative determinant. Since Stone's conjecture is true for matrices of order less than or equal to four, it follows that $A \notin \boldsymbol{Q}_0$. However, this can also be seen using chain condition. Note that PPT of A with respect to $\alpha = \{2, 4\}$ is given by

$$A = \begin{bmatrix} 0 & 0 & 3 & 1 \\ 2 & \frac{2}{3} & 0 & \frac{-1}{3} \\ 3 & 1 & 0 & -1 \\ 0 & \frac{1}{3} & 1 & \frac{1}{3} \end{bmatrix}$$

Note that each positive entry of first row of B leads to a chain. Hence, by Proposition 6, it follows that $B \notin \boldsymbol{Q}_0$. Since PPTs of $\boldsymbol{Q}_0$-matrices are in $\boldsymbol{Q}_0$, it follows that A does not belong to $\boldsymbol{Q}_0$.

Remark 8. A real square matrix is said to be an almost $\boldsymbol{P}_0$-matrix if all its proper principal minors are nonnegative and its determinant is negative. In the light of chain condition we make the following observations:

(i) A nonnegative matrix is a $\boldsymbol{Q}_0$-matrix if, and only if, it has no chains,

(ii) A triangular (upper or lower) matrix is in $\boldsymbol{Q}_0$ if, and only if, it has no chains,

(iii) An almost $\boldsymbol{P}_0$-matrix which is also in $\boldsymbol{E}_0^f$, can not be a $\boldsymbol{Q}_0$-matrix.

Item (ii) above can also be paraphrased as: if A is a triangular matrix, then A belongs to $\boldsymbol{Q}_0$ if, and only if, it satisfies Property (**) defined in Murthy, Parthasarathy and Sriparna [9] (A has Property (**) means the rows corresponding to nonpositive diagonal entries of A are nonpositive and this property holds for all PPTs of A). This is easy to establish using chain condition. Similarly, regarding item (iii), if A is an almost $\boldsymbol{P}_0$ and $\boldsymbol{E}_0^f$-matrix, then using sign structure of A^{-1} (see Theorems 3.2 and 3.3 of [7]) and using chain condition it can be shown that A^{-1} is not in $\boldsymbol{Q}_0$.

We now mention two other results that were very useful in establishing Stone's conjecture for 5 × 5 matrices.

Proposition 9. Suppose $A \in \boldsymbol{R}^{n\times n} \cap \boldsymbol{Q}_0$. If $A_{.j} = 0$, for some j, then $A_{\alpha\alpha} \in \boldsymbol{Q}_0$, where $\alpha = \{1, 2, \ldots, j-1, j+1, \ldots, n\}$.

Proposition 10. Let $A \in \boldsymbol{R}^{n\times n}$. Suppose i and j are such that $A_{i.} \leq 0$ and $a_{ij} < 0$. If there exists an $x \in \boldsymbol{R}_+^{n-1}$ such that $Bx < 0$, where B is obtained from A by dropping the ith row and j^{th} column, then $A \notin \boldsymbol{E}_0$.

Proof. Follows from the fact that if $A \in \boldsymbol{E}_0$, then $\{x \in \boldsymbol{R}_+^n : Ax < 0\} = \emptyset$.

Example 11. Repeated application of the above proposition proves that the following matrix A does not belong to $\boldsymbol{E}_0$.

$$A = \begin{bmatrix} 0 & 0 & - & * \\ - & 0 & * & * \\ 0 & 0 & 0 & - \\ \oplus & - & \oplus & \oplus \end{bmatrix}$$

We now come to the main result of this note.

Theorem 12. Suppose $\boldsymbol{R}^{5\times 5} \cap \boldsymbol{E}_0^f \cap \boldsymbol{Q}_0$. Then $A \in \boldsymbol{P}_0$.

Outline of the Proof. This is proved in an iterative manner. We first show that every 2 × 2 principal submatrix of A is in $\boldsymbol{P}_0$. Using this, then, we show that every 3 × 3 principal submatrix of A is also in $\boldsymbol{P}_0$. Similarly we show that every 4 × 4 principal submatrix of A is also in $\boldsymbol{P}_0$. It then follows from (iii) of Remark 8 that A is a $\boldsymbol{P}_0$-matrix.

A matrix A is said to be in $\boldsymbol{R}_0$ if $(0, A)$ has a unique solution. Aganagic and Cottle [1] showed that if $A \in \boldsymbol{P}_0$, then $A \in \boldsymbol{Q}$ if, and only if, $A \in \boldsymbol{R}_0$. Pang [11] showed that $\boldsymbol{E}_0 \cap \boldsymbol{R}_0 \subseteq \boldsymbol{Q}$. Jeter and Pye [6] showed that $\boldsymbol{R}^{4\times 4} \cap \boldsymbol{E}_0^f \cap \boldsymbol{Q} \subseteq \boldsymbol{R}_0$. This result is true in the case of 5 × 5 matrices.

Corollary 13. Suppose $A \in \boldsymbol{R}^{5\times 5} \cap \boldsymbol{E}_0^f$. Then $A \in \boldsymbol{Q}$ if, and only if, $A \in \boldsymbol{R}_0$.

Concluding Remarks.

The main interest of this note is to record that Stone's conjecture is true even in the case of 5 × 5 matrices. Aganagic and Cottle [2] characterized the $\boldsymbol{Q}_0$-matrices with nonnegative principal minors and showed that Lemke's algorithm processes LCP (q, A) if $A \in \boldsymbol{P}_0 \cap \boldsymbol{Q}_0$. Hence, if $A \in \boldsymbol{R}^{5\times 5} \cap \boldsymbol{E}_0^f \cap \boldsymbol{Q}_0$, then (q, A) can be processed by Lemke's algorithm for all $q \in \boldsymbol{R}^5$. of order 5 or less. Proposition 3 provides a useful necessary condition on $\boldsymbol{Q}_0$-matrices, particularly in the case of completely $\boldsymbol{Q}_0$-matrices. Similarly, Proposition 5 has been very useful in quickly identifying nonsemimonotone principal submatrices while establishing Theorem 12.

References

1. M. Aganagic and R. W. Cottle (1979) 'A note on $\boldsymbol{Q}$-matrices,' *Mathematical Programming* 16 pp.374-377.

2. M. Aganagic and R. W. Cottle (1987) 'A constructive characterization of $\boldsymbol{Q}_0$-matrices with nonnegative principal minors,' *Mathematical Programming* 37 pp.223-231.

3. Amit K. Biswas and G. S. R. Murthy, 'A note on $\boldsymbol{E}_0^f \cap \boldsymbol{Q}_0$-matrices,' Technical Report No.24, Indian Statistical Institute, Madras, India.

4. R. W. Cottle, J. S. Pang and R. E. Stone (1992) *The Linear Complementarity Problem*, Academic Press, Inc., 1992, Boston.

5. R. W. Cottle and R. E. Stone (1983) 'On the uniqueness of solutions to linear complementarity problems,' *Mathematical Programming* 27 191-213.

6. M. W. Jeter and W. C. Pye (1989) 'An example of nonregular semimonotone $\boldsymbol{Q}$-matrix,' *Mathematical Programming* 44 pp.351-356.

7. G. S. R. Murthy and T. Parthasarathy (1995) 'Some properties of fully-semimonotone $\boldsymbol{Q}_0$-matrices,' *SIAM J. MATRIX ANNL. APPL.*, 16 pp.1268-1286.

8. G. S. R. Murthy, T. Parthasarathy and G. Ravindran (1995) 'On copositive, semimonotone $\boldsymbol{Q}$-matrices,' *Mathematical Programming* 68 pp.187-203.

9. G. S. R. Murthy, T. Parthasarathy and Sriparna, 'Constructive characterization of Lipschitzian $\boldsymbol{Q}_0$-matrices,' to appear in *Linear Algebra and Its Applications*,

10. K. G. Murty (1988) *Linear Complementarity, Linear and Nonlinear Programming*, Heldermann Verlag, Berlin, Germany

11. J. S. Pang (1979) 'On $\boldsymbol{Q}$-matrices,' *Mathematical Programming* 17 pp.243-247.

12. T. D. Parsons(1970) 'Applications of principal pivoting,' Proceedings of the Princeton Symposium on Mathematical Programming 567-581.

13. R. E. Stone (1981) 'Geometric aspects of linear complementarity problem,' Ph.D. thesis, Department of Operations Research, Stanford University, Stanford, California.

A. K. Biswas
Indian Statistical Institute
110, Nelson Manickam Road
Aminjikarai, Madras 600 029
India

G. S. R. Murthy
Indian Statistical Institute
110, Nelson Manickam Road
Aminjikarai, Madras 600029
India

LINEAR COMPLEMENTARITY AND THE IRREDUCIBLE POLYSTOCHASTIC GAME WITH THE AVERAGE COST CRITERION WHEN ONE PLAYER CONTROLS TRANSITIONS [1]

S. R. Mohan S. K. Neogy and T. Parthasarathy

Abstract: *We consider the polystochastic game in which the transition probabilities depend on the actions of a single player and the criterion is the limiting average of the expected costs for each player. Using linear complementarity theory, we present a computational scheme for computing a set of stationary equilibrium strategies and the corresponding costs for this game with the additional assumption that under any choice of stationary strategies for the players the resulting one step transition probability matrix is irreducible. This work extends our previous work on the computation of a set of stationary equilibrium strategies and the corresponding costs for a polystochastic game in which the transition probabilities depend on the actions of a single player and the criterion is the total discounted expected cost for each player.*

1 Introduction

Zero-sum stochastic games with two players were introduced by Shapley in [17] as a generalization of matrix games. Nonzero-sum two person and n person stochastic games have been considered earlier by Fink [6] and Takahasi [19]. See also [14]. A nonzero-sum noncooperative polystochastic game is a repeated game which is defined by the objects

$$(\mathcal{N}, \mathcal{S}, \mathcal{N}_i(s), A_{ij}(s), p(t|s, a_1, a_2, \dots a_n), \ \forall\ s,\ t \in \mathcal{S}, a_i \in \mathcal{N}_i(s), i \neq j, i, j \in \mathcal{N}),$$

where $\mathcal{N} = \{1, 2, \dots, n\}$ denotes the set of players, $\mathcal{S} = \{1, 2 \dots, m\}$ denotes the set of states, the set $\mathcal{N}_i(s) = \{1, 2, \dots, m_i(s)\}$ denotes the set of actions available to Player i in state s, the matrix $A_{ij}(s) = ((a(\bar{a}_i, \bar{a}_j)))$ denotes the matrix of partial costs incurred by Player i depending on the actions $\bar{a}_i$ and $\bar{a}_j$, chosen by Players i and j, $i \neq j$, respectively and $p(t|s, a_1, a_2, \dots, a_n)$ is the probability that the game moves to the state t on day r given that the game is played in state s and Player i chooses the action $a_i \in \mathcal{N}_i(s)$ on day $r-1$, $1 \leq i \leq n$. Suppose now the vector $x_i^r(s) \in R^{m_i(s)}$ denotes the vector of probabilities over $\mathcal{N}_i(s)$ used by Player i as a mixed strategy on day r. Then the total expected cost incurred by Player i on day r is given by $(x_i^r(s))^t(\sum_{i \neq j} A_{ij}(s) x_j^r(s))$.

Let $\xi^i = \begin{bmatrix} x_i(1) \\ \vdots \\ x_i(m) \end{bmatrix}$ be a $\sum_{s=1}^{m} m_i(s) \times 1$ vector with m components, the sth component being a probability vector over the set $\mathcal{N}_i(s)$, whose $\tilde{a}_i^{th}$ coordinate $x_i(\tilde{a}_i \,|\, s)$ gives the probability that action $\tilde{a}_i \in \mathcal{N}_i(s)$ is chosen by Player i. Thus ξ^i specifies the probabilities of the choice of actions in each state, for Player i. A *mixed strategy* for Player i is a sequence

[1] Dedicated to Professor K. R. Parthasarathy on the occasion of his sixtieth birthday

$\{\xi^{ir}\}$, where ξ^{ir} specifies the probabilities of actions in different states on the rth day. By the sequence $\{\pi^r\}$, where $\pi^r = \begin{bmatrix} \xi^{1r} \\ \vdots \\ \xi^{nr} \end{bmatrix}$ we denote the mixed strategies of all players, over the infinite horizon. This is called a *game plan.* Given a game plan $\{\pi^r\}$, on day r the players use their mixed strategies ξ^{ir}. Suppose the state of the game is s on day r . Then the probability that the game moves to state t on day $(r+1)$ under $\{\pi^r\}$ is given by

$$\rho^{(r)}(t \mid s) = \sum_{\tilde{a}_1 \ldots, \tilde{a}_n} p(t \mid s, \tilde{a}_1, \ldots, \tilde{a}_n) \prod_{i=1}^{n} x_i^r(\tilde{a}_i | s)$$

where the sum is taken over all possible $(\tilde{a}_1, \ldots, \tilde{a}_n) \in \Pi_{i=1}^n \mathcal{N}_i(s)$. Let $\mathcal{Q}^{(r)}(\{\pi^r\}) = ((\rho^{(r)}(t|s)))$ be an $m \times m$ matrix whose s^{th} row gives the transition probabilities to various states from state s under π^r.

There are two types of criteria which have been considered in the literature for evaluating a given game plan of the players over the infinite horizon. The first criterion is to consider the sum over days, of the expected cost incurred by a player on day r, discounted by a factor β^r to obtain its present value. Each player then tries to minnimise his total discounted expected cost. The other type of criteria is called the undiscounted criteria introduced by Gillet [8] one of which is to consider the limit infimum of the average expected cost of the first N days as N tends to ∞. In this paper we consider the criterion to be the undiscounted limiting average expected cost in the above sense. Suppose $\{\pi_i^r\} = \{x_i^r(s), s \in \mathcal{S}\}$ is a sequence of mixed strategies used by Player i, $1 \leq i \leq n$ where $\{\pi^r\}$ denotes a game plan. One can then calculate the limit infimum of the expected average cost for Player i when the game is initiated in state $s \in \mathcal{S}$, over the infinite horizon using the transition probabilities $p(t|., ., , .)$ and the total expected cost on day r given that the game is played in state s on that day as follows.

Let $\phi_i(\{\pi^r\})$ denote the $m \times 1$ vector of limiting average expected costs where its sth coordinate $\phi_i(\{\pi^r\})(s)$ denotes the limiting average expected cost incurred by Player i under the game plan $\{\pi^r\}$. Then it is easy to see that

$$\phi_i(\{\pi^r\}) = \liminf_{L \to \infty} \frac{1}{L} \sum_{r=0}^{L} \prod_{k=1}^{r} \mathcal{Q}^{(k)}(\{\pi^r\})[\psi_i^r]$$

where $\psi_i^r(\pi^r)$ is the $m \times 1$ vector of expected immediate costs incurred by Player i on the rth day whose s^{th} coordinate $\psi_i^r(\pi^r)(s)$ is given by

$$\psi_i^r(\pi^r)(s) = \sum_{\tilde{a}_i \in \mathcal{N}_i(s)} x_i^r(\tilde{a}_i|s) (\sum_{k \neq i} \sum_{\tilde{a}_k \in \mathcal{N}_k(s)} a(\tilde{a}_i, \tilde{a}_k)(s) x_k^r(\tilde{a}_k|s)).$$

Suppose $\{\xi^{ir}\}$ is a mixed strategy for Player i. We say that $\{\xi^{ir}\}$ is *stationary* if $\xi^{ir} = \xi^{i1} = \xi^i$ for all r. We say that a game plan $\{\pi^r\}$ is *stationary* if each component ξ^{ir} of $\{\pi^r\}$ for $1 \leq i \leq n$ is stationary. We use the symbol π to denote a stationary game plan $\{\pi\}$. For a stationary game plan π the expression for $\phi_i(\{\pi\}) = \phi_i(\pi)$ simplifies to $\liminf_{L \to \infty} \frac{1}{L} \sum_{r=0}^{L} \mathcal{Q}^r(\pi)(\psi_i)(\pi)$ where $\mathcal{Q}^r(\pi)$ is the rth power of the matrix $((\rho^1(t|s)))$ and the sth coordinate $\psi_i(\pi)(s)$ of $\psi_i(\pi)$ is given by

$$\psi_i(\pi)(s) = \sum_{\tilde{a}_i \in \mathcal{N}_i(s)} x_i(\tilde{a}_i|s) (\sum_{k \neq i} \sum_{\tilde{a}_k \in \mathcal{N}_k(s)} a(\tilde{a}_i, \tilde{a}_k)(s) x_k(\tilde{a}_k|s)).$$

A stationary strategy $\xi^i = \begin{bmatrix} x^i(1) \\ \vdots \\ x^i(m) \end{bmatrix}$ for Player i is said to be a *pure stationary strategy* if for all s $\exists$ a $\tilde{a} \in \mathcal{N}_i(s)$ such that $x_i(\tilde{a}\,|s) = 1$ and $x_i(\bar{a}|s) = 0$ if $\tilde{a} \neq \bar{a} \in \mathcal{N}_i(s)$. A mixed strategy ξ^{ir} for Player i is called *a behavior strategy* if ξ^{ir} depends on the history of the strategies of all the players and resulting states in the past upto the $(r-1)^{th}$ day.

We say that a game plan $\{\bar{\pi}^r\}$ is an *equilibrium game plan* if

$$\phi_i(\{(\bar{\xi}^{1r} \ldots \bar{\xi}^{ir} \ldots \bar{\xi}^{nr})^t\}) \leq \phi_i(\{(\bar{\xi}^{1r} \ldots \xi^{ir} \ldots \bar{\xi}^{nr})^t\})$$

for all ξ^{ir} and for all i.

It is known that under certain restrictive assumptions on the game there is an equilibrium in stationary strategies. See [16]. For stationary strategies the above inequality simplifies (see [6]) to the following:

We use the symbol $\begin{bmatrix} \bar{\pi}_{-i} \\ \xi^{ir} \end{bmatrix}$ to denote a game plan whose i^{th} component is replaced by ξ^{ir}. In this notation note that $\bar{\pi} = \begin{bmatrix} \bar{\pi}_{-i} \\ \bar{\xi}^i \end{bmatrix}$. We say that $\bar{\pi}$ is a *stationary equilibrium strategy* if for each $1 \leq i \leq n$

$$\phi_i\left(\begin{bmatrix} \bar{\pi}_{-i} \\ \bar{\xi}^i \end{bmatrix}\right) \leq \phi_i\left(\left\{ \begin{matrix} \bar{\pi}^r_{-i} \\ \xi^{ir} \end{matrix} \right\}\right)$$

where $\{\xi^{ir}\}$ is a behavioral strategy for Player i.

The special case $n = 2$, i.e., a *bistochastic* nonzero-sum noncooperative game has earlier been considered. See [13]. In particular, the situation where the transition probabilities depend on the actions of a single player has been well studied. Parthasarathy and Raghavan [15] show that a single controller nonzero-sum noncooperative game with $n = 2$ and the limiting (liminf) average expected cost as the criterion has the *orderfield property* (i.e., a pair of Nash equilibrium costs and the corresponding equilibrium pair of strategies lie in the same ordered field as the data). See also Raghavan and Filar [16]. Nowak and Raghavan [13] reduce the problem of finding a pair of stationary equilibrium strategies in a bistochastic noncooperative game when the criterion is the limiting average expected cost with the additional assumption that under any stationary strategy, the transition probability matrix is irreducible, to the problem of computing a Nash equilibrium pair in a suitably constructed bimatrix game.

In this paper, we consider the polystochastic game in which the criterion is the limiting average expected cost and the transition probabilities depend on the actions of a single player, namely Player n, mainly from the point of view of computing a set of stationary equilibrium strategies for the players and the corresponding equilibrium cost vector. We extend the methodology and algorithm of our earlier work [11] on the problem of computing a set of stationary Nash equlibirum strategies and the corresponding costs for the game with the total discounted expected costs as the criterion to the case where the criterion is the limiting average expected costs to the players with the additional assumption that the one step transition probability matrix under any stationary strategy is irreducible. As in our earlier work [11], we do this by formulating this problem as a linear complementarity problem that can be solved using the algorithm of Lemke [9]. We may note that the approach of Nowak and Raghavan [13] for the case $n = 2$ with the assumption of one player control transitions involve computing for each possible combination of pure stationary strategies, the corresponding limiting average expected costs and then constructing the associated bimatrix game and finally applying the algorithm of Lemke-Howson [10] to the associated linear complementarity problem. In contrast, the procedure suggested in the next section uses Lemke's algorithm on a suitable linear complementarity problem formulated with the given data only without requiring any additional computation. The complementarity formulation is presented in Section 2. The main result on the processability by Lemke's algorithm is proved in Section 3. In the Appendix we present some of the basic definitions and results from the theory of linear complementarity useful to us.

2 The Complementarity Formulation

In what follows we use the symbol e^{is} to denote a column vector with $m_i(s)$ coordinates each of which is 1. The symbol e^{*k} denotes a column vector with k coordinates each of which is 1. We denote a stationary game plan by $\{\pi\}$ which is a sequence and use $\phi_i(\pi)$ to denote the corresponding $m \times 1$ vector of limiting average of expected costs of Player i when the game plan $\pi = (\xi^1, \xi^2, \ldots, \xi^n)$ is being used. The ith component ξ^i of π is a stationary strategy of Player i. In what follows, we shall assume that the transition probabilities depend only on the actions of Player n. This means that

$$p(t|s, \tilde{a}_1, \tilde{a}_2, \ldots, \tilde{a}_n) = p(t|s, \tilde{a}_n).$$

Further we shall also assume that under any stationary strategy $(x_n(s), 1 \leq s \leq m)$ the transition probability matrix is irreducible. As an immediate consequence of this assumption we have the following lemma.

LEMMA 2.1 *Let π be any stationary strategy. Then the vector $\phi_n(\pi) = ge^{*m}$ where g is a real number. (In other words, $\phi_n(\pi)(s) = g \; \forall \; s \in \mathcal{S}$.)*

Proof. This follows from our assumption that under any stationary strategy, the one step transition probability matrix is irreducible. ∎

LEMMA 2.2 *Let $\phi_i(\pi)$ be the vector of limiting average expected costs for player $1 \leq i \leq n$ with the matrices of immediate partial costs as $A_{ij}(s)$. The set of Nash equilibria of the noncooperative game with cost vectors as $\psi_i(\pi), 1 \leq i \leq n-1$ and $\phi_n(\pi)$ is the same as the set of Nash equilibria of the noncooperative game with costs as $\phi_i(\pi)$, $1 \leq i \leq n$. (i.e.,any Nash equilibrium of the game in which the criterion for players other than n is the first day's cost is also a Nash equilibrium of the original game and conversely.)*

Proof. Let SG1 be the *auxiliary game* in which the criterion for Player $i, i \neq n$ is $\psi_i(\pi)(s)$ for $i \leq n$ and $\phi_n(\pi)(s)$ for Player n when the game is initiated in state s. Let π^* be an equilibrium plan. Let $\xi^{*i} = \{x_i(s), s \in \mathcal{S}\}$ be the ith component of π^* which specifies the vectors of probabilities over $\mathcal{N}_i(s)$ used by Player i as a mixed strategy. Note that ξ^{*n} determines the one step transition probability matrix $\mathcal{Q}(\pi^*)$. By the Nash equilibrium conditions applied to SG1, we have,

$$\psi_i\left(\begin{bmatrix} \pi^*_{-i} \\ \xi^{*i} \end{bmatrix}\right) \leq \psi_i\left(\begin{bmatrix} \pi^*_{-i} \\ \xi^{i} \end{bmatrix}\right), \; 1 \leq i \leq n-1$$

for any mixed strategy ξ^i of Player i.

Now since the one step transition probability matrix $\mathcal{Q}(\pi^*)$ is the same under both the stationary strategies π^* and $\begin{bmatrix} \pi^*_{-i} \\ \xi^{i} \end{bmatrix}$ for any fixed Player $i, i \neq n$ by multiplying both sides of the above equation by $Q(\pi^*)^r$ where r is any nonnegative integer, summing over r upto N terms and dividing by N we see that the above inquality continues to hold. Now taking limit we note that $\liminf \frac{1}{N}\sum_{r=0}^{N}(Q(\pi^*)^r\psi_i\left(\begin{bmatrix}\pi^*_{-i}\\ \xi^{*i}\end{bmatrix}\right) \le \liminf \frac{1}{N}\sum_{r=0}^{N}(Q(\pi^*)^r\psi_i\left(\begin{bmatrix}\pi^*_{-i}\\ \xi^{i}\end{bmatrix}\right)$.

It follows from here that the Nash equilibrium conditions in the original game hold for all the Players $i, i < n$. For Player n, the Nash equilibrum condition in SG1 is the same as the one in the original stochastic game. This concludes one part of the assertion. To complete the proof we have to show that any Nash equilibrium point of the original stochastic game is a Nash equilibrium point of SG1. So let $((x_i^*(s)), (\phi_i(\pi)(s)),\ 1 \le s \le m, 1 \le i \le n)$ be a stationary equilibrium point of the original stochastic game. It is then known that there is a sequence $\{\beta_N\} \uparrow 1$ and a sequence $\{x_i^{*\beta_N}(s)\}$ such that $\{x_i^{*\beta_N}(s)\}$ converges to $x_i^*(s), \forall\ i$. Further if $\{\pi^{*\beta_N}\}$ denotes a stationary game plan with the strategies as $(x_i^{*\beta_N})$ then $(x_i^{*\beta_N}(s), i \in \mathcal{N},\ s \in \mathcal{S})$ with $(\phi_i^{\beta_N}(\pi^{*\beta_N})(s),\ i \in \mathcal{N}, s \in \mathcal{S})$ forms a stationary Nash equilibrium point of the stochastic game in which the strategies are evaluated by the sum of β_N discounted expected daily costs [15]. For such a β_N discounted game it is known that $x_i^{*\beta_N}$ along with the expected first day's costs for players other than n and $\pi_n^{\beta_N}(s)$ form a stationary equilibrium point for an auxiliary game in which for players other than n the criterion is the expected first day's cost and for Player n it is $\phi_n^{\beta_N}$. See [15] and [11]. Note also that $(1-\beta_N)\phi_n^{\beta_N}(\pi^{*\beta_N})(s)$ converges to $\phi_n(\pi)(s)$. Thus we have for the first day expected costs $\psi_i(s)'s,\ i \ne n$,

$$\psi_i^{\beta_N}\left(\begin{bmatrix}\pi^{*\beta_N}_{-i}\\ \xi^{*\beta_N i}\end{bmatrix}\right) \le \psi_i^{\beta_N}\left(\begin{bmatrix}\pi^{*\beta_N}_{-i}\\ \xi^{i}\end{bmatrix}\right),$$

for any arbitrary stationary strategy ξ^i of Player i. Taking limit $\{\beta_N\} \uparrow 1$ as $N \to \infty$ we see that the conditions of Nash equilibrium for $i \ne n$ in the game SG1 are satisfied by π^*. For $i = n$ we have,

$$\phi_n^{\beta_N}\left(\begin{bmatrix}\pi^{*\beta_N}_{-n}\\ \xi^{*\beta_N n}\end{bmatrix}\right) \le \phi_n^{\beta_N}\left(\begin{bmatrix}\pi^{*\beta_N}_{-n}\\ \xi^{n}\end{bmatrix}\right),$$

for any arbitrary stationary strategy ξ^n of Player n. Multiplying both sides of the above inequality by $(1-\beta_N)$ and taking limits as $N \to \infty$ it is seen that the Nash equilibrium conditions with respect to Player n hold in the game SG1 for π. This completes the proof. ■

Let $v_1, v_2, \ldots, v_{n-1}$ be the vector of first day expected costs for player 1 through $(n-1)$ corresponding to an equilibrium set of stationary strategies $(\xi^{*1}, \ldots, \xi^{*n})$ and ge^{*m} be the vector of limiting average of expected cost of Player n corresponding to the same equilibrium set of stationary strategies.

THEOREM 2.1 *$\{x_i(s), 1 \le s \le m,\ 1 \le i \le n\}$, $\{v_i(s), 1 \le s \le m, 1 \le i \le n-1\}$ and g, a real number form a set of equilibrium strategies for players $1, \ldots, n$ and the corresponding equilibrium costs if and only if there exist real numbers $\mu(s)$, $1 \le s \le m$ such that along with $(\{x_i(s), 1 \le s \le m, 1 \le i \le n\}$, $\{v_i(s), 1 \le s \le m, 1 \le i \le n-1\})$ and g they satisfy the following system of inequalities and equations:*

$$w_i(s) = \sum_{j \ne i} A_{ij}(s)x_j(s) - v_i(s)e^{is} \ge 0, 1 \le s \le m,\ 1 \le i \le (n-1) \tag{2.1}$$

$$w_n(s) = \sum_{j=1}^{n-1} A_{nj}(s)x_j(s) + P(s)\mu - \mu(s)e^{ns} - ge^{ns} \ge 0 \tag{2.2}$$

$$\sum_{\bar{a}\in\mathcal{N}_i(s)} x_i(\bar{a}|s) = 1,\ 1 \le s \le m,\ 1 \le i \le n \tag{2.3}$$

$$x_i(\bar{a}|s) \ge 0\ \forall\ i, s,\ and\ \bar{a} \in \mathcal{N}_i(s),\ 1 \le s \le m, 1 \le i \le n \tag{2.4}$$

$$(w_i(s))^t x_i(s) = 0,\ 1 \le s \le m,\ 1 \le i \le n \tag{2.5}$$

where μ is the vector $(\mu(1), \dots, \mu(m))^t$ and $P(s)$ is the $m_n(s) \times m$ matrix whose $(\bar{a}, t)^{th}$ entry is $p(t|s, \bar{a})$ where $\bar{a} \in \mathcal{N}_n(s)$.

Proof. Observe that the set of inequalities in (2.1) along with the complementarity conditions in (2.5) for $i \le n-1$ and appropriate normalising conditions in (2.3) are equivalent to the Nash equilibrium conditions in the game SG1 for players other than n. Hence we need to verify that a solution that also satisfies (2.2), (2.3) with $i = n$ and the complementarity condition $(w_n(s))^t x_n(s) = 0$ satisfies the Nash equilibrium conditions of Player n for the game SG1. We proceed as in [18]. We have

$$(x_n(s))^t w_n(s) = (x_n(s))^t (\sum_{j=1}^{n-1} A_{nj}(s) x_j(s) + P(s)\mu - \mu(s)e^{ns} - ge^{ns}) = 0.$$

This gives the equation

$$g + \mu(s) = \psi_n(s)^1 + \alpha_1 \tag{2.6}$$

where $\psi_n(s)^1$ gives the first day's expected cost of Player n, α_1 is given by $x_n(s)^t P(s)\mu$ which is $\sum_{t=1}^{m} \rho(t|s, x_n(s))\mu(t)$. We now substitute

$$\mu(t) = \psi_n(t) + \sum_{s'=1}^{m} p(s'|t, x_n(t))\mu(s') - g$$

where $\psi_n(t)$ denotes $(x_n(t))^t(\sum_{j=1}^{n-1} A_{nj}(t) x_n(t))$, in equation (2.6) to obtain

$$2g + \mu(s) = \psi_n(s)^1 + \psi_n(s)^2 + \alpha_2 \tag{2.7}$$

where $\psi_n(s)^2 = \sum_{t=1}^{m} \rho(t|s, x_n(s))\psi_n(t)$ which is the second day expected cost of Player n under π and α_2 is given by $\sum_{t=1}^{m} \rho(t|s, x_n(s))[\sum_{s'=1}^{m} \rho(s'|t, x_n(t))\mu(s')]$. Iterating N times, we note that

$$Ng + \mu(s) = \sum_{r=1}^{N} \psi_n(s)^r + \alpha_N \tag{2.8}$$

Dividing by N and allowing $N \to \infty$ we note that ge^{*m} gives the vector of limiting average expected costs for Player n. To complete the proof, suppose $\hat{\xi}^n = \begin{bmatrix} y_n(1) \\ \vdots \\ y_n(m) \end{bmatrix}$ is any other strategy for Player n while the strategy of Player $i, \forall \ i \neq n$ remains the same as $\xi^i = \begin{bmatrix} x_i(1) \\ \vdots \\ x_i(m) \end{bmatrix}$. Multiplying equation (2.2) by $(y_n(t))^t$, we obtain

$$(y_n(t))^t A_{nj}(s) x_j(s) + \sum_{t=1}^{m} \rho(t|s, y_n(s))\mu(t) \geq g + \mu(s) \ \forall s \in \mathcal{S} \tag{2.9}$$

We shall use the notation $\psi_n(s)^{*1}$ for $(y_n(t))^t A_{nj}(s) x_j(s)$ and $\psi_n(s)^{*r}$ for $(Q(\pi^*)^r \psi_n(s)^{*1}$. Then as before substituting for $\mu(t)$ in equation (2.9) using the inequality $\psi_n(t)^{*1} + Q^*\mu - ge^{*m} \geq \mu(t)$, we obtain, $\psi_n(s)^{*1} + \psi_n(s)^{*2} + (Q^*)^2\mu \geq 2g + \mu(s)$ Iterating this N times, dividing by N and taking limit as $N \to \infty$ we obtain $\phi_n\left(\begin{bmatrix} \pi^*_{-i} \\ \xi^n \end{bmatrix}\right) \geq g$.

To complete the proof we need to show that if $((x_i^*(s), v_i(s), g)$, $1 \leq i \leq n, 1 \leq s \leq m)$ is a Nash equilibrium point of the game SG1, then $(x_i^*(s)$, $v_i(s), g)$ satisfy the system of inequalities and equations given by (2.1) through (2.5). Let $(x_i^{*\beta_N}(s))$ be a set of Nash equilibrium strategies of the corresponding discounted game with discount factor β_N so that as $\{\beta_N\} \uparrow 1$, $x_i^{*\beta_N}(s) \to x_i^*(s)$. It is known that $(x_i^{*\beta_N}(s), 1 \leq i \leq n, 1 \leq s \leq m)$ along with $v_i^{\beta_N}(s)$ and $\phi_n^{\beta_N}(\pi^{*\beta_N})(s)$ satisfy the following system of inequalities and equations:

$$w_i(s) = \sum_{j \neq i} A_{ij}(s) x_j^{*\beta_N}(s) - v_i^{\beta_N}(s) e^{is} \geq 0, 1 \leq s \leq m, \ 1 \leq i \leq (n-1) \tag{2.10}$$

$$w_n(s) = \sum_{j=1}^{n-1} A_{nj}(s) x_j^{*\beta_N}(s) + \beta_N P(s) \phi_n^{\beta_N} - \phi_n^{\beta_N}(s) e^{ns} \geq 0 \tag{2.11}$$

$$\sum_{\bar{a} \in \mathcal{N}_i(s)} x_i^{\beta_N}(\bar{a}|s) = 1, \ 1 \leq s \leq m, \ 1 \leq i \leq n \tag{2.12}$$

$$x_i(\bar{a}|s) \geq 0 \ \forall \ i, s, \text{ and } \bar{a} \in \mathcal{N}_i(s), \ 1 \leq s \leq m, 1 \leq i \leq n \tag{2.13}$$

$$(w_i(s))^t x_i(s) = 0, \ 1 \leq s \leq m, \ 1 \leq i \leq n \tag{2.14}$$

where $P(s)$ is the $m_n(s) \times m$ matrix whose $(\bar{a}, t)^{th}$ entry is $p(t|s, \bar{a})$, $\bar{a} \in \mathcal{N}_n(s)$, $t \in \mathcal{S}$. See [11]

Now by taking limit as $N \to \infty$ and noting that $x_i^{*\beta_N}(s) \to x_i^*(s)$, $\forall \ i \leq n, 1 \leq s \leq m$ and that for $1 \leq i \leq n-1$, since $(1-\beta_N)\phi_i^{\beta_N}(s) = v_i^{\beta_N}(s)$, $\lim_{N\to\infty} v_i^{\beta_N}(s) = v_i(s)$, it is seen that $(x_i^*(s), v_i(s))$, $\forall \ i \leq n-1, 1 \leq s \leq m$ satisfy (2.1) as well as (2.3), (2.4) and (2.5). For $i = n$ we note that $x_n^{*\beta_N}(s) \to x_n^*(s)$ and $\beta_N P(s)\phi_n^{\beta_N} - \phi_n^{\beta_N}(s)e^{ns} = \beta_N P(s)[\phi_n^{\beta_N} - \phi_n^{\beta}(1)e^{ns}] + \beta_N[\phi_n^{\beta_N}(1) - \phi_n^{\beta_N}(s)]e^{ns} - (1-\beta_N)\phi_n^{\beta_N}(s)e^{ns}$. Now it is known (see [18]) that for any fixed state (say, 1) if its mean recurrence time under any stationary strategy of Player n is finite then for any state s, $\phi_n^{\beta_N}(s) - \phi_n^{\beta_N}(1) \to \mu(s)$ for some real number $\mu(s)$. Further $(1-\beta_N)\phi_n^{\beta_N}(s) \to \phi_n(s)$. Making use of these and taking limit as $N \to \infty$ in (2.2) it is easy to see that $((x_i^*), (v_i(s), g)$, $1 \leq i \leq n$, $1 \leq s \leq m)$ satisfy (2.1) through (2.5). This completes the proof. ∎

We shall now formulate the problem of finding a solution to the above system of linear inequalities and equations (2.1) through (2.5) as a linear complementarity problem. For a given a square matrix $M \in R^{\ell \times \ell}$ and a vector $q \in R^{\ell}$ the linear complementarity problem (denoted by LCP(q, M)) is to find vectors $w,\ z \in R^{\ell}$ such that

$$w - Mz = q, \ \ w \geq 0, \ z \geq 0 \tag{2.15}$$

$$w^t z = 0. \tag{2.16}$$

A pair (w, z) of vectors satisfying (2.15) and (2.16) is called a solution to the LCP(q, M). This problem is well studied in the literature over the years. For the literature on the linear complementarity problem and for Lemke's algorithm to solve LCP(q, M) see [9]. For a recent book on this topic see Cottle, Pang and Stone [2]. Some of the basic definitions and results from the literature on this problem that are needed in this paper are presented in the Appendix.

Given a one player control polystochastic game

$$(\mathcal{N}, \mathcal{S}, \mathcal{N}_i(s), A_{ij}(s), p(t|s, a_n), \ \forall\ s,\ t \in \mathcal{S}, a_i \in \mathcal{N}_i(s), i \neq j, i, j \in \mathcal{N}),$$

we shall assume without loss of generality that all the partial cost matrices $A_{ij}(s)$ are strictly positive. Under this assumption note that all daily expected costs and the limiting average expected costs are positive. Let B, a square matrix of order $\sum_{i=1}^{n}\sum_{s=1}^{m} m_i(s)$ be defined as $B = \begin{bmatrix} \bar{B} \\ \tilde{B} \end{bmatrix}$ where $\bar{B}$ is a matrix of order $\sum_{i=1}^{n-1}\sum_{s=1}^{m} m_i(s) \times \sum_{i=1}^{n}\sum_{s=1}^{m} m_i(s)$ given by

$$\begin{bmatrix}
0 & A_{12}(1) & \cdots & \cdot & A_{1n}(1) & 0 & 0 & \cdots & 0 & 0 & \cdots & 0 & 0 & \cdots & 0 \\
0 & 0 & \cdots & 0 & 0 & 0 & A_{12}(2) & \cdots & A_{1n}(2) & 0 & \cdots & 0 & 0 & \cdots & 0 \\
\vdots & \vdots & & & & & & & \vdots & & \vdots & & & \vdots & \vdots \\
0 & 0 & 0 & 0 & 0 & 0 & 0 & 0 & 0 & 0 & 0 & 0 & A_{12}(m) & \cdots & A_{1n}(m) \\
\vdots & \vdots & & & & & & & \vdots & & \vdots & & & \vdots & \vdots \\
A_{n-11}(1) & \cdot & \cdots & 0 & A_{n-1n}(1) & 0 & 0 & \cdots & 0 & 0 & \cdots & 0 & 0 & \cdots & 0 \\
\vdots & \vdots & & & & & & & \vdots & & \vdots & & & \vdots & \vdots \\
0 & 0 & 0 & 0 & 0 & 0 & 0 & 0 & 0 & 0 & 0 & A_{n-11}(m) & \cdots & \cdots & A_{n-1n}(m)
\end{bmatrix}$$

and $\tilde{B}$ is the matrix of order $\sum_{s=1}^{m} m_n(s) \times \sum_{i=1}^{n}\sum_{s=1}^{m} m_i(s)$ given by

$$\begin{bmatrix}
A_{n1}(1) & A_{n2}(1) & \cdots & A_{nn-1}(1) & 0 & 0 & 0 & \cdots & 0 & 0 & \cdots & 0 & 0 & \cdots & 0 \\
0 & 0 & \cdots & 0 & 0 & A_{n1}(2) & A_{n2}(2) & \cdots & A_{nn-1}(2) & 0 & \cdots & 0 & 0 & \cdots & 0 \\
\vdots & \vdots & & & & & & & \vdots & & \vdots & & & \vdots & \vdots \\
0 & 0 & 0 & 0 & 0 & 0 & 0 & 0 & 0 & 0 & 0 & A_{n1}(m) & A_{n2}(m) & \cdots & 0
\end{bmatrix}$$

Let $\bar{E}$ be a matrix of order $\sum_{i=1}^{n-1}\sum_{s=1}^{m} m_i(s) \times m(n-1)$ which is of the form

$$\bar{E} = \begin{bmatrix}
-e^{11} & 0 & \cdots & 0 \\
0 & -e^{12} & \cdots & 0 \\
\vdots & \cdots & \cdots & \vdots \\
0 & \cdots & \cdots & -e^{(n-1)m}
\end{bmatrix}.$$

Let E be a matrix of order $(nm+1) \times \sum_{i=1}^{n}\sum_{s=1}^{m} m_i(s)$ which is of the form

$$E = \begin{bmatrix} (e^{11})^t & 0 & \dots & 0 \\ 0 & (e^{12})^t & \dots & 0 \\ \vdots & \dots & \dots & \vdots \\ 0 & \dots & \dots & (e^{nm})^t \\ 0 & \dots & f^1 & f^2 \end{bmatrix}$$

where e^{is} is as defined earlier, and $f^t = ((f^1)^t, (f^2)^t)$ is a row vector of order $\sum_{s=1}^{m} m_n(s)$ whose coordinates are all equal to 1. Let Q be the matrix of order $\sum_{s=1}^{m} m_n(s) \times m$ given by

$$Q = \begin{bmatrix} Q^1 \\ \vdots \\ Q^m \end{bmatrix}$$ where Q^s is an $m_n(s) \times m$ matrix which is

$$Q^s = \begin{bmatrix} p(1|s,1) & \dots & -1+p(s|s,1) & \dots & p(m|s,1) \\ p(1|s,2) & \dots & -1+p(s|s,2) & \dots & p(m|s,2) \\ \vdots & \vdots & \vdots & \vdots & \vdots \\ p(1|s,m_n(s)) & \dots & -1+p(s|s,m_n(s)) & \dots & p(m|s,m_n(s)) \end{bmatrix}$$

Let M be a square matrix of order $\sum_{i=1}^{n}\sum_{s=1}^{m} m_i(s) + nm + 1$ which is given by

$$M = \begin{bmatrix} \bar{B} & \bar{E} & 0 & 0 \\ \tilde{B} & 0 & Q & -f \\ E & 0 & 0 & 0 \end{bmatrix}.$$

Let the vector q be defined as $q = \begin{bmatrix} q^1 \\ q^2 \\ q^3 \end{bmatrix} = \begin{bmatrix} 0 \\ -e^{*mn} \\ -m \end{bmatrix}$

where q^1 is of order $\sum_{i=1}^{n}\sum_{s=1}^{m} m_i(s) \times 1$, q^2 is of order $mn \times 1$, q^3 is of order 1 and e^{*mn} is (as has been defined earlier) a vector of order $mn \times 1$ of 1's.

With these notations we have the following theorem.

THEOREM 2.2 *Consider the LCP(q, M) where q and M are defined as above. Any solution to the LCP(q, M) is a solution to the system of inequalities and equations (2.1) through (2.5) and hence yields a stationary equilibrium point of the auxiliary game SG1. Conversely, given any stationary equilibrium point $((x_i(s), v_i(s), ge^{*m})i \in \mathcal{N},\ s \in \mathcal{S})$ of SG1 there exist nonnegative numbers $\mu^*(s)$, $1 \le s \le m$, such that $((x_i(s), v_i(s), ge^{*m}, \mu^*)i \in \mathcal{N},\ s \in \mathcal{S})$ yields a solution to the LCP(q, M)*

Proof. Suppose (w, z) is a solution to the LCP(q, M). Note that the inequality $q + Mz \ge 0$ gives us the inequalities in (2.1) (2.2) and (2.4) where z^t can be identified as

$$((x_1(1))^t, \dots, (x_n(m))^t, v_1(1), \dots, v_1(m), \dots, v_{n-1}(m), \mu_1, \dots \mu_m, g). \tag{2.17}$$

The vector w^t can be identified as

$$((w_1(1))^t, \ldots, (w_n(1))^t, \ldots, (w_n(m))^t, u_1(1), \ldots, u_1(m), \ldots, u_n(m), u_{n+1}). \qquad (2.18)$$

where $w_i(s)$ is a vector of order $m_i(s) \times 1$ defined earlier, $u_i(s)$ are real numbers given by

$$u_i(s) = \sum_{\bar{a} \in \mathcal{N}_i(s)} x_i(\bar{a}|s) - 1 \qquad (2.19)$$

and u_{n+1} is a real number given by

$$u_{n+1} = \sum_{s=1}^{m} \sum_{\bar{a} \in \mathcal{N}_i(s)} x_i(\bar{a}|s) - m. \qquad (2.20)$$

We note that $w_i(s) \geq 0$, $x_i(s) \geq 0$ and further, $\sum_{\bar{a} \in \mathcal{N}_i(s)} x_i(\bar{a}|s) \geq 1$. Now note that $u_i(s)v_i(s) = 0$. This implies that if $u_i(s) > 0$ then $v_i(s) = 0$. However, since $x_i(s)$ for all i and s are nonnegative with at least one coordinate positive and as A_{ij}'s are strictly positive matrices, it follows that for $i \leq (n-1)$, if $v_i(s) = 0$, then $w_i(s) > 0$. This implies that $(w_i(s))^t x_i(s) > 0$ and contradicts the hypothesis that (w, z) solves the LCP(q, M). Hence it follows that $u_i(s) = 0, i \neq n$ and hence $\sum_{\bar{a} \in \mathcal{N}_i(s)} x_i(\bar{a}|s) = 1$, $\forall\, i \leq (n-1)$. We now proceed to show that the same conclusion holds also for $i = n$. Note that since $\sum_{\bar{a} \in \mathcal{N}_n(s)} x_n(\bar{a}|s) \geq 1$ it follows as earlier that for each s there is at least one equality in the block of inequalities of (2.2). Now suppose for s, the $\bar{a}_s^{th}$ inequality of (2.2) holds as an equality. We thus have,

$$\sum_{\bar{a}_{ks} \in \mathcal{N}_k(s)} a(\bar{a}_{ns}, \bar{a}_{ks})(s) x_k(\bar{a}_{ks}|s) + \sum_{t=1}^{m} p(t|s, \bar{a}_s)\mu_t - \mu_s - g = 0 \qquad (2.21)$$

This leads to the equation

$$\Theta = -\mathcal{R}\mu + ge^{*m}$$

where $\mathcal{R}$ is a row representative submatrix of the vertical block matrix $\mathcal{Q}$ and Θ is a positive column vector of order $m \times 1$. For an explanation about vertical block matrices and row represetative matrices see [12]. Note that $-\mathcal{R}$ is an irreducible $Z \cap P_0$-matrix. It follows that the above equation cannot hold if $g = 0$. Hence $g > 0$ and by complementarity $u_{n+1} = 0$ and hence it follows that $\sum_{s=1}^{m} \sum_{\bar{a} \in \mathcal{N}_n(s)} x_n(\bar{a}|s) = m$. It follows that $\sum_{\bar{a} \in \mathcal{N}_n(s)} x_n(\bar{a}|s) = 1\ \forall s$. Thus, $(x_i(s), v_i(s), w_i(s)\mu(s))\ \forall s \in \mathcal{N})$ and g solve the LCP(q, M). Conversely, given an equilibrium point $((x_i(s), v_i(s)\ \forall i,\ s), ge^{*m})$, by Theorem 2.1 there exist real numbers $\mu(s)$ that satisfy (2.1) through (2.5). Let $\mu^*(s) = \mu(s) + \theta$ where θ is a positive real number large enough so that $\mu^*(s) > 0$ for all s. Now define $u_i(s) = 0$, $\forall i$, s and take w and z as defined in (2.6) and (2.7). It is easy to see that (w, z) solves LCP(q, M). This yields a solution to the LCP(q, M). ■

REMARK 2.1 Thus the problem of finding a set of stationary equilibrium strategies and corresponding costs for a one player control irreducible polystochastic game with the limiting average expected cost as the criterion can be formulated as a problem of solving a linear complementarity problem. We shall show in the next section that a modification of this formulation will yield an LCP which can be solved by Lemke's algorithm.

REMARK 2.2 Note that the matrix Q is a vertical block Z-matrix. Z-matrices have been introduced and studied in [7]. Vertical block matrices have been considered in the literature on generalized complementarity problem since 1970. For the definition and other details see [1], [4] and [5]. Vertical block Z matrices have been studied recently in [12].

3 Lemke's Algorithm for Computing an Equilibrium

Lemke's algorithm when applied to the LCP(q, M) formulated in the previous section may terminate in a *secondary ray*. However, we shall show that a slight modification leads to an LCP$(q, \bar{M})$ where $\bar{M}$ is in the class defined by Eaves [3]. Appendix A presents the definition of secondary ray and the class $\mathcal{L}$ class of matrices. For more explanations see also [2].

As in [11] we first replace the matrix M by $\bar{M}$ which is obtained by replacing $B = \begin{bmatrix} \bar{B} \\ \tilde{B} \end{bmatrix}$ in M by $B + \Lambda = C$ where Λ is a square matrix of order $\sum_{i=1}^{n}\sum_{s=1}^{m} m_i(s)$ each of whose entries is 1. Thus

$$\bar{M} = \begin{bmatrix} \bar{C} & \bar{E} & 0 & 0 \\ \tilde{C} & 0 & Q & f \\ E & 0 & 0 & 0 \end{bmatrix}$$

where $\bar{C}$ and $\tilde{C}$ are the row partition of C induced by the partition of B as $\begin{bmatrix} \bar{B} \\ \tilde{B} \end{bmatrix}$. We have the following lemma.

LEMMA 3.1 *Consider the LCP$(q, \bar{M})$. If $(\bar{w}, \bar{z})$ solves LCP$(q, \bar{M})$ then (w^*, z^*) solves LCP(q, M) where $w^* = \bar{w}$, $z_r^* = \bar{z}_r$, for $1 \le r \le \sum_{i=1}^{n}\sum_{s=1}^{m} m_i(s)$, $z_r^* = \bar{z}_r - mn$ for $\sum_{i=1}^{n}\sum_{s=1}^{m} m_i(s) < r \le \sum_{i=1}^{n}\sum_{s=1}^{m} m_i(s) + (n-1)m$ $z_r^* = \bar{z}_r$ for $\sum_{i=1}^{n}\sum_{s=1}^{m} m_i(s) + (n-1)m < r \le \sum_{i=1}^{n}\sum_{s=1}^{m} m_i(s) + nm$. and $z_r^* = \bar{z}_r - mn$ for $r = \sum_{i=1}^{n}\sum_{s=1}^{m} m_i(s) + nm + 1$.*

Proof. This is easy to verify. ∎

THEOREM 3.1 *LCP$(d, \bar{M})$ has a unique solution when $d > 0$, $d \in R^{m^*}$ where $m^* = \sum_{i=1}^{n}\sum_{s=1}^{m} m_i(s) + mn$.*

Proof. The proof is similar to the proof in the case of the matrix obtained for the discounted game See [11]. We shall show that $\bar{M} \in \mathcal{L}_1$, the class introduced by Eaves [3] by verifying its defining condition. So, suppose $0 \ne x = \begin{bmatrix} \xi_1 \\ \xi_2 \\ \xi_3 \\ \xi_4 \end{bmatrix} \ge 0$ is given where $x \in R^{m^*}$ and the partition of x is induced by the partition in $\bar{M}$. Now $\bar{M}x = \begin{bmatrix} \bar{C}\xi_1 + \bar{E}\xi_2 \\ \tilde{C}\xi_1 + Q\xi_3 + f\xi_4 \\ E\xi_1 \end{bmatrix}$.

If $\xi_4 \neq 0$, or if $\xi_4 = 0$ and $\xi_3 \neq 0$ then there exists a $r > \sum_{i=1}^{n}\sum_{s=1}^{m} m_i(s) + m(n-1)$ such that $\bar{x}_r > 0$ and $(\bar{M}x)_r \geq 0$ as $E\xi_1 \geq 0$. If $\xi_4 = \xi_3 = 0$ and $\xi_2 \neq 0$, then there exists a r, $\sum_{i=1}^{n}\sum_{s=1}^{m} m_i(s) < r < \sum_{i=1}^{n}\sum_{s=1}^{m} m_i(s) + m(n-1)$ such that $\bar{x}_r > 0$ and $(M\bar{x})_r \geq 0$ as $E\xi_1 \geq 0$. If $\xi_2 = \xi_3 = \xi_4 = 0$ then $\xi_1 \neq 0$ and since $\bar{C}\xi_1 > 0$ and $\tilde{C}\xi_1 > 0$ it follows that $\exists$ a $r \leq \sum_{i=1}^{n}\sum_{s=1}^{m} m_i(s)$ such that $x_r > 0$ and $(\bar{M}x)_r > 0$.

Thus it follows that $\bar{M} \in \mathcal{L}_1$. Now from Lemma (3) (page 620) in [3] it follows that LCP$(d, \bar{M})$ has a unique solution for each $d > 0, d \in R^{m^*}$. This completes the proof. ∎

However unlike in the case of the discounted game, LCP$(0, \bar{M})$ has a nontrivial solution. We shall first characterize the nontrivial solutions of LCP$(0, \bar{M})$.

THEOREM 3.2 *Suppose $((x_i(s), v_i(s), \mu_s, g), 1 \leq i \leq n,\ 1 \leq s \leq m)$ is a solution to LCP$(0, \bar{M})$ then $x_i(s) = 0, v_i(s) = 0,\ \forall\ i, s$ and $g = 0$ and $\mu = ce^{*m}$ where c is a positive number.*

Proof. Suppose there is a $(\bar{w}, \bar{z})$ such that

$$\bar{w} - \bar{M}\bar{z} = 0,\ \bar{w} \geq 0,\ \bar{z} \geq 0, \text{ and } \bar{w}^t\bar{z} = 0.$$

It follows that $\bar{z}^t\bar{M}\bar{z} = 0$. Let $X = \begin{bmatrix} \bar{E} & 0 & 0 \\ 0 & Q & -f \end{bmatrix}$. Also let $E^t = \begin{bmatrix} Y_{11} & Y_{12} \\ Y_{21} & Y_{22} \end{bmatrix}$ be the partition of E as in X, since X and E^t are of the same order. Note that $\bar{M} = \begin{bmatrix} C & X \\ E & 0 \end{bmatrix}$ and let $\bar{z} = \begin{bmatrix} \eta_1 \\ \eta_2 \end{bmatrix}$ be the partition of $\bar{z}$ induced by this partition of $\bar{M}$. We then have

$$\bar{z}^t\bar{M}\bar{z} = \eta_1^tC\eta_1 + \eta_1^t(E^t + X)\eta_2 = 0. \tag{3.1}$$

Noting that $C > 0$ and $E^t + X \geq 0$, we see that equation (3.1) implies that $\eta_1^tC\eta_1 = 0$ and $\eta_1^t(E^t + X)\eta_2 = 0$. Now $\eta_1^tC\eta_1 = 0 \Rightarrow \eta_1 = 0$. Noting that $\bar{M}\bar{z} \geq 0$, we conclude that $X\eta_2 \geq 0$. Let $\eta_2 = \begin{bmatrix} \bar{\eta}_2 \\ \tilde{\eta}_2 \\ \eta_2^* \end{bmatrix}$ be the partition of η_2 induced by the partition of X as $\begin{bmatrix} \bar{E} & 0 & 0 \\ 0 & Q & -f \end{bmatrix}$. We then have $\bar{E}\bar{\eta}_2 \geq 0$ which implies that $\bar{\eta}_2 = 0$ as $\bar{E} \leq 0$. Now consider the inequality $Q\tilde{\eta}_2 + (-f)\eta_2^* \geq 0$. Note that Q is a matrix of order $\sum_{s=1}^{m} m_n(s) \times m$ which has the row partition $\begin{bmatrix} Q^1 \\ \vdots \\ Q^m \end{bmatrix}$ defined earlier. As noted earlier this a vertical block matrix each of whose row representative submatrices is an irreducible singular $Z \cap P_0$ matrix and hence the above inequality can be satisfied only as an equation and any η_2 satisfying the equation $Q\tilde{\eta}_2 + (-f)\eta_2^* = 0$ has to be of the form ce^{*m}. This concludes the proof of the theorem. ∎

REMARK 3.1 Thus unlike in the discounted case with the limiting average cost as the criterion even when the game is irreducible the LCP formulation given above with the matrix as $\bar{M}$ we are unable to solve the problem by an application of Lemke's algorithm.

We shall now further modify the problem to obtain an LCP which has a unique solution for any positive vector d and for the null vector 0. Let M^* be the matrix of order $\sum_{i=1}^{n}\sum_{s=1}^{m} m_i(s) + nm$ obtained from $\bar{M}$ by omitting its $(\sum_{i=1}^{n}\sum_{s=1}^{m} m_i(s) + (n-1)m+1)^{th}$ column, which is the column corresponding to $\mu(1)$ and the corresponding row which is the row corresponding to the equation $\sum_{a\in\mathcal{N}_n(1)} x_i(a|1) = 1$. Let $q*$ of order $\sum_{i=1}^{n}\sum_{s=1}^{m} m_i(s) + nm$ be defined as $q^* = \begin{bmatrix} q^{*11} \\ q^{*12} \\ q^{*2} \\ q^3 \end{bmatrix} = \begin{bmatrix} 0 \\ \theta(Q)_{.1} \\ -e^{*(mn-1)} \\ -m \end{bmatrix}$ where θ is a large fixed positive number, $(Q)_{.1}$ denotes the first column of the vertical block matrix Q, q^{*11} is of order $\sum_{i=1}^{n}\sum_{s=1}^{m} m_i(s) + (n-1)m \times 1$ and q^{*12} is of order $m \times 1$.

THEOREM 3.3 *The only solution to LCP(d, M^*) is the solution $w = d$; $z = 0$ and the only solution to LCP$(0, M^*)$ is the trivial solution $w = 0$; $z = 0$.*

Proof. Since M^* is a prinicpal submatrix of $\bar{M}$ it follows that M^* is also in the class $\mathcal{L}_1$. It follows from here that the LCP(d, M^*) has a unique solution for any positive vector d. Since any nontrivial solution to LCP$(0, \bar{M})$ assigns positive weights to all the columns of Q it follows from the arguments of the proof of the earlier theorem that the only solution to LCP$(0, M^*)$ is the trivial solution. ■

THEOREM 3.4 *There is a real number θ^* such that for $\theta \geq \theta^*$, a solution to LCP$(q^*(\theta), M^*)$ obtained by Lemke's algorithm yields a set of Nash equilibrium strategies for the players.*

Proof. From the previous theorem it follows that for any q, LCP(q, M^*) has a unique solution and that this can be computed by applying Lemke's algorithm to it initiating it with any positive vector d. See [3]. It is also known that the original game with the limiting average cost as the criterion has a Nash equilibrium point $(x_i^*(s)), (\phi_(i)(s), ge^{*m})$, $1 \leq i \leq n$, $1 \leq s \leq m)$. See [15]. By Theorem 2.1 $(x_i^*(s), 1 \leq i \leq n,\ 1 \leq s \leq m)$ along with the corresponding expected first day's costs $v_i(s)$ for players 1 through $(n-1)$, the real number g and a set real numbers $\mu(s)$, $1 \leq s \leq m$ satisfy the system of inequalities and equations given in (2.1) through (2.5). It follows from here that $((x_i^*(s)), (v_i(s), ge^{*m})$, $1 \leq i \leq n$, $1 \leq s \leq m)$ along with $\mu(s) + \theta^*$ yields a solution to the LCP(q, M), where θ^* is as defined in the second part of the proof of Theorem 2.2. By Lemma 3.1 it follows that a corresponding solution can be obtained for LCP$(q, \bar{M})$. Since any solution to LCP$(q, \bar{M})$ is also a solution to the LCP$(q^*(\theta), M^*)$ with $\bar{\mu} > \mu^*$, the theorem follows. ■

The case $n = 2$ is discussed more often in the literature. We therefore state the following corollary.

COROLLARY 3.1 *Lemke's algorithm processes the modified linear complementarity problem LCP$(q^*(\theta), M^*)$ associated with the problem of finding a pair of equilibrium strategies and the corresponding limiting average costs of a two person nonzero-sum stochastic game in which transition probabilities depend on the actions of a single player, under any stationary strategies for the players the resulting transition probability matrix is irreducible and the criterion is the limiting average of expected costs.*

REMARK 3.2 The formulation of the problem of computing a stationary set of Nash equilibrium strategies for the Player n control irreducible polystochastic game with the limiting average of expected daily costs as the criterion given in this paper can be extended to the case when under any stationary strategy of Player n there is a fixed state, say state 1, such that this state is visited with positive probability whatever be the initial state of the game and that the mean recurrence time of it is finite. This assumption will ensure that under any stationary strategy there is only one positive recurrent class and that all the states not included in this class are transient and are eventually absorbed into this class. It is easy to see that in this case also $\phi_n(\pi) = ge^{*m}$ for some real number g and that the matrix Q is a vertical block matrix each of whose row representative matrix is a $Z \cap P_0$-matrix with rank $(m-1)$. The proof of our formulation of LCP$(q^*(\theta), M^*)$ goes through and a set of stationary equilibrium strategies can be computed by solving the LCP$(q^*(\theta), M^*)$ using Lemke's algorithm for a sufficiently large value of θ.

REMARK 3.3 It is clear from the linear complementarity formulation of the problem of finding a set of stationary equilibrium strategies and corresponding limiting average expected cost vectors for all the players of a polystochastic game considered here, that such a game has a set of rational equilibrium strategies and corresponding rational limiting average expected costs, if all the data are rational. Thus we have the *orderfield property* for a polystochastic game with limiting average expected cost as a criterion in which the transition probabilities depend on the actions of a single player and under any stationary strategy of this player the resulting Markov Chain is irreducible.

EXAMPLE 3.1 Consider the stochastic game with $n = 2$ players and $m = 2$ states for which a Nash equilibrium point is computed in [13]. The cost matrices $A_{ij}(s)$'s are as follows:

$$A_{12}(1) = \begin{bmatrix} 9 & 6 & 4 \\ 5 & 7 & 9 \end{bmatrix} \quad A_{12}(2) = \begin{bmatrix} 2 & 9 & 8 \\ 9 & 0 & 8 \end{bmatrix} \quad A_{21}(1) = \begin{bmatrix} 4 & 7 \\ 7 & 5 \\ 8 & 3 \end{bmatrix} \quad A_{21}(2) = \begin{bmatrix} 9 & 3 \\ 1 & 9 \\ 2 & 1 \end{bmatrix}$$

Let the transition probabilities be given as follows:

$$q(1|1,1) = 0.5 \quad q(2|1,1) = 0.5$$
$$q(1|1,2) = 0.0 \quad q(2|1,2) = 1.0$$
$$q(1|1,3) = 0.8 \quad q(2|1,3) = 0.2$$
$$q(1|2,1) = 0.5 \quad q(2|2,1) = 0.5$$
$$q(1|2,2) = 1.0 \quad q(2|2,2) = 0.0$$
$$q(1|2,3) = 0.2 \quad q(2|2,3) = 0.8$$

M and q are given by

$$\begin{bmatrix}
0 & 0 & 0 & 0 & 9 & 6 & 4 & 0 & 0 & 0 & -1 & 0 & 0 & 0 & 0 \\
0 & 0 & 0 & 0 & 5 & 7 & 9 & 0 & 0 & 0 & -1 & 0 & 0 & 0 & 0 \\
0 & 0 & 0 & 0 & 0 & 0 & 0 & 2 & 9 & 8 & 0 & -1 & 0 & 0 & 0 \\
0 & 0 & 0 & 0 & 0 & 0 & 0 & 9 & 0 & 8 & 0 & -1 & 0 & 0 & 0 \\
4 & 7 & 0 & 0 & 0 & 0 & 0 & 0 & 0 & 0 & 0 & 0 & -0.5 & 0.5 & -1 \\
7 & 5 & 0 & 0 & 0 & 0 & 0 & 0 & 0 & 0 & 0 & 0 & -1.0 & 1.0 & -1 \\
8 & 3 & 0 & 0 & 0 & 0 & 0 & 0 & 0 & 0 & 0 & 0 & -0.2 & 0.2 & -1 \\
0 & 0 & 9 & 3 & 0 & 0 & 0 & 0 & 0 & 0 & 0 & 0 & 0.5 & -0.5 & -1 \\
0 & 0 & 1 & 9 & 0 & 0 & 0 & 0 & 0 & 0 & 0 & 0 & 1.0 & -1.0 & -1 \\
0 & 0 & 2 & 1 & 0 & 0 & 0 & 0 & 0 & 0 & 0 & 0 & 0.2 & -0.2 & -1 \\
1 & 1 & 0 & 0 & 0 & 0 & 0 & 0 & 0 & 0 & 0 & 0 & 0 & 0 & 0 \\
0 & 0 & 1 & 1 & 0 & 0 & 0 & 0 & 0 & 0 & 0 & 0 & 0 & 0 & 0 \\
0 & 0 & 0 & 0 & 1 & 1 & 1 & 0 & 0 & 0 & 0 & 0 & 0 & 0 & 0 \\
0 & 0 & 0 & 0 & 0 & 0 & 0 & 1 & 1 & 1 & 0 & 0 & 0 & 0 & 0 \\
0 & 0 & 0 & 0 & 1 & 1 & 1 & 1 & 1 & 1 & 0 & 0 & 0 & 0 & 0
\end{bmatrix} \text{ and } \begin{bmatrix} 0 \\ 0 \\ 0 \\ 0 \\ 0 \\ 0 \\ 0 \\ 0 \\ 0 \\ 0 \\ -1 \\ -1 \\ -1 \\ -1 \\ -2 \end{bmatrix}.$$

We fix $\mu(1) = 10.0$ and obtain a solution to LCP$(q^*(10), M^*)$ as $\bar{w}_6 = 0.648870$, $\bar{w}_8 = 4.525670$, $\bar{z}_1 = 0.689938$, $\bar{z}_2 = 0.310062$ $\bar{z}_3 = 0.809035$ $\bar{z}_4 = 0.190965$, $\bar{z}_5 = 0.55556$, $\bar{z}_7 = 0.444444$, $\bar{z}_9 = 0$, $\bar{z}_{10} = 1.0$, $\bar{z}_{11} = 10.777780$, $\bar{z}_{12} = 12.0$, $\bar{z}_{13} = 8.398356$ and $\bar{z}_{14} = 8.129363$.

The solution for the auxiliary game is given below:

The strategies of the players are as follows:

$$x_1(1) = \begin{bmatrix} 0.689938 \\ 0.310062 \end{bmatrix}, \; x_1(2) = \begin{bmatrix} 0.809035 \\ 0.190965 \end{bmatrix}, \; x_2(1) = \begin{bmatrix} 0.555556 \\ 0.0 \\ 0.444444 \end{bmatrix}, \; x_2(2) = \begin{bmatrix} 0.0 \\ 0.0 \\ 1.0 \end{bmatrix}.$$

The corresponding costs of the auxiliary game are $v_1(1) = 6.777780$, $v_1(2) = 8.0$, $\phi_2(1) = \phi_2(2) = 4.129363$.

The solution for the original game is obtained as follows:
Strategies are the same as in the auxiliary game and costs for players $1 \leq i \leq n-1$ of the original game are obtained by multiplying the corresponding first day expected costs of the game by $\lim_{N\to\infty} \mathcal{Q}(\pi)^N$ which is the same as $\lim_{N\to\infty} \frac{1}{N}\sum \mathcal{Q}(\pi)^N$.

The costs of the original game are $\phi_1(1) = \phi_1(2) = 7.5686$,
$\phi_2(1) = \phi_2(2) = 4.129363$

REMARK 3.4 As a concluding remark we wish to note that the problem of computing a set of stationary equilibrium strategies and the corresponding average expected costs for the more general non-zero sum one player control polystochastic game when the criterion is the limiting average expected cost still remains open.

Acknowledgement: We wish to thank Professor T. E. S. Raghavan, University of Illinois at Chicago for many discussions on this subject. We are also thankful to an annonymous referee whose helpful comments led to this improved version.

References:

1. R. W. Cottle and G.B. Dantzig (1970) "A generalization of the linear complementarity problem", Journal of Combinatorial Theory 8 pp. 79-90.

2. R. W. Cottle, J. S. Pang, and R. E. Stone (1992) *The Linear Complementarity Problem* Academic Press, New York.

3. B. C.Eaves (1971) "The linear complementarity problem", Management Science 17 pp. 612-634.

4. A. A. Ebiefung and M. M. Kostreva (1991) "Z-matrices and the generalized linear complementarity problem", Technical Report #608, Department of Mathematical Sciences, Clemson University, South Carolina.

5. A. A. Ebiefung and M. M. Kostreva (1993) "Generalized P_0 and Z-matrices", Linear Algebra and Its Applications 195 pp.165-179.

6. A. M. Fink (1964) "Equilibrium in a stochastic n-person game", Journal of Science of Hiroshima University, Series A-I 28 pp.89-93.

7. M. Fiedler and V. Pták (1962) "On matrices with non-positive off-diagonal elements and positive principal minors", Czechoslovak Mathematical Journal 12 pp.382-400.

8. D. Gilette (1957) "Stochastic games with zero stop probabilities", In: A. W. T. M. Dresher and P. Wolfe (eds), Contributions to the theory of games, Princeton University Press, Annals of Mathematical Studies 39 (1957).

9. C. E. Lemke (1965) "Bimatrix equilibrium points and mathematical programming", Management Science 11 pp.681-689.

10. C. E. Lemke and J. T. Howson (1964) "Equilibrium points of bimatrix games", SIAM Journal on Applied Mathematics 12 pp.413-423.

11. S. R. Mohan, S. K. Neogy and T. Parthasarathy (1997) "Linear complementarity and discounted polystochastic game when one player controls transitions", in *Complementarity and Variational Problems* eds. M. C. Ferris and Jong -Shi Pang, SIAM, Philedelphia, pp 284-294.

12. S. R. Mohan and S. K. Neogy (1996) "Algorithms for the generalized linear complementarity problem with a vertical block Z-matrix", SIAM Journal On Optimization 6 pp. 994-1006.

13. A. S. Nowak and T. E. S. Raghavan (1993) "A finite step algorithm via a bimatrix game to a single controller non-zerosum stochastic game", Mathematical Programming 59 pp. 249-259.

14. T. Parthasarathy and T. E. S. Raghavan (1971). *Some topics in two-person games*, American Elsevier Publishing Company, Inc., New York.

15. T. Parthasarathy and T. E. S. Ragavan (1981) "An orderfield property for stochastic games when one player controls transition probabilities", Journal of Optimization theory and Applications 33 pp.375-392.

16. T. E. S. Raghavan and J. A. Filar (1991) "Algorithms for stochastic games, a survey", Zeitschrift für Operations Research 35 pp. 437-472.

17. L. S. Shapley (1953) " Stochastic games", Proceedings of the National Academy of Sciences 39 pp.1095-1100.

18. M. A. Stern (1975) "On stochastic games with limiting average payoff", Ph. D. thesis in Mathematics submitted to the Graduate College of the University of Illinois, Chicago, Illinois.

19. M. Takahasi (1964) "Equilibrium points of stochastic noncooperative n-person game", Journal of Science of Hiroshima University, Series A-I 28 pp.95-99.

A Appendix: Results from linear complementarity

Given a square matrix M of order n and a vector $q \in R^n, q \neq 0$ consider the linear complementarity problem $LCP(q, M)$ of determining a solution to equations (2.15) and (2.16). Let (w, z) be a solution to the $LCP(q, M)$. The pair (w_j, z_j) is called a *complementary pair of variables.* The pair of columns $(I_{.j}, -M_{.j})$ is called a *complemenatry pair of column vectors.*

Lemke's complementary pivoting algorithm proceeds by introducing an artificial positive column vector d to be referred to as the *covering vector* and a corresponding artificial variable z_0. The system of equations and inequalities considered is:

$$w - Mz - dz_0 = q; \ (w, z, z_0) \geq 0. \tag{A.1}$$

We say that a solution (w, z, z_0) is *almost complementary* if $z_0 > 0$ and $w^t z = 0$, and *complementary* if $w^t z = 0$ and $z_0 = 0$. A square submatrix B of order n of $(I, -M)$ is said to be a *complementary matrix* if $I_{.j}$ is a column of B implies that $-M_{.j}$ is not a column of B. A square submatrix B of $(I, -M, -d)$ with n columns is said to be an *almost complementary matrix* if $-d$ is a column of B and $I_{.j}$ is a column of B implies that $-M_{.j}$ is not a column of it. An almost complementary matrix B is said to *correspond* to an almost complementary solution (w, z, z_0) if the columns corresponding to the positive components of (w, z, z_0) are a subset of the columns of the matrix B. An almost complementary solution (w, z, z_0) is said to be an *almost complementary basic feasible solution* if there is a nonsingular almost complementary matrix corresponding to it. Suppose (w, z, z_0) is an almost complementary basic feasible solution to (A.1). Suppose there is a vector $(w^*, z^*, z_0^*) \geq 0$ such that $(w, z, z_0) + \lambda(w^*, z^*, z_0^*)$ solves (A.1) for all λ positive. Then we say that the unbounded edge $(w, z, z_0) + \lambda(w^*, z^*, z_0^*)$ of the set of feasible solutions to (A.1) is an *almost complementary ray* and that (w^*, z^*, z_0^*) *generates an almost complementary ray* at (w, z, z_0).

Lemke's algorithm starts with an almost complementary initial basic feasible solution to (A.1) and generates a sequence of adjacent almost complementary basic feasible solutions until either a complementary basic feasible solution is found or an almost complementary ray is found. When $q \not\geq 0$, the initial basic feasible solution to the above system is taken as $w^0 = q + \theta d \geq 0$ and $z^0 = 0$, $z_0^0 = \theta$ where θ is chosen as $min\{\frac{-q_i}{d_i} | q_i < 0\}$. This produces for the initial solution (w^0, z^0, z_0^0) at least one complementary pair of variables (w_α^0, z_α^0) such that $w_\alpha^0 = z_\alpha^0 = 0$. Let k be an index such that $w_k^0 = z_k^0 = 0$. Of the pair, one, namely w_k has been driven out of the basis in the initial pivot operation that has made z_0 a basic variable. By the complementary rule, its complement, namely z_k is now chosen to be included in the basis. As in the simplex method for linear programming, the variable to be excluded from the basis is determined using the minimum ratio criterion for feasibility. In general at any iteration there is exactly one nonbasic pair of complementary column vectors one of which has been removed from the basis at the previous iteration. Its complement is chosen to be included in the basis at the next iteration. The iterations continue until either the variable z_0 is removed from the basis by the minimum ratio criterion or the algorithm terminates in a secondary ray. Suppose at an iteartion $(I_{.k}, -M_{.k})$ is the pair of nonbasic complementary columns, B is the almost complementary basis matrix and $A_{.k}$ where $A_{.k}$ is either $-M_{.k}$ or $I_{.k}$. If $B^{-1}A_{.k} \leq 0$ then it is easy to see that at the almost complementary solution corresponding to B an almost complementary ray is encountered and the algorithm terminates without finding a solution to the problem.

The following classes matrices considered in the literature on the linear complementarity problem are of interest to us.

DEFINITION A.1 *We say that a square matrix M is an R_0 matrix or $M \in R_0$ if there is no nontrivial solution to $LCP(0, M)$.*

DEFINITION A.2 *We say that a square matrix M is a semimonotone matrix if $z \neq 0$, $z \geq 0$, $\Rightarrow \exists$ an index $i, 1 \leq i \leq n$, such that $z_i > 0$ and $(Mz)_i \geq 0$.*

The class of semimonotone matrices has been introduced by B. C. Eaves. It is denoted as $\mathcal{L}_1$. The following results due to Eaves are easy to verify. Their proofs are available in [3].

THEOREM A.1 *A square matrix M is a semimonotone matrix if and only if the $LCP(d, M)$ has only the trivial solution $w = d, z = 0$ for any positive vector d.*

THEOREM A.2 *Suppose M is a semimonotone matrix which is also an R_0 matrix. Then Lemke's algorithm with the covering vector as d where d is any positive vector, applied to the problem $LCP(q, M)$ where q is any given vector in R^n terminates with a solution to the problem.*

The class of matrices satisfying the hypothesis of the above theorem is denoted as $\mathcal{L}_1^*$ in [3].

Definition A.3 *A square matrix is called an $\mathcal{L}_2$ matrix if (w, z) is a solution to LCP$(0, M)$, $z \neq 0, \rightarrow \exists(\hat{w}, \hat{z})$, $\hat{z} \neq 0$ such that $\hat{w} = -M^t\hat{z}$, $w \geq \hat{w} \geq 0$, $z \geq \hat{z} \geq 0$.*

Definition A.4 *A square matrix is said to be an $\mathcal{L}$ matrix if $M \in \mathcal{L}_1 \cap \mathcal{L}_2$.*

Theorem A.3 *Let M be a square matrix and let $M \in \mathcal{L}$. Suppose Lemke's algorithm with a positive covering vector d terminates in a secondary ray for some q. Then there is no nonnegative solution to $w - Mz = q$.*

For a proof and for more details see [3] and [2].

S. R. Mohan
Indian Statistical Institute
New Delhi-110016, India

S. K. Neogy
Indian Statistical Institute
New Delhi-110016, India

T. Parthasarathy
Indian Statistical Institute
New Delhi-110016, India

ON THE LIPSCHITZ CONTINUITY OF THE SOLUTION MAP IN SOME GENERALIZED LINEAR COMPLEMENTARITY PROBLEMS

Roman Sznajder and Seetharama Gowda[1]

Abstract: *This paper investigates the Lipschitz continuity of the solution map in the settings of horizontal, vertical, and mixed linear complementarity problems. In each of these cases, we show that the solution map is (globally) Lipschitzian if and only if the solution map is single-valued. These generalize a similar result of Murthy, Parthasarathy, and Sabatini proved in the LCP setting.*

1 Introduction

This paper is a continuation of our recent efforts to understand the Lipschitzian behavior of the solution map arising from piecewise affine equations. For the linear complementarity problem (LCP), see Section 4, corresponding to a matrix $M \in I\!R^{n\times n}$, Murthy, Parthasarathy, and Sabatini [8] have shown that the solution mapping

$$q \mapsto \mathcal{S}(q) := \{x : x \geq 0,\ Mx + q \geq 0, \quad \text{and} \quad x^T(Mx+q) = 0\}$$

is (globally) Lipschitzian on $I\!R^n$ if and only if $\mathcal{S}$ is single-valued (equivalently, M is a **P**-matrix). The main aim of this paper is to show that a similar result is valid in the contexts of horizontal, vertical, and mixed linear complementarity problems, see Section 4 for definitions. Unlike [8] (where the analysis, though elementary, is based on LCP ideas), our approach is via piecewise affine functions. In [6] Gowda and Sznajder showed that a piecewise affine function $f : I\!R^n \to I\!R^n$ is surjective and the inverse map f^{-1} is Lipschitzian on $I\!R^n$ if and only if f is open (or equivalently, coherently oriented); moreover, when the branching number of f is less than or equal to four, these conditions are equivalent to f being a homeomorphism. While this result can be immediately applied to the LCP (via the mapping $f(x) := Mx^+ - x^-$) and more generally to the affine variational inequality problem (AVI) (via the normal map) [6], it cannot be applied directly to the horizontal, vertical, and mixed LCPs. However, as we see below, simple transformations will allow us to rewrite these problems as piecewise affine equations where the above result could be applied.

2 Preliminaries

Throughout this paper, $\boldsymbol{B}$ denotes the closed unit ball in the space under consideration. We define $x \wedge y$, $x \vee y$, and $\langle x, y\rangle (= x^T y)$ as, respectively, the componentwise minimum, componentwise maximum, and the usual inner product of vectors x and y. Also, $x^+ := x \vee 0$ and $x^- := (-x) \vee 0$.

For a comprehensive treatment of piecewise affine functions, see [2] or [13]. Formally, a continuous function $f : I\!R^n \to I\!R^m$ is called *piecewise affine* if there exists a set of triples

[1]Research supported by the National Science Foundation Grant CCR-9307685

(Ω_j, A_j, a_j) $(j = 1, 2, \ldots, K)$ such that each Ω_j is a polyhedral set in $\mathbb{R}^n$ with nonempty interior, $A_j \in \mathbb{R}^{m \times n}$, $a_j \in \mathbb{R}^m$, and

(a) $\mathbb{R}^n = \cup_{i=1}^K \Omega_i$;

(b) For $i \neq j$, $\Omega_i \cap \Omega_j$ is either empty or a proper common face of Ω_i and Ω_j. In particular, $\text{int } \Omega_i \cap \text{int } \Omega_j = \emptyset$ for $i \neq j$;

(c) $f(x) = A_i x + a_i$ for $x \in \Omega_i$, $i = 1, 2, \ldots, K$.

We shall refer to A_i $(i = 1, 2, \ldots, K)$ as the matrices of f (or matrices defining f). The collection $\{\Omega_i, i = 1, 2, \ldots, K\}$ is said to be a *polyhedral subdivision of* $\mathbb{R}^n$ corresponding to f.
The *branching number* of this polyhedral subdivision (or simply that of f) is the maximal number of Ωs that have a common face of dimension $(n-2)$.

When $m = n$, we say that f is *coherently oriented* if all the (square) matrices corresponding to f have the same nonzero determinantal sign.

Piecewise affine functions can also be described equivalently [13] as follows. A continuous function $f : \mathbb{R}^n \to \mathbb{R}^m$ is piecewise affine if there exist affine functions $f_1, f_2, \ldots, f_J$ from $\mathbb{R}^n$ to $\mathbb{R}^m$ such that

$$f(x) \in \{f_1(x), f_2(x), \ldots, f_J(x)\} \quad \text{for all} \quad x \in \mathbb{R}^n.$$

This formulation is particularly useful in studying examples.

We shall say that a multivalued function $G : \mathbb{R}^m \to \mathbb{R}^n$ with the domain $\text{dom}\, G$ is *Lipschitzian* if there exists a positive number γ such that

$$G(y) \subseteq G(z) + \gamma \|y - z\| \mathbb{B} \quad \text{for all} \quad y, z \in \text{dom}\, G,$$

The above condition implies that G is *lower semicontinuous* on $\text{dom}\, G$ where we define lower semicontinuity of G on a set $Y \subseteq \text{dom}\, G$ as follows: for each sequence $\{y^k\}$ in Y converging to $\bar{y} \in Y$, and for any $\bar{x} \in G(\bar{y})$, there exists a sequence $\{x^k\}$ in $\text{ran}\, G$ such that $x^k \in G(y^k)$ for each k and $\{x^k\}$ converges to $\bar{x}$. When G is polyhedral (that is, the graph of G is a finite union of polyhedral sets) whose domain is convex (or more generally, Lipschitz path-connected), lower semicontinuity turns out to be equivalent to the Lipschitzian property [7]. With specific applications in mind, we shall restrict our attention to the case when G is the inverse of a piecewise affine function. The following results from [7] are crucial for our analysis.

Theorem 1 *Suppose that $f : \mathbb{R}^n \to \mathbb{R}^n$ is piecewise affine and the range of f has nonempty interior. If f^{-1} is lower semicontinuous on the range of f, then the matrices corresponding to f are nonsingular.*

Theorem 2 *Assume $f : \mathbb{R}^n \to \mathbb{R}^n$ is a piecewise affine function. Then the following conditions are equivalent:*

(a) f is surjective and f^{-1} is lower semicontinuous on $\mathbb{R}^n$.

(b) f is surjective and f^{-1} is Lipschitzian on $\mathbb{R}^n$.

(c) f is coherently oriented.

Moreover, when the branching number of f is less than or equal to four, these conditions are equivalent to

(d) f is a homeomorphism.

We should note here that a piecewise affine function from $\mathbb{R}^n$ into itself is a homeomorphism if and only if it is injective, and coherently oriented if and only if it is an open map, see Thm. 2.3.1 and Prop. 2.3.1 in [13]. Also, the equivalence of (c) and (d) holds under conditions (involving the so called *k-th branching number*) weaker than what is stated here, see [13] Thm. 2.3.7.

3 The main result

We see from the equivalence of (a) and (d) in Theorem 2 that lower semicontinuity of f^{-1} on all of $\mathbb{R}^n$ guarantees the unique solvability of the equation $f(x) = q$ for all $q \in \mathbb{R}^n$. We may ask whether such a result is valid if we replace $\mathbb{R}^n$ by a subspace of $\mathbb{R}^n$. To be precise, let $f : \mathbb{R}^n \to \mathbb{R}^n$ be piecewise affine, Y be a subspace of $\mathbb{R}^n$, $f^{-1}(q) \neq \emptyset$ for all $q \in Y$, and f^{-1} is lower semicontinuous on Y. Does it follow that $f(x) = q$ has a unique solution for all $q \in Y$? Even under the branching number condition, this question does not seem to have a simple and clearcut answer. The Lipschitzian behavior of the solution map arising in horizontal, vertical, and mixed linear complementarity problems is related to this question. Fortunately, the extra structure available in the formulations of these problems allows us to apply Theorem 2 in an appropriate way.

We now present our main result. Applications of this to various complementarity problems will be discussed in the next section.

Let $\psi : \mathbb{R}^n \times \mathbb{R}^m \to \mathbb{R}^k$ be a function with the following properties:

(a) ψ is piecewise affine, onto,

(b) branching number of ψ is less than or equal to four, and

(c) there exist matrices $P \in \mathbb{R}^{n\times k}$ and $Q \in \mathbb{R}^{m\times k}$ such that

$$\psi(x, y) = r \iff \psi(x - Pr, y - Qr) = 0 \quad \text{for every } r.$$

A simple example of such a function is $\psi(x, y) = x \wedge y$.

Now consider the piecewise affine function $H : \mathbb{R}^n \times \mathbb{R}^m \to \mathbb{R}^l \times \mathbb{R}^k$ defined by

$$H(x, y) = \begin{pmatrix} Mx + Ny \\ \psi(x, y) \end{pmatrix} \tag{1}$$

where $M \in \mathbb{R}^{l\times n}$, $N \in \mathbb{R}^{l\times m}$. It is clear that H is piecewise affine and the branching number of H is less than or equal to four.

For a given $q \in \mathbb{R}^l$, we consider the equation

$$H(x, y) = \begin{pmatrix} q \\ 0 \end{pmatrix}$$

and let $\mathcal{S}(q)$ denote the solution set of this equation. We have the following result characterizing the Lipschitzian behavior of $\mathcal{S}$.

Theorem 3 *Consider the above H with $n+m=k+l$. Then the following are equivalent:*

(i) $\mathcal{S}(q) \neq \emptyset$ for all $q \in \mathbb{R}^l$ and the map $q \mapsto \mathcal{S}(q)$ is Lipschitzian on $\mathbb{R}^l$.

(ii) $\mathcal{S}(q) \neq \emptyset$ for all $q \in \mathbb{R}^l$ and the map $q \mapsto \mathcal{S}(q)$ is lower semicontinuous on $\mathbb{R}^l$.

(iii) $|\mathcal{S}(q)| = 1$ for all $q \in \mathbb{R}^l$.

(iv) H is coherently oriented.

Proof. The implication $(i) \Longrightarrow (ii)$ is obvious. Assume (ii). For any $q \in \mathbb{R}^l$ and $r \in \mathbb{R}^k$, it follows easily from property (c) of ψ that

$$H^{-1}\begin{pmatrix} q \\ r \end{pmatrix} = (Pr, Qr) + \mathcal{S}(q - MPr - NQr). \tag{2}$$

¿From this equality we easily verify that the piecewise affine function H is onto and H^{-1} is lower semicontinuous. Since the branching number of H is less than or equal to four, from Theorem 2, we see that H is a homeomorphism. By putting $r = 0$ in the above equality, we see that $|\mathcal{S}(q)| = 1$ for all $q \in \mathbb{R}^l$. This is (iii). Now suppose (iii) holds. Then $|\mathcal{S}(q - MPr - NQr)| = 1$ for all q and r. By the equality (2), H is one-to-one, i.e., it is a homeomorphism. By Theorem 2, H is coherently oriented, thus proving (iv). Finally when (iv) holds, by Theorem 2, H is surjective and H^{-1} is Lipschitzian on $\mathbb{R}^l \times \mathbb{R}^k$. Restricting H^{-1} to $\mathbb{R}^l \times \{0\}$, we see that $\mathcal{S}$ is Lipschitzian on $\mathbb{R}^l$. Thus we have (i). ∎

Theorem 4 *Let $n+m=k+l$. Suppose that $\mathcal{S}(q) \neq \emptyset$ for all q in some open subset $\mathcal{E}$ of $\mathbb{R}^l$. If the mapping $\mathcal{S} : q \mapsto \mathcal{S}(q)$ is lower semicontinuous on the domain of $\mathcal{S}$, then the matrices that define H are all nonsingular.*

Proof. Under the given assumption on $\mathcal{S}$, it follows from (2) that $H^{-1}(p)$ will be nonempty for all p in some open set, moreover H^{-1} is lower semicontinuous on $\operatorname{ran} H$. Now the conclusion follows from Theorem 1. ∎

4 Applications

In this section we specialize the previous two results to horizontal, vertical, and mixed linear complementarity problems.

To begin with, recall that the linear complementarity problem LCP(M,q) [1] is to find a vector x such that

$$x \geq 0, \quad Mx + q \geq 0, \quad \text{and} \quad x^T(Mx+q) = 0 \tag{3}$$

where $M \in \mathbb{R}^{n\times n}$ and $q \in \mathbb{R}^n$. This problem is equivalent to solving the piecewise equation $x \wedge (Mx+q) = 0$ or the piecewise equation $Mx^+ - x^- = -q$.

4.1 The horizontal linear complementarity problem

Given a pair of matrices $A, B \in \mathbb{R}^{m\times n}$ and a vector $q \in \mathbb{R}^m$, the horizontal linear complementarity problem, HLCP(A,B,q) [14], [15] is to find vectors x and y in $\mathbb{R}^n$ such that

$$Ax - By = q$$
$$x \wedge y = 0.$$

This problem can be formulated as a piecewise linear equation $H(x,y) = \begin{pmatrix} q \\ 0 \end{pmatrix}$ where

$$H(x,y) = \begin{bmatrix} Ax - By \\ x \wedge y \end{bmatrix}. \tag{4}$$

As before, $\mathcal{S}(q)$ denotes the solution set of $H(x,y) = \begin{pmatrix} q \\ 0 \end{pmatrix}$. Note that this H is like the one given in (1) with $\psi(x,y) = x \wedge y$. Clearly this ψ is piecewise affine, onto, and $\psi(x,y) = r$ implies that $\psi(x-r, y-r) = 0$. The polyhedral subdivision corresponding to this ψ is given by $\{\Omega_\alpha : \alpha \subseteq \{1,\dots,n\}\}$ where

$$\Omega_\alpha = \{(x,y) \in I\!R^n \times I\!R^n : x_\alpha \geq y_\alpha, \; x_\beta \leq y_\beta\} \text{ for } \alpha \subseteq \{1,\dots,n\} \text{ and } \beta := \alpha^c.$$

It is easily seen that the branching number of ψ is less than or equal to four. Thus Theorem 3 is applicable.

Theorem 5 *Consider the horizontal LCP corresponding to the matrix pair (A,B). Assume that A and B are square. Then the following are equivalent:*

(a) (A,B) is a Q-pair (that is, for every $q \in I\!R^n$, $\mathcal{S}(q) \neq \emptyset$) and the solution map $q \mapsto \mathcal{S}(q)$ is Lipschitzian.

(b) (A,B) is a Q-pair and the solution map $q \mapsto \mathcal{S}(q)$ is lower semicontinuous.

(c) $|\mathcal{S}(q)| = 1 \;\; \forall q \in I\!R^n$.

(d) All the column representative matrices of (A,B) have the same nonzero determinantal sign.

We recall that an $n \times n$ matrix C is a *column representative* of (A,B) if for each j, the jth column of C is either the jth column of A or the jth column of B.

Proof. The equivalence of (a), (b), and (c) follow immediately from Theorem 3. We complete the proof by showing that (d) is nothing but the coherence property of H: on the polyhedral set Ω_α described above,

$$H(x,y) = \begin{bmatrix} Ax - By \\ \begin{pmatrix} y_\alpha \\ x_\beta \end{pmatrix} \end{bmatrix} = \begin{bmatrix} A & -B \\ \begin{bmatrix} 0 & 0 \\ 0 & I_\beta \end{bmatrix} & \begin{bmatrix} I_\alpha & 0 \\ 0 & 0 \end{bmatrix} \end{bmatrix} \begin{pmatrix} x \\ y \end{pmatrix}.$$

By the Schur determinantal formula [1] (p. 76), [10], the determinant of the matrix defining H on Ω_α is $\det[A_{\cdot\alpha} \; B_{\cdot\beta}]$ which is precisely the determinant of the column representative of (A,B) corresponding to the index set α. Thus the coherence property of H is condition (d) of the theorem. ■

Some comments regarding the above theorem are in order. The above result can also be derived using Theorem 19 in [15] by reducing the HLCP problem to the classical linear complementarity problem and then applying the theorem of Murthy, Parthasarathy, and Sabatini [8] mentioned in the Introduction. At the same time, it is possible to deduce this result of Murthy, Parthasarathy, and Sabatini from Theorem 5. We shall omit the details.

At this stage, one may ask whether Theorem 5 is valid for non square matrices. It is known that uniqueness can be achieved in the HLCP only when A and B are square [3]. How about the Lipschitzian property of the solution map? The following proposition and example pertain to this question.

Proposition 1 *Assume that* (A, B) *is a* Q*-pair where* $A, B \in R^{m\times n}$ *and the solution map* $q \mapsto \mathcal{S}(q)$ *is Lipschitzian. Then* $m \leq n$.

Proof. Suppose, if possible, that $m > n$. Then HLCP (A, B, q) can be written as

$$\begin{aligned} x \wedge y &= 0 \\ A_1 x - B_1 y &= q_1 \\ A_2 x - B_2 y &= q_2 \end{aligned}$$

where $A_1, B_1 \in \mathbb{R}^{n\times n}$, $A_2, B_2 \in \mathbb{R}^{(m-n)\times n}$, $q_1 \in \mathbb{R}^n$, and $q_2 \in \mathbb{R}^{m-n}$. Obviously, (A_1, B_1) is a Q-pair. Let $(x^*, y^*) \in \mathcal{S}(A_1, B_1, q_1)$. Since the solution map for the pair (A, B) is Lipschitzian, we have

$$(x^*, y^*) \in \mathcal{S}\left(A, B, \begin{pmatrix} q_1 \\ A_2x^* - B_2y^* \end{pmatrix}\right) \subseteq$$

$$\mathcal{S}\left(A, B, \begin{pmatrix} \bar{q}_1 \\ \bar{q}_2 \end{pmatrix}\right) + \gamma \left\| \begin{pmatrix} q_1 \\ A_2x^* - B_2y^* \end{pmatrix} - \begin{pmatrix} \bar{q}_1 \\ \bar{q}_2 \end{pmatrix} \right\| \mathbf{B}.$$

Now let $\bar{q}_1$ be arbitrary and $\bar{q}_2 = A_2x^* - B_2y^*$. Then $(x^*, y^*) \in \mathcal{S}(A_1, B_1, \bar{q}_1) + \gamma\|q_1 - \bar{q}_1\|\mathbf{B}$. It follows that the solution map $\bar{q}_1 \mapsto \mathcal{S}(A_1, B_1, \bar{q}_1)$ is Lipschitzian and hence Theorem 5 shows that the problem HLCP $(A_1, B_1, \bar{q}_1)$ has a unique solution. For a given $\bar{q}_1 \in \mathbb{R}^n$, take $q_2 \neq A_2x^* - B_2y^*$ with $(x^*, y^*) \in \mathcal{S}(A_1, B_1, \bar{q}_1)$. Then, HLCP $\left(A, B, \begin{pmatrix} \bar{q}_1 \\ q_2 \end{pmatrix}\right)$ has no solution, contradicting the assumption that (A, B) is a Q-pair. Hence $m \leq n$. ■

In the following example, m is less than n, the matrices A and B form a Q-pair, and the solution map is Lipschitzian, yet the corresponding HLCP has more than one solution.

Example. Let

$$\begin{aligned} [I\ 0]x - [I\ 0]y &= q \\ x \wedge y &= 0 \end{aligned}$$

where I denotes the $m \times m$ identity matrix, x and y are in $\mathbb{R}^n$. An easy inspection shows that $\forall q \in \mathbb{R}^m$,

$$\mathcal{S}(q) = ((q^+, 0), (q^-, 0)) + L$$

where $L := \{((0, u), (0, v)) : u \wedge v = 0\}$. Evidently, $([I\ 0], [I\ 0])$ is a Q-pair, and the corresponding solution map is Lipschitzian, yet $|\mathcal{S}(q)| > 1$.

Here is an application of Theorem 4.

Theorem 6 *Let* $n = m$. *Suppose that* HLCP (A, B, q) *has nonempty solution set for every* $q \in \mathcal{E} \subseteq \mathbb{R}^n$ *with* int $\mathcal{E} \neq \emptyset$. *Also, assume that the solution map* $q \mapsto \mathcal{S}(q)$ *is lower semicontinuous on the domain of* $\mathcal{S}$. *Then all column representative matrices of* (A, B) *are nonsingular.*

Proof. We saw in the proof of Theorem 5 that the determinants of column representative matrices of (A, B) are nothing but the determinants of the matrices defining H (given by (4)). The equality (2) shows that H^{-1} is lower semicontinuous on the range of H. The same equality shows that if $\bar{q} \in \text{int } \mathcal{E}$, then $\begin{pmatrix} \bar{q} \\ 0 \end{pmatrix}$ belongs to the interior of the range of H. To complete the proof, we need only quote Theorem 1. ■

4.2 The vertical linear complementarity problem

Given

$$\mathbf{M} = (M_1, M_2, \ldots, M_k) \quad \text{and} \quad \mathbf{q} = (q_1, q_2, \ldots, q_k),$$

where each M_j is an $n \times n$ matrix and q_j is an n-vector, the VLCP$(\mathbf{M}, \mathbf{q})$ [5], [14], [15] is to solve the piecewise affine equation

$$(M_1 x + q_1) \wedge (M_2 x + q_2) \wedge \cdots \wedge (M_k x + q_k) = 0. \tag{5}$$

We shall write $\mathbf{\Phi}(\mathbf{q})$ for the solution set of this equation.

By introducing the variables $y^j = M_j x + q_j$, we can write the above equation as

$$F(x, y^1, \ldots, y^k) = \begin{pmatrix} q_1 \\ q_2 \\ \vdots \\ q_k \\ 0 \end{pmatrix} \tag{6}$$

where

$$F(x, y^1, y^2, \ldots, y^k) := \begin{bmatrix} y^1 - M_1 x \\ y^2 - M_2 x \\ \vdots \\ y^k - M_k x \\ y^1 \wedge y^2 \ldots \wedge y^k \end{bmatrix}$$

with y^j denoting the jth vector (and not the jth coordinate).

Let $\mathcal{S}(\mathbf{q})$ denote the solution set of (6). Note that the mappings $\mathbf{\Phi}$, $\mathcal{S}$, and F^{-1} have similar lower semicontinuity (Lipschitzian) behavior. This can be easily seen by the equalities

$$F^{-1}\begin{pmatrix} q_1 \\ q_2 \\ \vdots \\ q_k \\ r \end{pmatrix} = (0, r, r, \ldots, r) + F^{-1}\begin{pmatrix} q_1 - r \\ q_2 - r \\ \vdots \\ q_k - r \\ 0 \end{pmatrix} \tag{7}$$

and

$$F^{-1}\begin{pmatrix} q_1 \\ q_2 \\ \vdots \\ q_k \\ 0 \end{pmatrix} = \{(x, M_1 x + q_1, M_2 x + q_2, \ldots, M_k x + q_k) : x \in \mathbf{\Phi}(\mathbf{q})\}. \tag{8}$$

For $\boldsymbol{l} = (l_1, \ldots, l_i, \ldots, l_n)$ with $i \in \{1, \ldots, n\}$ and $l_i \in \{1, \ldots, k\}$, we put

$$\Omega_{\boldsymbol{l}} = \bigcap_{i=1}^{n} \bigcap_{j \neq l_i} \{ (x, y^1, \ldots, y^k) \in I\!R^n \times I\!R^n \times \cdots \times I\!R^n : \ (y^j)_i \geq (y^{l_i})_i \}. \tag{9}$$

Certainly, $\{\Omega_{\boldsymbol{l}}\}$ forms a polyhedral subdivision associated with the piecewise linear mapping F. With $y = (y^1, \ldots, y^k)$, and $\psi(x, y) := y^1 \wedge y^2 \wedge \ldots \wedge y^k$, the above F looks like H defined in (1). Since ψ has branching number less than or equal to four, Theorem 3 is applicable.

Theorem 7 *Consider the vertical LCP corresponding to* $\mathbf{M}$. *Then the following are equivalent:*

(a) $\mathbf{M}$ *is of* type Q *(that is, for every* $\mathbf{q} \in \mathbb{R}^n \times \cdots \times \mathbb{R}^n$, $\mathbf{\Phi}(\mathbf{q}) \neq \emptyset$*), and the solution map* $\mathbf{q} \mapsto \mathbf{\Phi}(\mathbf{q})$ *is Lipschitzian.*

(b) $\mathbf{M}$ *is of* type Q, *and the solution map* $\mathbf{q} \mapsto \mathbf{\Phi}(\mathbf{q})$ *is lower semicontinuous.*

(c) $|\mathbf{\Phi}(\mathbf{q})| = 1$ *for all* $\mathbf{q} \in \mathbb{R}^n \times \cdots \times \mathbb{R}^n$.

(d) All row representative matrices of $\mathbf{M}$ *have the same nonzero determinantal sign.*

The equivalence of the first three items follows (via the mapping F) immediately from Theorem 3. Only item (d) requires an explanation. By definition, an $n \times n$ matrix C is a *row representative* of $\mathbf{M}$ if for each index j, the jth row of C belongs to the set consisting of jth rows of matrices $M_1, M_2, \ldots, M_k$. It can be shown that the determinant of a row representative of $\mathbf{M}$ is the determinant of a matrix that appears in the piecewise affine formulation of F and conversely. Theorem 3 now gives the equivalence of (c) and (d). The equivalence of (c) and (d) also follows from Theorem 17 in [5].

The following result is an analogue of Theorem 6.

Theorem 8 *Suppose that* VLCP$(\mathbf{M}, \mathbf{q})$ *has nonempty solution set for every* $\mathbf{q} \in \mathcal{E}$ *with* int $\mathcal{E} \neq \emptyset$. *If the solution map* $\mathbf{q} \mapsto \mathbf{\Phi}(\mathbf{q})$ *is lower semicontinuous on the domain of* $\mathbf{\Phi}$, *then all row representative matrices of* $\mathbf{M}$ *are nonsingular.*

Proof. In view of equalities (7) and (8), the lower semicontinuity of $\mathbf{\Phi}$ implies the lower semicontinuity of F^{-1} on the range of F. To complete the proof, we need only show that the range of F has nonempty interior. This is easily seen since for $\bar{\mathbf{q}} \in \mathcal{E}$, the element $\begin{pmatrix} \bar{\mathbf{q}} \\ 0 \end{pmatrix}$ belongs to the interior of the range of F. ■

4.3 The mixed linear complementarity problem

Given matrices $A \in \mathbb{R}^{n \times n}$, $B \in \mathbb{R}^{n \times m}$, $C \in \mathbb{R}^{m \times n}$, and $D \in \mathbb{R}^{m \times m}$, and vectors $a \in \mathbb{R}^n$ and $b \in \mathbb{R}^m$, the mixed linear complementarity problem [4] is to find vectors $x \in \mathbb{R}^n$ and $y \in \mathbb{R}^m$ such that

$$\begin{aligned} Ax + By + a &= 0\,, \\ u &= Cx + Dy + b\,, \\ u \wedge y &= 0\,. \end{aligned}$$

Let $\mathcal{S}(a,b)$ denote the solution set of the above MLCP.

Theorem 9 *The following are equivalent.*

(1) For all $(a,b) \in \mathbb{R}^n \times \mathbb{R}^m$, $|\mathcal{S}(a,b)| \neq \emptyset$, *and the solution map* $(a,b) \mapsto \mathcal{S}(a,b)$ *is Lipschitzian.*

(2) For all $(a,b) \in \mathbb{R}^n \times \mathbb{R}^m$, $|\mathcal{S}(a,b)| \neq \emptyset$, *and the solution map* $(a,b) \mapsto \mathcal{S}(a,b)$ *is lower semicontinuous.*

(3) For all $(a,b) \in \mathbb{R}^n \times \mathbb{R}^m$, $|\mathcal{S}(a,b)| = 1$.

(4) A *is invertible and* $D - CA^{-1}B$ *is a* $\mathbf{P}$*-matrix.*

Proof. Define the following piecewise affine function

$$F(x,y,u) := \begin{bmatrix} Ax+By \\ u-(Cx+Dy) \\ u\wedge y \end{bmatrix}. \tag{10}$$

Observe that $F : \mathbb{R}^n \times \mathbb{R}^m \times \mathbb{R}^m \mapsto \mathbb{R}^n \times \mathbb{R}^m \times \mathbb{R}^m$ is like H described in (1): for $\alpha \subseteq \{1,\dots,m\}$ and $\beta := \alpha^c$, define

$$\Omega_\alpha := \{(x,y,u) \in \mathbb{R}^n \times \mathbb{R}^m \times \mathbb{R}^m : y_\alpha \geq u_\alpha,\ y_\beta \leq u_\beta\}.$$

The family $\{\Omega_\alpha\}$ forms a polyhedral subdivision of $\mathbb{R}^n \times \mathbb{R}^m \times \mathbb{R}^m$. For any $(x,y,u) \in \Omega_\alpha$ we have

$$F(x,y,u) = \begin{bmatrix} Ax+By \\ u-(Cx+Dy) \\ \begin{pmatrix} u_\alpha \\ y_\beta \end{pmatrix} \end{bmatrix} = \begin{bmatrix} A & B & 0 \\ -C & -D & I \\ 0 & E_1 & E_2 \end{bmatrix} \begin{pmatrix} x \\ y \\ u \end{pmatrix} \tag{11}$$

where

$$E_1 = \begin{bmatrix} 0 & 0 \\ 0 & I_\beta \end{bmatrix} \text{ and } E_2 = \begin{bmatrix} I_\alpha & 0 \\ 0 & 0 \end{bmatrix}.$$

Also,

$$(x^*,y^*) \in \mathcal{S}(a,b) \text{ if and only if } F(x^*,y^*,u^*) = \begin{pmatrix} -a \\ b \\ 0 \end{pmatrix}$$

where $u^* = Cx^* + Dy^* + b$. The equivalence of (*1*), (*2*) and (*3*) follows from Theorem 3. The equivalence of (*3*) and (*4*) is given in Proposition 2 of [11]. ■

We point out that under the Lipschitzian assumption, Pang [12] proved that matrix A is nonsingular, in which case the MLCP problem can be transformed to the standard LCP, and then we can apply the result of Murthy, Parthasarathy, and Sabatini [8]. Again, our approach is consistent with Theorem 2.

We now state an analogue of Theorem 4.

Theorem 10 *Suppose that* MLCP(A,B,C,D,a,b) *has nonempty solution set for every* $(a,b) \in \mathcal{E} \subseteq \mathbb{R}^n \times \mathbb{R}^m$ *with* int $\mathcal{E} \neq \emptyset$. *Assume also that the solution map* $(a,b) \mapsto \mathcal{S}(a,b)$ *is lower semicontinuous on the domain of* $\mathcal{S}$. *Then* A *is invertible and* $D - CA^{-1}B$ *is nondegenerate (that is, every principal minor is nonzero).*

Let $(a,b) \in \text{int}\,\mathcal{E}$. It is easily seen that $(-a,b,0)^T \in \text{int ran}\,F$. Also, F^{-1} is lower semicontinuous on the range of F. By Theorem 1, matrices of F are nonsingular. The lemma below shows how algebraic manipulations involving the Schur determinantal formula lead to the desired conclusion.

Lemma 1 *Let the matrices* A,B,C,D *be as above. Then* A *is invertible and for any index set* $\alpha \subseteq \{1,\dots,n\}$,

$$\det \begin{bmatrix} A & B & 0 \\ -C & -D & I \\ 0 & E_1 & E_2 \end{bmatrix} = (-1)^m \det A \cdot \det S_{\alpha\alpha}$$

where $S := D - CA^{-1}B$.

Proof of Lemma 1. Let $\varepsilon > 0$ be small, so that $A_\varepsilon := A + \varepsilon I$ is invertible. Then

$$\det \begin{bmatrix} A_\varepsilon & B & 0 \\ -C & -D & I \\ 0 & I & 0 \end{bmatrix} = \det A_\varepsilon \det \begin{bmatrix} -S_\varepsilon & I \\ I & 0 \end{bmatrix} = (-1)^m \det A_\varepsilon$$

by the Schur determinantal formula, where $S_\varepsilon := D - CA_\varepsilon^{-1}B$. Letting $\varepsilon \to 0$, we see that $\det N = (-1)^m \det A$, where

$$N = \begin{bmatrix} A & B & 0 \\ -C & -D & I \\ 0 & I & 0 \end{bmatrix}.$$

Since N is a matrix that appears (for $\alpha = \emptyset$) in the definition of F, we see that $\det A \neq 0$. Now, assume that $\alpha \subseteq \{1, \ldots, n\}$ is arbitrary. Then

$$\det \begin{bmatrix} A & B & 0 \\ -C & -D & I \\ 0 & E_1 & E_2 \end{bmatrix} = \det A \det \begin{bmatrix} -S & I \\ E_1 & E_2 \end{bmatrix} =$$
$$\det A \det (-SE_2 - E_1) = (-1)^m \det A \det (E_1 + SE_2)$$

where the first equality comes from the Schur determinantal formula, the second equality holds because the matrices E_1 and E_2 commute. Also,

$$\det (E_1 + SE_2) = \det \begin{bmatrix} S_{\alpha\alpha} & 0 \\ S_{\beta\alpha} & I \end{bmatrix} = \det S_{\alpha\alpha}.$$

■

5 Concluding Remarks

In this paper we dealt with the global Lipschitzian behavior of the solution map in each of the settings of the horizontal, vertical, and the mixed linear complementarity problem. It is possible to describe the (local) pseudo-Lipschitzian behavior of the solution map at a given solution point for these problems following the results in [7].

References

[1] R.W. Cottle, J.-S. Pang and R.E. stone(1992) *The linear complementarity problem*, Academic Press, Boston .

[2] B.C. Eaves and U.G. Rothblum(1990) *Relationships of properties of piecewise affine maps over ordered fields*, Linear Algebra and Its Applications, 132 pp. 1-63.

[3] M.S. Gowda (1996) *On the extended linear complementarity problem*, Mathematical Programming 72 pp. 33-50.

[4] M.S. Gowda and J.-S. Pang(1994) *Stability analysis of variational inequalities and nonlinear complementarity problems, via the mixed linear complementarity problem and degree theory*, Mathematics of Operations Research, 19 pp. 831-879.

[5] M.S. Gowda and R. Sznajder (1994) *The generalized order linear complementarity problem*, SIAM J. Matrix Analysis and Applications, 15 pp. 779-795.

[6] M.S. Gowda and R. Sznajder(1996) *On the Lipschitzian properties of polyhedral multifunctions,* Mathematical Programming, 74 pp. 276-278.

[7] M.S. Gowda and R. Sznajder(1997)*On the pseudo-Lipschitzian behavior of the inverse of a piecewise affine function,* to appear in the Proceedings of the International Conference on Complementarity Problems and Their Applications; M.C. Ferris and J.-S. Pang, eds. SIAM Publications, 1997.

[8] G.S.R. Murthy, T. Parthasarathy, and M. Sabatini(1996) *Lipschitzian* **Q**-*matriceso are* **P**-*matrices*, Mathematical Programming, 74 .

[9] G.S.R. Murthy, T. Parthasarathy, and B. Sriparna, *Constructive characterization of Lipschitzian Q_0-matrices*, Linear Algebra and Its Applications, *forthcoming.*

[10] D.V. Ouellette, *Schur complements and Statistics* (1981) Linear Algebra and Its Applications, 36 pp. 187-295.

[11] J.-S. Pang(1990) *Newton's method for B-differentiable equations*, Mathematics of Operations Research, 15 pp. 311-341.

[12] J.-S. Pang (1993) *A degree-theoretic approach to parametric nonsmooth equations with multivalued perturbed solution sets*, Mathematical Programming, 62 pp. 359-383.

[13] S. Scholtes (1994) *Introduction to piecewise differentiable equations*, Preprint 53/1994, Institute für Statistik und Mathematische Wirtschaftstheorie, Universität Karlsruhe, 7500 Karlsruhe, Germany, May.

[14] R. Sznajder (1994) *Degree-theoretic analysis of the vertical and horizontal linear complementarity problem*, Ph.D. thesis, University of Maryland Baltimore County, Baltimore, Maryland, May.

[15] R. Sznajder and M.S. Gowda(1995) *Generalizations of* $\mathbf{P}_0$ *and* **P**-*properties; extended vertical and horizontal LCPs*, Linear Algebra and Its Applications, 223/224 pp. 695-715.

Roman Sznajder
Department of Natural Sciences
and Mathematics
Bowie State University

Bowie, Maryland 20715

M. Seetharama Gowda
Department of Mathematics
and Statistics
University of Maryland baltimore County

Baltimore, Maryland 21228

PARI MUTUEL AS A SYSTEM OF AGGREGATION OF INFORMATION

Guillermo Owen

Abstract: *A bettor is faced with the problem of choosing optimal bets, given his (subjective) probabilities of outcomes of an experiment (horse race), and the payoff odds on these outcomes. Conversely, a bookie (pari-mutuel system), faced by several bettors with different subjective probabilities, has the problem of choosing payoff odds so as to avoid the risk of loss. It is shown that, under certain reasonably broad conditions on the several bettors' utility functions and subjective probabilities, such an equilibrium set of payoff odds always exists. Some examples are worked out in detail.*

1. The Individual Bettor's Problem

Let us consider the problem faced by a bettor at a race track. There are n horses, with (subjective) probabilities $p_1, p_2, \ldots, p_n$ of winning. The race track (bookie) has posted odds $q_1, q_2, \ldots, q_n$ on the several horses. Bettor has A dollars to bet on these horses, and a bet of x dollars on horse j will return a payoff of x/q_j dollars if that horse wins (this includes the bettor's original x dollar bet). It is assumed that the p_j and q_j satisfy the standard conditions

$$\sum_{j=1}^{n} p_j = 1; \quad p_j \geq 0 \tag{1}$$

$$\sum_{j=1}^{n} q_j = 1; \quad q_j \geq 0. \tag{2}$$

Condition (1) means that the bettor's subjective probabilities are consistent; condition (2), that the payoff odds are fair. (Though in fact they seldom are, and most bookies will normally post odds such that $\sum q_j > 1$, i.e. they pay less than a fair system would.)

The bettor has a utility function u for money; we assume that u is monotone non-decreasing and continuous. The bettor's problem, then, is to choose his bets, $x_1, x_2, \ldots, x_n$, so as to maximize his expected utility, given by

$$F(x_1, x_2, \ldots, x_n) = \sum_{j=1}^{n} p_j \, u(A - B + x_j/q_j) \tag{3}$$

where

$$B = \sum_{j=1}^{n} x_j \tag{4}$$

subject to

$$B \leq A \tag{5}$$

$$x_j \geq 0. \tag{6}$$

The first thing to notice here is that, assuming the fairness condition (2) holds, the bettor might as well set $B = A$, i.e. bet all his available funds. In fact, suppose we had $B < A$. We could then set $\epsilon = A - B$, and

$$x'_j = x_j + \epsilon q_j , \qquad j = 1, \ldots , n$$

This would give us a new vector of bets, x'_j, with $B' = A$, and it is not difficult to see that

$$F(x'_1, x'_2, \ldots, x'_n) = F(x_1, x_2, \ldots, x_n).$$

Thus the bettor can do at least as well with bets giving $B = A$ as with $B < A$.

This simplifies the problem, then, to one of maximizing

$$F(x_1, x_2, \ldots, x_n) = \sum_{j=1}^{n} p_j \, u(x_j/q_j) \tag{7}$$

subject to

$$\sum_j x_j = A \tag{8}$$

$$x_j \geq 0. \tag{9}$$

Assuming the differentiability of u, the first-order conditions for optimality will be

$$p_j \, u' (x_j/q_j) = \lambda q_j \qquad \text{if} \quad x_j > 0 \tag{10a}$$

$$p_j \, u' (x_j/q_j) \leq \lambda q_j \qquad \text{if} \quad x_j = 0 \tag{10b}$$

where λ is a Lagrange multiplier representing the marginal utility of money.

[In case u is not differentiable at the point x_j/q_j, conditions (10) must be modified, in terms of the right-hand and left-hand derivatives of u, to give

$$p_j \, u' (x_j/q_j +) \leq \lambda q_j \leq p_j \, u' (x_j/q_j -).] \tag{11}$$

Now, in the simplest case, all x_j are positive, so that (10a) holds for all j. Adding with respect to j, we have, by (2),

$$\lambda = \sum p_j \, u' (x_j/q_j) \tag{12}$$

so that λ is simply the expected value of u'.

More generally, of course, (10a) does not hold for all j, and so we can only state that λ is at least equal to the expected value of u'.

In case u is concave, the first-order conditions are sufficient for optimality. We rewrite these as

$$u'(x_j/q_j) = \lambda q_j/p_j \qquad \text{if} \quad x_j > 0 \tag{14a}$$

$$u'(0) \leq \lambda q_j/p_j \qquad \text{if} \quad x_j = 0 \tag{14b}$$

Using the fact that u' is monotone non-increasing (for concave u), we obtain the interesting result

$$x_j/q_j \geq x_k/q_k \quad \text{whenever} \quad p_j/q_j > p_k/q_k. \tag{15}$$

with the stronger result that, for strictly concave u, (15) holds even if the second inequality is loose.

Thus, a discrepancy between the bettor's subjective probability and the payoff odds leads the bettor to bet so that his conditional winnings will be greater for horses for which the ratio p_j/q_j is greater, and conversely.

Conditions (13) and (14) are meaningful if both p_j and q_j are positive. In case $p_j = 0$, $q_j > 0$, it is easily seen that optimality requires $x_j = 0$, i.e. never bet on a horse which has no chance of winning. It is not clear what happens if $q_j = 0$, though in practice it is difficult to imagine a situation in which infinite odds are offered. In case $p_j = q_j = 0$, we imagine the bettor will still set $x_j = 0$; in case $p_j > q_j = 0$, however, we seem to reach some sort of contradiction. We note then that $q_j = 0$ leads to contradictions which would best be avoided; among other things, the payoff functions are discontinuous or fail to exist here.

In case u is strictly concave, we may use the inverse function $w = (u')^{-1}$, and (13)-(14) now take the form

$$x_j = q_j\, w(\lambda q_j/p_j) \quad \text{if} \quad w(\lambda q_j/p_j) > 0 \tag{16}$$

$$x_j = 0 \quad \text{if} \quad w(\lambda q_j/p_j) \leq 0 \tag{17}$$

Condition (8) can be restated as

$$A = \sum{}' q_j\, w(\lambda q_j/p_j) \tag{18}$$

where the prime on the summation symbol means that it should include only those j such that (16) holds, i.e. such that

$$\lambda q_j < p_j\, u'(0) \tag{19}$$

The right side in (18) is seen to be a monotone non-decreasing function of λ and thus (18) can be solved, numerically or analytically, for λ. This solves the single bettor's problem.

2. The Equilibrium Odds

In general, bookies are risk-averse and seek to set payoff odds in such a way as to eliminate the possibility of loss. Of course, a bookie is not bound by the fairness condition (2), so that, in practice, the sum of the q_j is greater than 1. If (2) were to be enforced, however (perhaps under cutthroat competition among bookies), the bookie could only eliminate the risk of loss if the amounts bet on the several horses were proportional to the q_j, i.e. if

$$b_j = q_j C \tag{20}$$

where b_j is the total amount bet on horse j (by all bettors) and C is the total amount of all wagers. If there is only one bettor, it is easy to see that this can be accomplished by setting $q_j = p_j$. For then $x_j = q_j$ A will; satisfy conditions (10) (with $\lambda = u'(A)$). In case u is strictly concave, moreover, this will be the bettor's unique optimum, so that $q_j = p_j$ gives an equilibrium. (Clearly, with one bettor, $b_j = x_j$ and $C = A$.)

In case there are two or more bettors, with different subjective probabilities, the bookie must look for some way of combining the several bettors' beliefs so as to avoid loss. At a race track, this is normally accomplished by a pari mutuel system, which simply sets $q_j = b_j/C$, so that (20) is automatically achieved, after the amounts of the bets are known. In effect, the players bet against each other, with the track as intermediary. This has the disadvantage–from the players' point of view–that bets are made with incomplete knowledge of the payoff odds. Thus, a player may well feel he would have changed his bets, had he known the true payoff odds in advance. Of course, such a change would in turn cause the q_j to change, leading to a further change in bets, et sic ad infinitum or at least until some equilibrium is reached. The question is whether such an equilibrium exists.

Assume, then, m bettors. Bettor i ($i = 1, \ldots, m$) has a subjective probability distribution $(p_{i1}, p_{i2}, \ldots, p_{in})$ satisfying $p_{ij} \geq 0$, and

$$\sum_{j=1}^{n} p_{ij} = 1$$

This same bettor has a sum of money, A_i, available for betting, and a utility function for money, u_i. If the odds are posted as $(q_1, q_2, \ldots, q_n)$, then each bettor will choose his bets $(x_{i1}, x_{i2}, \ldots, x_{in})$ so as to maximize his expected utility, as discussed above. Total bets on horse j are then

$$b_j = \sum_{i=1}^{m} x_{ij} \tag{21}$$

and the total amount bet on all horses is

$$C = \sum_{i=1}^{m} A_i = \sum_{j=1}^{n} b_j \tag{22}$$

There will be an equilibrium if (20) holds for all j.

To simplify the proofs, we will assume that all bettors have capital equal to at least 1 unit. This represents no loss of generality since, in the first place, bettors with zero capital cannot affect the outcome of the process. Since all (active) bettors have capital $A_i > 0$, we can simply change the unit of currency so that the poorest of them all has at least one unit.

As we mentioned above, difficulties arise if $q_j = 0$ for any j. We will therefore try to avoid this, and will specifically rule out such equilibria. We make, then, the following assumption:

Assumption Z. For every j, there is some i such that $p_{ij} > 0$.

We will prove the existence of an equilibrium under the further assumption that the utility functions are strictly concave. Our proof uses a fixed-point theorem. Some care must however be taken to avoid the possibility that the fixed point lies on the boundary of the simplex of bets.

Theorem 1 Suppose Assumption Z holds, and suppose moreover that all the utility functions are strictly concave. Then there is an equilibrium n-tuple of payoff odds, $q_j^* > 0$.

Proof: Let Q be the unit n-simplex, i.e. the set of all vectors $(q_1, \dots q_n)$ satisfying (2). Let Q° be the interior of Q (the set of Q with all components positive), and let ∂Q be the boundary of Q (the set of all q with at least one $q_j = 0$).

For $q \in Q^\circ$, consider bettor i's optimal choice of bets. As discussed above, it cannot be optimal for him to bet on a horse with no (subjective) chance of winning, and so his bets must satisfy, not just (8) and (9), but also the condition $x_{ik} = 0$ whenever $p_{ik} = 0$.

Restricted to that set, bettor i's expected utility,

$$F_i(x_i, q) = \sum_{j=1}^{n} p_{ij}\, u_i(x_{ij}/q_j)$$

is strictly concave, and so has a unique maximizing vector, $x_i^*(q)$, Since F is continuous for all x_i and all $q \in Q^\circ$, it will follow that $x_i^*(q)$ is continuous for all $q \in Q^\circ$.

Let, now,

$$b^*(q) = \sum_{i=1}^{m} x_i^*(q).$$

Then b^* is a continuous mapping from Q° into R^n. Let, finally,

$$J(q) = \{j \mid b_j^*(q)/q_j = \max_k [b_k^*(q)/q_k]\}$$

Clearly, J assigns to each $q \in Q^\circ$ a non-empty subset of $N = \{1, 2, \dots, n\}$. By the continuity of b^*, J is upper semi-continuous (if restricted to Q°).

Next, for $q \in \partial Q$, define

$$J(q) = \{j \mid q_j = 0\}$$

Since $q \in \partial Q$, $J(q)$ is non-empty here also. Trivially, it is upper semi-continuous if restricted to ∂Q.

In this way, the mapping J is defined over the entire simplex Q. We wish to show it is upper semi-continuous, i.e. if $q \to \hat{q}$ and $j \in J(q)$, then $j \in J(\hat{q})$.

In this, we can dispense with the case in which $\hat{q} \in Q^\circ$, since such $\hat{q}$ can only be approached through $q \in Q^\circ$, and we know J, restricted to Q°, is upper semi-continuous. Similarly, we can dispense with the case in which $q \to \hat{q}$ with all q and $\hat{q} \in \partial Q$, since we know that J, restricted to ∂Q, is semi-continuous.

It remains to consider the case in which $q \to \hat{q}$ with $q \in Q^\circ$ and $\hat{q} \in \partial Q$. Let $K = K(\hat{q})$. We must show that, for q sufficiently close to $\hat{q}$, $J(q) \subset K$.

Take some fixed $k \in K$: we have $\hat{q}_k = 0$. By Assumption Z, there is some bettor, h, with $p_{hk} > 0$. Keeping h fixed, let $L(q)$ be the set of all j which maximize p_{hj}/qj. Suppose

$\hat{q}_j > 0$. As $q \to \hat{q}$, the ratio p_{hk}/q_k increases without bound, whereas $p_{\hat{h}j}/q_{\hat{j}}$ approaches the finite limit, $p_{hj}/\hat{q}_j$. Thus for q sufficiently close to $\hat{q}$, $j \in L(q)$, and we conclude that there exists $\epsilon_1 > 0$ such that, if $|q - \hat{q}| < \epsilon_1$, $L(q) \subset K$.

Let r be the minimum of all the positive components $\hat{q}_j$. Let $\epsilon_2 = r/2$. Then, for all q such that $|q - \hat{q}| < \epsilon_2$ and all $j \in K$, we will have $q_j > r/2$.

Let s be the minimum of all p_{hj} such that $\hat{q}_j = 0$ and $p_{hj} > 0$. Since u_h is strictly increasing, we know $u'_h(y+) > 0$ for all y. Set then,

$$\epsilon_3 = \frac{rs\, u'_h(2C/r)}{2\, u'_h(1/2)}$$

Finally, let $\epsilon_4 = r/4nC$.

Let now ϵ be the smallest of ϵ_1, ϵ_2, ϵ_3, and ϵ_4. Assume $q \in Q^\circ$, with $|q - \hat{q}| < \epsilon$. We will show $J(q) \subset K$.

Let $x^*_h(q)$ be bettor h's optimal response to the payoff odds q, and suppose $x^*_{hj} > q_j/2$ for some $j \in K$. Let $1 \in L(q)$. Since $|q - \hat{q}| < \epsilon_1$, $1 \in K$. Moreover, $j \in L(q)$, so

$$\frac{p_{h1}}{q_1} > \frac{p_{hj}}{q_j}$$

and hence, by (15),

$$\frac{x^*_{h1}}{q_1} \geq \frac{x^*_{hj}}{q_j}.$$

Thus x^*_{h1} and x^*_{hj} are both positive, and so we can apply (13) to get

$$u'_h(x^*_{\hat{h}j}/q_1) = \frac{q_1\, p_{hj}}{q_j\, p_{h1}}\, u'_h(x^*_{hj}/q_j)$$

Now, since $|q - \hat{q}| < \epsilon$, we will have $q_j > r/2$, $q_1 < \epsilon_3$, and $p_{h1} \geq s$.

Also, $p_{hj} \leq 1$. Finally, $x^*_{hj}/q_j > 1/2$, so by monotonicity $u'_h(x^*_{hj}/q_j) \leq u'_h(1/2)$. Therefore

$$u'_h(x^*_{h1}/q_1) < \frac{2\epsilon\, u'_h(1/2)}{rs}$$

and, using the definition of ϵ_3,

$$u'_h(x^*_{h1}/q_1) < u'_h(2C/r)$$

By the monotonicity of u', this gives us

$$\frac{x^*_{h1}}{q_1} \geq \frac{2C}{r}$$

Clearly, $b^*_1 \geq x^*_{h1}$, and so $b^*_1/q_1 \geq 2C/r$.

On the other hand, for any $k \in K$, $b^*_k \leq C$ and $q_k > r/2$. Thus

$$b^*_k/q_k < 2C/r \leq b^*_1/q_1$$

and we see that $k \in J(q)$. We conclude that $J(q) \subset K$.

Suppose, on the other the hand, there is no j, $j \epsilon K$, with $x^*_{hj} > q_j/2$. In this case,

$$\sum_{j \epsilon K} x^*_{hj} \leq 1/2 \sum_{j \epsilon K} q_j \leq 1/2$$

and so

$$\sum x^*_{hj} > A_h - 1/2 \geq 1/2.$$

Thus there is some $\ell \in K$ with $x^*_{h\ell} > 1/2n$.

Since $|q - \hat{q}| < \epsilon_4$ and $\hat{q}_\ell = 0$, we have $q_\ell < \epsilon_4$ and so

$$b^*_\ell / q_\ell \geq x^*_{h\ell}/q_\ell > 1/2n\,\epsilon_4 = 2C/r.$$

Once again $b^*_j/q_j < 2C/r$ for all $j \epsilon \epsilon K$, and so $b^*_j/q_j < b^*_\ell/q_\ell$. Thus we conclude once again that $J(q) \subset K$.

We see then that J is an upper semi-continuous mapping, assigning a non-empty subset of N to each $q \epsilon Q$. Define, now, for $S \subset N$,

$$\Phi(S) = \{q \mid q \epsilon Q, q_j = 0 \quad \text{for all} \quad j \epsilon S\}$$

Now, Φ is an upper semi-continuous mapping from the collection of all subsets of N to Q. The composition $\Omega = \Phi \circ J$. is then an upper semi-continuous mapping assigning a non-empty, closed convex subset of Q to each $q \epsilon Q$ By the Kakutani fixed point theorem, such a mapping must have a fixed point, i.e. there is $q^* \epsilon Q$, such that $q^* \epsilon \Omega(q^*)$.

Clearly $q^* \notin \partial Q$ since, for $q \epsilon \partial Q$, $J(q)$ consist of those indices j with $q_j = 0$, and so $\Omega(q)$ will consist of those vectors $z \epsilon Q$ such that $z_j = 0$ whenever $q_j > 0$. Thus $q^* \epsilon Q^\circ$. But, if $q^* \epsilon Q^\circ$, the only $S \subset N$ such that $q \epsilon \Phi(S)$ is N itself, i.e. $J(q^*) = N$. This means that $b^*_j(q^*)/q^*_j$ is equal for all j and this in turn means that

$$b^*_j(q^*) = q^*_j C$$

This means that q^* is the desired equilibrium odds vector.

The hypothesis of Theorem 1 –namely, strict concavity–is overly restrictive. We can weaken it to require only concavity together with strict monotonicity of the utility functions.

Assume then, that the u_i are concave, strictly increasing functions of money. For $t \geq 0$, define

$$w_i(x;\, t) = u_i(x) - te^{-x}.$$

Then, $w_i(x, 0) = u_i(x)$. We have, however,

$$w''_i(x;\, t) = u''_i(x) - te^{-x}$$

where the double-primes denote second derivatives with respect to x. We thus find that w_i is strictly concave and monotone in x for each $t > 0$.

Suppose, then, that each bettor's utility function u_i is replaced by the strictly concave $w_i(x;\, t)$. For each $t > 0$, there is an equilibrium n-tuple $q^*(t)$. As $t \to 0$, these $q^*(t)$ will have an accumulation point q^{**}. The only difficulty is that, although all $q^*(t) \epsilon Q^\circ$, q^{**} could conceivable belong to ∂Q.

Theorem 2 below shows that this will not happen.

<u>Theorem 2.</u> Assume all the function u_i are concave and strictly monotone increasing, and suppose that Assumption Z holds. Then there exists an equilibrium n-tuple $q^{**} \epsilon Q^\circ$.

Proof:
As discussed above, let $q^*(t)$ be the equilibrium obtained by using the function $w_i(x;\, t)$. Let $t \to 0$. By the compactness of Q, the points $q^*(t)$ will have an accumulation point $q^{**} \in Q$. We must show $q^{**} \in Q^o$.

Suppose that $\hat{q} \in \partial Q$. We will show that, if $|q - \hat{q}|$ and t are sufficiently small, $q \neq q^*(t)$.

In fact, if $q = q^*(t)$, then $J_t(q) = N$, where $J_t(q)$ si the set J, as described above, with the u_i replaced by $w_i(x;\, t)$.

Define ϵ_1, ϵ_2, ϵ_3, ϵ_4 as in the proof of Theorem 1, and let $\epsilon_5 = \epsilon_3/2$. Let $0 < \delta < u_i'(1/2)$.

Suppose, next, that ϵ is the minimum of ϵ_1, ϵ_2, ϵ_3, ϵ_4 and ϵ_5, that $|q - \hat{q}| < \epsilon$, and that $0 < t < \delta$. We have, now,

$$w_h'(2C/r;\, t) = u_h'(2C/r) + t\, e^{-2C/r} > u_h'(2C/r)$$

$$w_h'(1/2;\, t) = u_h'(1/2) + t\, e^{-1/2} < u_h'(1/2) + \delta < 2u_h'(1/2)$$

and so

$$\epsilon_5 = \frac{\epsilon_3}{2} = \frac{rs}{4}\,\frac{u_h'(2C/r)}{u_h'(1/2)} < \frac{rs}{2}\,\frac{w_h'(2C/r;\, t)}{2\,w_h'(1/2;\, t)}$$

The proof then proceeds as in Theorem 1, leading to the conclusion that

$$J_t(q) \subset K \neq N$$

and so $q \neq q^*(t)$. Thus $\hat{q} \neq q^{**}$, i.e.$q^{**} \in Q^o$ as desired.

It is easy to see that if $q^*(t)$ is an equilibrium or each t, then q^{**} will be an equilibrium for $t = 0$, i.e. for the functions u_i.

It is possible to weaken the conditions on the utility functions further, so that they are not strictly increasing, but only if we strengthen Assumption Z. Consider then

Assumption Y. For all i and all j, $p_{ij} > 0$.

Theorem 3. Assume all the utility functions are concave and non-decreasing, and suppose Assumption Y holds. Then there is an equilibrium $q^{**} \in Q^o$.

Proof: We consider first the trivial case in which, for each bettor i, u_i is maximized at $x = A_i$ (or less). In this case it is clear that, for any $q \in Q^o$, the bets $x_{ij} = A_i\, q_j$ will be optimal for bettor i, and so every $q \in Q^o$ will give an equilibrium.

Suppose, next, that for at least one bettor, say bettor 1, u_1 is not maximized at A_1. Then $u_1'(A_1 +) > 0$, and we will set

$$r = \frac{u_1'(A_1 +)}{u_1'(A_1 -)}$$

Clearly $0 < r \leq 1$.

As before, we replace the utility functions $u_i(x)$ by the functions $w_i(x;\, t) = u_i(x) - te^{-x}$, and consider the equilibria $q^*(t)$ obtained in this manner. (Theorem 1 guarantees their existence for all $t > 0$.) As $t \to 0$, these $q^*(t)$ will have an accumulation point $q^{**} \notin \partial Q$. If $q^{**} \in Q^o$, it is the desired equilibrium. We must show $q^{**} \notin \partial Q$. In fact , assume $\hat{q} \in \partial Q$, and let $K = J(\hat{q})$. Let $\epsilon > 0$ be such that, if $|q - \hat{q}| < \epsilon$,

$$\frac{p_{ij}}{q_j} < \frac{r}{2}\,\frac{p_{ik}}{q_k}$$

for all i, all $k \in K$, and all $j \notin K$. (This can be done since all $p_{ik} > 0$.)

By (15), and since $r \leq 1$, we have

$$\frac{x_{ik}}{q_k} \leq \frac{x_{ij}}{q_j} \tag{24}$$

for all i, and all $k \in K$, and all $j \notin K$.

Since (24) holds for all i, the equilibrium condition can hold only if (24) holds as an equation throughout. This will mean

$$\frac{x_{1j}}{q_j} = A_1 \quad \text{for} \quad j = 1, 2, \ldots, n$$

Then, for $k \in K$, $j \in K$

$$\frac{p_{ij}}{q_j} < \frac{r}{2}\,\frac{p_{1k}}{q_k}$$

Now, since $p_{1k}/q_k = p_{1j}/q_j = A_1$, we have, by (11),

$$\frac{p_{ij}}{q_j} < w_1'(A_1+;t) \leq \frac{p_{1k}}{q_k}\, w_1'(A_1-;t)$$

and we have, then

$$\frac{w_1'(A_1+;t)}{w_1'(A_1-;t)} < \frac{r}{2}$$

or

$$\frac{u_1'(A_1+)te^{-A}}{u_1'(A_1-)+te^{-A}} < \frac{r}{2}$$

so that, since $r \leq 1$,

$$\frac{u_1'(A_1+)}{u_1'(A_1-)} < \frac{r}{2}$$

and this is a contradiction. Thus q cannot be $q^*(t)$ for any $t > 0$, and therefore $q^{**} \neq \hat{q}$. We conclude that $q^{**} \in Q^{\circ}$, and this is the desired equilibrium.

3. Examples

We consider here several examples. The first three deal with some "reasonable" utility functions; the last shows that the conditions of Theorems 2 and 3 cannot be further weakened.

(a) Logarithmic utility

Assume that each bettor has a utility function

$$u_i(x) = \log(K_i + x)$$

where K_i is a parameter, representing perhaps bettor i's reserves. In this case, the optimality conditions (10)-(11) take the form

$$\frac{p_{ij}}{k_i q_j + x_{ij}} = \lambda_i \quad \text{if} \quad x_{ij} > 0$$

or equivalently,

$$x_{ij} = p_{ij}/\lambda_i - K_i q_j \qquad \text{if positive} \tag{25}$$

$$x_{ij} = 0 \quad \text{if} \quad p_{ij} < K_i q_j \lambda_i. \tag{26}$$

In the simplest case, when all $x_{ij} > 0$, this will give us

$$A_i = \sum_{j=1}^{n} x_{ij} = 1/\lambda_i - K_i.$$

And so

$$\lambda_i = \frac{1}{A_i + K_i}$$

Substituting in (25), this gives us

$$x_{ij} = (A_i + K_i) p_{ij} - K_i q_j$$

To look for the equilibrium, we have, from (20)-(21),

$$\sum_{i=1}^{m} x_{ij} = q_j^* C.$$

Thus

$$\sum_i (A_i + K_i) p_{ij} - q_j^* \sum_i K_i = q_j^* C$$

so

$$q_j^* = \frac{\sum_i (A_i + K_i) p_{ij}}{C + \sum_i K_i} \tag{27}$$

But $C = \sum A_i$, and we see that q_j^* is then simply a weighted average, with weights $A_i + K_i$, of the several bettors' subjective probabilities p_{ij}.

(b) Exponential utilities

Consider, next, the exponential utility functions

$$u_i(x) = -\exp\{-\alpha_i x\}$$

where α_i is a parameter, measuring, in some sense, bettor i's risk aversion. (Essentially, we can think of $1/\alpha_i$ as representing the sum of money which i would be "hurt" by losing.)

In this case, the optimality conditions (10)-(11) take the form

$$\alpha_i p_{ij} \exp\{\alpha_i x_{ij}/q_j\} = \lambda_i q_j \quad \text{if} \quad x_{ij} > 0$$

which reduces to

$$\alpha_i x_{ij} = q_j (\log \alpha_i - \log \lambda_i) + q_i \log (p_{ij}/q_j) \tag{28}$$

if this right-hand side is positive, and $x_{ij} = 0$ otherwise (i.e., $x_{ij} = 0$ if $\alpha_i p_j < \lambda_i q_j$).

Assuming, once again, that all $x_{ij} > 0$, summation with respect to j gives us

$$\alpha_i A_i = \log \alpha_i - \log \lambda_i + \sum_j q_j \log (p_{ij}/q_j)$$

so that, substituting in (28),

$$\frac{x_{ik}}{q_k} = \frac{1}{\alpha_i}\left[\log \frac{p_{ik}}{q_k} - \sum_j q_j \log \frac{p_{ij}}{q_j}\right] + A_i$$

To obtain the equilibrium odds, we add with respect to i, obtaining

$$C = \sum_i \frac{1}{\alpha_i}\left[\log \frac{p_{ik}}{q_k} \sum_j q_j \log \frac{p_{ij}}{q_j}\right\} + C$$

or, setting $\beta_i = 1/\alpha_i$,

$$\sum_i \beta_i \log \frac{p_{ik}}{q_k} = \sum_i \sum_j \beta_i q_j \log \frac{p_{ij}}{q_j}$$

The right-hand side of this last expression is independent of k, and so we can write

$$\sum_i \beta_i (\log p_{ik} - \log q_k) = \gamma$$

where γ is independent of k. Then

$$\log q_k^* = \frac{\sum \beta_i \log p_{ik} - \gamma}{\sum \beta_i}$$

or

$$q_k^* = \Gamma \; \{\Pi_{i=1}^n p_{ik} \beta_i\}^{(\beta_1 + \ldots + \beta_n)^{-1}}$$

where Γ is a constant. Thus in this case q_k^* is proportional to a weighted geometric mean, with weights β_i, of the subjective probabilities p_{ik}.

(c) Linear utility

Yet another possibility is to equate utility with money. This case was treated in detail by Gale and Eisenber (1959), and so we will merely refer the reader to that interesting article.

(d) A counter-example

Let us consider a two-horse, two-bettor situation. Bettor 1 is certain that horse 1 will win, while bettor 2 believes the race is a toss-up. They each have capital equal to 1 unit. We thus have

$$p_{11} = 1 \qquad p_{12} = 0$$

$$p_{21} = 1/2 \qquad p_{22} = 1/2$$

$$A_1 = A_2 = 1$$

Bettor 1's utility function does not matter; he will bet all his money on horse 1. Bettor 2 has a utility function

$$u_2(x) = \begin{cases} x & \text{if } x \leq 3/2 \\ 3/2 & \text{if } x \geq 3/2 \end{cases}$$

It is easily seen that, if $q_1 < 1/2$, then bettor 2 will choose

$$x_{21} = 3q_1/2, \qquad x_{22} = 1 - 3q_1/2$$

whereas, if $q_1 > 1/2$, then 2 will choose

$$x_{21} = 1 - 3q_2/2, \qquad x_{22} = 3q_2/2.$$

For $q_1 = q_2 = 1/2$, bettor 2 can choose any bet with $1/4 \leq x_{21} \leq 3/4$, $x_{22} = 1 - x_{21}$.

Since $x_{11} = 1$, $x_{12} = 0$ whatever q is. we will then have

$$b_1^*(q) = 1 + 3q_1/2 \qquad \text{if } q_1 < 1/2$$

$$b_1^*(q) = 2 - 3q_2/2 = 1/2 + 3q_1/2 \quad \text{if } q_1 > 1/2$$

$$5/4 \leq b_1^*(q) \leq 7/4 \qquad \text{if } q_1 = 1/2$$

For an equilibrium, we must have $b_1^*(q) = 2q_1$ (since $C = 2$). But, from the above, we see that this holds only if $q_1 = 1$. But his leads to the undesirable discontinuity on the boundary of the simplex, and we must conclude that there is no equilibrium for this situation. Briefly speaking: bettor 2 will win everything if horse 2 wins, so long as he (the bettor) bets a positive amount on this horse (no matter how small). Thus 2 wants to bet as little as possible, subject to a positive bet. This is of course impossible.

APPENDIX

Concavity and Convexity

A function f from $\Re^m$ to $\Re$ is said to be concave if for every $x, y \in \Re^m$, and $0 \leq \lambda \leq 1$,

$$f(\lambda x + (1 - \lambda)y) \geq \lambda f(x) + (1 - \lambda) f(y) \tag{30}$$

It is strictly concave if strict inequality holds in (30) whenever $x \neq y$ and $0 < \lambda < 1$. A function f is convex [strictly convex] if $-f$ is concave [strictly concave].

A set $S \subset \Re^m$ is convex if, for any $x, y \in S$ and $0 \leq \lambda \leq 1$,

$$\lambda x + (1 - \lambda) y \in S.$$

Generally, if f is a concave function then for any q, the set

$$S_q = \{x \mid f(x) \geq q\}$$

is convex. In particular, the set of all x which maximize $f(x)$ is convex (though it may be empty).

If a function is strictly concave, it need not have a maximum. If there is a maximum, however, then the maximizing point is unique.

If $f : \Re \rightarrow \Re$ is concave, it will be differentiable almost everywhere in its domain. Even when not differentiable, however, f has both right and left derivatives, $f'(x+)$ and $f'(x-)$. The derivative is monotone non-increasing, satisfying

$$f'(x+) \leq f'(x-) \qquad \text{for all x} \qquad (31a)$$

$$f'(x-) \leq f'(x+) \qquad \text{if} \quad y < x. \qquad (31b)$$

If f is strictly concave, its derivative is strictly decreasing, satisfying 31(b) with strict inequality.

Upper Semi-Continuity

Let X, Y be topological spaces. A set-valued mapping from X to Y is a mapping Φ which assigns, to each $x \in X$, a subset $\Phi(x) \subset Y$. It is a correspondence if $\Phi(x) \neq \phi$ for each $x \in X$.

The set-valued mapping Φ is upper semi-continuous if, whenever $x_n - x^*$, $y_n \in \Phi(x_n)$, and $y_n \rightarrow y^*$, then $y^* \in \Phi(x^*)$.

Theorem. Let f be a continuous real-valued function defined on the product space $X \times Y$. Define, for $x \in X$,

$$\Phi(x) = \{y \,|\, f(x, y) = \max_{t \in Y} f(x, t)\}$$

Then Φ is an upper semi-continuous set-valued mapping from X to y. If Y is compact and non-empty, then Φ is also a correspondence.

Kakutani's fixed-point theorem.

Let X be a compact convex subset of $\Re^n$, and let Φ be an upper semi-continuous mapping from X to X such that, for each $x \in X$, $\Phi(x)$ is compact and convex. Then there exists some $x^* \in X$ such that $x^* \in \Phi(x^*)$.

References

1. Eisenberg, E., and D. Gale (1959) "Consensus of Individual Probabilities the Pari-Mutuel Approach", Annals Math Stat 165-168.

2. Kakutani, S (1941) "A Generalization of Brouwer's Fixed-Point Theorem." Duke Math J. 457-458.

3. Rockafellar, R. T.(1970) Convex Analysis, Princeton University Press, Princeton, New Jersey

Guillermo Owen
Department of Mathematics
Naval Postgraduate School
Monterey, CA 93943

Genetic Algorithm for Finding the Nucleolus of Assignment Games

Hubert H. Chin

Abstract *This paper describes a heuristic approach to finding the nucleolus of assignment games using genetic algorithms. The method consists of three steps, as follow. The first step is to maintain a set of possible solutions of the core, called population. With the concept of nucleolus, the lexicographic order is the function of fitness. The second step is to improve the population by a cyclic three-stage process consisting of a reproduction (selection), recombination (mating), and evaluation (survival of the fittest). Each cycle is called a generation. Generation by generation, the selected population will be a set of vectors with the higher fitness values. A mutation operator changes individuals that may lead to a high fitness region by performing an alternative search path. The last step is to terminate the loop by setting an acceptable condition. The highest fitness individual presents the nucleolus. The discussion includes an outline of the processing pseudocode.*

1 Introduction

Assignment game with side payments is a model of certain two-sided markets[1]. It is known that prices which competitively balance supply and demand correspond to elements in the core. The nucleolus[2], lying in the lexicographic center of the non-empty core, has the additional property of satisfying each coalition as much as possible. The corresponding prices favor neither the sellers nor the buyers, hence providing some stability for the market. The practical methods to find the nucleolus are based on linear programming techniques[3], which do not seem to be well-suited, because the combinatorial structures of the lexicographic order involve the NP-hard problem. For example, Kohlberg[4] solved the nucleolus of general cooperative games leading to extremely large linear programming. Owen[5] improved this method by solving it as a single minimization problem, but it still causes some serious numerical difficulties. Solymosi and Raghavan[6] gave an algorithm based on a geometric approach to locate the nucleolus. When players increase greatly, the method to compute the exact location of the nucleolus is an NP-complete problem[7, 8, 9]. This paper exploits a heuristic approach based on the genetic algorithm to search for the nucleolus.

The Genetic Algorithm(GA) paradigm has been proposed to generate solutions to a wide range of problems [10]. Serial implementations have presented empirical evidence of its effectiveness on a combinatorial optimization problem. These include control systems[11], function optimizations[12], and combinatorial problems[13]. In all cases, a population of solutions to the problem at hand is maintained and successive "generations" are produced with new solutions replacing some of the older ones. The population is typically kept at a fixed size. Most new solutions are formed by mating two previous ones; this is done with a "recombination" operator and probability of mixing their genes. There are two mechanisms leading to success. First, the better fit solutions are more likely to recombine and hence propagate.

It is important to realize that the GA approach is inherently sequential. It follows a "trajectory" of the best solution to a local maximum. Second, there may be many local maximums to be considered, and the mutation operation jumps forward when a high performance region has been identified. At end, an ad hoc termination condition is used, and the best remaining solution (or the best ever seen) is reported to be the nucleolus.

This paper is organized as follows. Section II describes the task of integrating domain knowledge into the GA algorithm. Section III presents the pseudocode, including the framework of reproductive plans and genetic operators. Section IV discusses an example of the real estate market and its results. Section V concludes the paper.

2 Domain Knowledge and Formulations

In this section, the focus is on the incorporation of domain knowledge of Assignment Games(AGs) into the traditional GAs, as an exploratory tool to identify the nucleolus. The conventional notions of AGs and formulation of GAs are introduced as follows.

2.1 Domain Knowledge

AGs consist of two types of players named row players, $M = \{r_1, r_2, \ldots, r_m\}$, and column players, $N = \{c_1, c_2, \ldots, c_n\}$. When a transaction between $r_i \in M$ and $c_j \in N$ takes place, a certain profit $a_{ij} \geq 0$ occurs. Then $A = (a_{ij})_{(m,n)}$ is called an augmented gain matrix. Also, (S, T) stands for a coalition of $S \subseteq M$ and $T \subseteq N$. The worth of a coalition, $\nu(S, T)$, is to maximize the total profit of an assignment of players within the coalition, (S, T). Put symbolically, $\nu(S,T) = Max\{\sum_{(i,j)} a_{ij} : i \text{ induces } c_i \in S \text{ and } j \text{ induces } r_j \in T\}$. An (S,T)-matching, μ, is a matching between the players of S and T, that is, $\mu = \{(r_i, c + j) : r_i \in S$ and j induces $c_j \in T\}$. This means that the matched players share the profit they can make, but an unmatched player receives nothing.

The characteristic functions state the worth $\nu(S, T)$ for every possibility of coalitions S and T. It is obvious that $\nu(S) = 0$ if $|S| = 0$ or 1, and $\nu(T) = 0$ if $|T| = 0$ or 1. Because no player can make any worth (profits) without help from another, the cases of one-sided coalitions are "flat", that is, $\nu(S) = 0$ if $S \subseteq M$, or $\nu(T) = 0$ if $T \subseteq N$. In other words, only a mixed coalition can ever hope to assure a trade. A larger coalition can split up into separate trading pairs and pool the profit. The trading activity determines ν by the augmented gain matrix $(a_{ij})_{(m,n)}$. In fact, n may be characterized as the smallest super-additive set function on $M \cup N$ satisfying character function requirements. The evaluation of $\nu(S, T)$ is commonly called the "optimal assignment problem" or simply "assignment game."

An imputation is a vector $\xi = (r_1, r_2, \ldots, r_n, c_1, c_2, \ldots, c_m)$ and the set of imputations is denoted by $I = I(N \cup M; \nu)$, such that $\xi \in I$. The core set $C = C(N \cup M; \nu)$ of imputations, that is, $C = \{\xi : \nu(\xi) \geq \nu(U)$ for all $U \subset N \cup M\}$. The core of the AG rarely consists of just a single imputation,$\{\xi\}$. Shapley and Shubik[1] showed that for assignment games, the core is never empty and is a closed, convex polyhedral set. The dimension of the core is typically equal to $MIN(m, n)$, but it may be less in the presence of degeneracy, i.e., special arithmetical relations among the a_{ij}. Note that the dimension of the imputation space in which the core is situated is $(m+n-1)$, considerably larger than $MIN(m, n)$. The conclusion is that the set of imputations is always a non-empty set.

Schmeidler[2] introduced the nucleolus solution concept and showed that every game possesses a single point. That is, for each $\xi \in I(N \cup M; \nu)$, let $Lex(\xi) = \{x_1, x_2, \ldots, x_k\}$, where $k = 2^{N \cup M}$, denote the $2^{N \cup M}$-vectors whose components are arranged in non-decreasing order (i.e., $x_i \leq x_j$ whenever $i \leq j$). Let $\geq_L$ denote the usual lexicographic order of vectors, that is, $\xi >_L \psi$, if $Lex(\xi) = \{x_1, x_2, \ldots, x_k\}$ and $Lex(\psi) = \{y_1, y_2, \ldots, y_k\}$,where $x_j \geq y_j \exists j$, and $x_i = y_i$, for $\forall i \leq j$. The weak form, $\geq_L$, is that $\xi \geq_L \psi$, if $\xi >_L \psi$ or $\xi = \psi$. The nucleolus of an AG are the imputations which lexicographically maximize the vector over the set of imputations. That is, the nucleolus is denoted by $\{\xi : Lex(\xi) \geq_L Lex(\psi)$, for $\forall \psi \in I(N \cup M; \nu)\}$.

Linear Programming (LP) is used as a tool for solving the core set. Consider the game of coalition of all players, i. e., the problem of determining $\nu(N \cup M)$. Introduce $(m \times n)$ non-negative real variables $x_{ij}, \forall i \in M, \forall j \in N$, and impose on them the $(m+n)$ constraints $\sum_{i \in M} x_{ij} \leq 1$ and $\sum_{j \in N} x_{ij} \leq 1$. The LP problem is then to maximize the following objective function: $z = \sum_{i \in M} \sum_{j \in N} a_{ij} x_{ij}$. It can shown that the maximum value z_{max} is attained with all $x_{ij} = 0$ or 1. Thus, the fractions or probabilities artificially introduced disappear from the solution, and the LP problem is effectively equivalent to the assignment problem, so that it is $z_{max} = \nu(M \cup N)$. The LP problem can be transposed into a dual form; the solutions of the two problems are intimately bound. In the present case, the dual has $(m+n)$ non-negative real variables, $r_1, r_2, \ldots, r_n; c_1, c_2, \ldots, c_n$, subject to the $(m \times n)$ constraints, $r_i + c_j \geq a_{ij}, \forall r_i \in M, \forall c_i \in N$, and the objective is to minimize the sum: $\sum_{i \in M} \sum_{j \in N} (r_i + c_j)$. The fact is that the core of an AG is precisely the set of solutions of the LP dual of the corresponding assignment problem.

2.2 Formulations

A population is a subset of vectors that are in the core. An individual of the population is called a genotype, and the component values at each position of a genotype are called alleles. The population is initialized to select a set of random genotypes which are the core members. These members are located in a closed, convex polyhedral set. One trick would be to employ a heuristic selection preference for individuals with extreme positions of the convex set. This preference takes the heuristic adjustment to random selections of the initial population. The size of the initial population is based upon the extreme points of the convex set. The population size is allowed to grow to a constant size (within physical memory limitations).

Recombination is the primary means of generating plausible new genotypes for addition to the population. Traditionally, mates are selected at random. The random mating is implemented by shuffling the population and mating pairs from the top. The heuristic mating is to select individuals with maximum Euclidean distance. In processing, let two genotypes, $\xi = \{r_1, r_2, \ldots, r_n; c_1, c_2, \ldots, c_n\}$ and $\psi = \{r'_1, r'_2, \ldots, r'_n; c'_1, c'_2, \ldots, c'_n\}$, be selected from the population. The process takes some probabilities $\pi = \{p_1, p_2, \ldots, p_{n+m}\}$ from a random generator to combine alleles $r''_i = r_i p_i + r'_i(1 - p_i)$, where $i = 1, 2, ..., N$, and $c''_j = c_j p_{n+j} + c'_i(1 - p_{n+j})$, where $j = 1, 2, \ldots, M$, producing a complete genotype, $\zeta = \{r''_1, r''_2, \ldots, r''_n; c''_1, c''_2, \ldots, c''_n.\}$. The computation applies only to r''_i, because the values of each c''_j are based upon the μ-matching function. The matching is to match each seller to a buyer in each population.

Evaluation performs a search for the least fit. For ξ, ψ, and ζ, assume that the order relation is $\xi \leq_L \psi \leq_L \zeta$. The least fit ξ is replaced. Since the population is weighted towards higher-fitness genotypes, eventually new genotypes will survive and rejoin the population; then the population is said to have "converged."

A reproduction operator has the opportunity to flourish or perish depending on its fitness. It also includes a "background" mutation operator. In a typical implementation, the mutation operator provides a chance for any allele to be changed to a highly fit individual. Since recombination redistributes corresponding alleles, the mutation operator guarantees generation of a new genotype, which may be better or worse. If the mutation rate is too low, a possibly critical genotype missing from the initial random distribution has only a small chance of getting back to the population. However, if the parameter space is steadily lost to random changes, the performance of the algorithm suffers.

3 Pseudocode of the Algorithm

The software procedure implementing this algorithm is GA_for_AG. The algorithm maintains a set of "current best" solutions and tries to improve them. The set of possible solutions is called a population. This population is improved by a cyclic three-step process consisting of reproduction (select the best individuals), recombination (mate two individuals), and evaluation (survival of the fittest). An outline of the structure for the program is given as follows:

```
Program GA_for_AG
    begin
        Initialize population;
        Select population(0);
        Evaluate population(0);
        t = 1;
        repeat
            Reproduce population(t) from population(t − 1);
            Recombine population(t);
            Evaluate population(t);
        until (termination condition true);
    end.
```

Procedures of the algorithm are described as follows:

1. Initialize population

This procedure defines the data structure of the program, including: genotypes = imputations; alleles = components. It invokes: a Linear Program that is implemented by the Hungarian method to determine the Core elements.

2. Select population

This procedure selects the initialized population at iteration 0. Since the geometry of the core is a polyhedron, selection must cover the most extreme forms of the core.

3. Evaluate population

This procedure evaluates the fitness of the selected population by calling the Fitness function. The evaluation is to guarantee that every individual is in the core. It invokes: a lexicographic order function to return a fitness vector of a given individual. The fitness function compares vector quantities ordered by lexicographic order. An outline of the structure for the Fitness function is described as follows.

```
function Fitness(A: Augmented Gain Matrix, X, Y: Lexicographic Vectors)
    begin
        Compute satisfaction elements of A;
        Call Lexicographic Order (X);
        Call Lexicographic Order (Y);
```

```
        Compare(X, Y);
        If (X Lexicon greater than Y) then
             Return (X);
        else
             Return (Y);
end;
```

4. Reproduce population

This procedure organizes a new population according to the evaluation results. The actual number of offspring attributed to the population is directly proportional to the algorithm's performance. The mutation is randomly generated by this procedure for checking the convergent condition. It also checks the termination condition for ending the repeat loop and making a final report. An outline of the structure for the Mutation function is described as follows:

```
function Mutation(ξ: individual, t: iteration, k: real number)
     begin
          if (rand is greater than k)
               Get ξ = imputation, back to the population;
               t = t + 1;
     end;
```

5. Recombine population

This procedure mates two randomly selected individuals, ξ and ψ, where $\xi = \{r_1, r_2, \ldots, r_m; c_1, c_2, \ldots, c_n\}$ and $\psi = \{r'_1, r'_2, \ldots, r'_m; c'_1, c'_2, \ldots, c'_n\}$. If there are S sellers and T buyers, a random generation produces m random variables between 0 and 1, i.e., $\pi = \{p_1, p_2, \ldots, p_n\}$. Therefore, the mating is $r''_i = r_i p_i + r'_i(1 - p_i)$, where $i = 1, 2, ..., n$. The matching function, μ, takes care of c_n's values based on the (S, T)-matching.

4 Example

An example of a realty market AG is formulated as follows: Let there be m houses in the market and n prospective purchasers referred to simply as sellers and buyers respectively. Each buyer is allowed to buy one house and each house sells to only one buyer. For demonstration purposes, the transfer can be summed up in the one simple observation: There are equal number of buyers and sellers. The ith seller values his house at least worth k_i dollars, while the jth buyer values the same house at most h_{ij} dollars. If $h_{ij} \geq k_i$, then a price favorable to both parties exists; otherwise, there is no deal between them. If s_i sells his house to b_j for a sale price q_{ij} dollars, then the ith seller's final profit is $(q_{ij} - k_i)$ and the jth buyer's final bid saves $(h_{ij} - q_{ij})$. The AG is to find the optimal solution of sale prices (q_{ij} for $\forall i, j$) of these houses.

4.1 Computation

There are three sellers (s_1, s_2, s_3) and three buyers (b_1, b_2, b_3), and their formulation is as follows: The first seller values his house at \$190K; the second seller values his house at \$230K; and the third seller values his house at \$250K. The first buyer wants to offer for the first house \$230K, the second house \$270K, and the third house \$300K. The second buyer wants to offer for the first house \$250K, the second house \$260K, and the third house \$280K. The third buyer wants to offer for the first house \$210K, the second house \$240K, and the third house \$200K. By the definition, the augmented gain matrix $A = (a_{ij})$, where $a_{ij} = (h_{ij} - k_i)$. Let player set $M \cup N = \{s_1, s_2, s_3, b_1, b_2, b_3\}$ and tableaux as follows:

Seller set: M	**House set**	**Buyer set N:** b_1	b_2	b_3
$s_1 \geq 190K$	house$_1$:	$\leq 230K$	$\leq 250K$	$\leq 210K$
$s_2 \geq 230K$	house$_2$:	$\leq 270K$	$\leq 260K$	$\leq 240K$
$s_3 \geq 250K$	house$_3$:	$\leq 300K$	$\leq 280K$	$\leq 200K$

Therefore, these data lead to the following $A = (a_{ij})_{(3,3)}$ matrix: (units of thousand dollars)

$$A = \begin{bmatrix} 40 & \mathbf{60} & 20 \\ 40 & 30 & \mathbf{10} \\ \mathbf{50} & 30 & 0 \end{bmatrix}$$

The results (shown in bold) of unique optimal assignment are computed by LP. By definition, we obtain worth, v(S), as follows:

$v(\{s_1\}) = v(\{s_2\}) = v(\{s_3\}) = v(\{b_1\}) = v(\{b_2\}) = v(\{b_3\}) = 0$
$v(\{s_1, b_1\}) = 40, v(\{s_1, b_2\}) = 60, v(\{s_1, b_3\}) = 20,$
$v(\{s_2, b_1\}) = 40, v(\{s_1, b_2\}) = 30, v(\{s_1, b_3\}) = 10,$
$v(\{s_3, b_1\}) = 50, v(\{s_1, b_2\}) = 30, v(\{s_1, b_3\}) = 0,$
....
$v(\{s_1, s_2, b_1, b_2\}) = MAX(|40, 60|, |40, 30|) = 60 + 40 = 100$, where
$v(\{s_1, s_2, b_1, b_2\})$ selects the maximum sum of the matrix of
the first and second rows and columns.
$v(\{s_1, s_2, s_3, b_1\}) = MAX(|40|, |40|, |50|) = 50$, where
$v(\{s_1, s_2, s_3, b_1\})$ selects the maximum value of the first column
of matrix A.
$v(P) = v(\{s_1, s_2, s_3, b_1, b_2, b_3\}) = MAX(A) = 120.$

By solving the following equations, the nucleolus = $[s_1, s_2, s_3, b_1, b_2, b_3]$ that satisfy the following constraints:

$s_1 + s_2 + s_3 + b_1 + b_2 + b_3 = 120, s_1 + b_1 \geq 40, s_2 + b_1 \geq 40, s_3 + b_1 \geq 20,$
$s_1 + b_2 \geq 40, s_2 + b_2 \geq 40, s_3 + b_2 \geq 40, s_1 + b_3 \geq 40, s_2 + b_3 \geq 40, s_3 + b_3 \geq 40,$

The answer shows that the nucleolus = [30, 10, 20, 30, 30, 0].

The optimal assignments for the three house values are indicated as follows: The final price of the first house is \$220K, the price of the second house is \$240K, and the price of the third house is \$270K.

4.2 Results

This example is computed by the GA method to search a nucleolus with an initial population of 8 individuals and mutation rates of .01. An average performance of population is defined as follows:

$\Phi = [\sum(\text{all individuals})(\sum(\text{all components})\ |x_i - y_i|/6)/8]$, where
$[y_1, y_2, y_3, y_4, y_5, y_6] = [30, 10, 20, 30, 30, 0] =$ ***nucleolus***, and
$[x_1, x_2, x_3, x_4, x_5, x_6]$ is an individual in the ith generation.

Recalling that the extreme vectors are $[60, 10, 50, 0, 0, 0]$ and $[0, 0, 0, 60, 50, 10]$, performance then ranges from 0 to 240. The following chart shows the average performance of the 500 generations experiment:

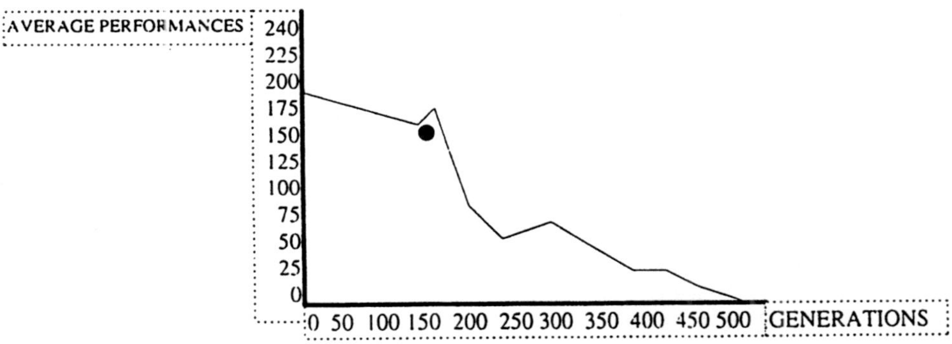

During an early generation (150 iterations), a dangerous property is discovered of premature convergence to an extreme individual in core. The true nucleolus is assumed known. After this observation, the mutation substitutes "middling" individuals for the extreme individuals. The mate-selection heuristic proves superior to the random mating. This suggests that maximum Euclidean distance is good for improving the performance in this case.

The outputs of the algorithm written in C programming language are summarized as follows: The initial population is :

$$\left\{ \begin{array}{cccc} (0,0,0,60,50,10) & (60,10,50,0,0,0) & (60,0,0,0,50,10) & (0,10,0,60,50) \\ (0,0,50,60,10,0) & (60,10,0,0,0,50) & (60,0,50,0,10,0) & (0,10,50,60,0,0) \end{array} \right\}$$

of size 8. The genotype length is 6. Lexicographic order length is 8. The total generation is equal to 500. The terminal condition is accepted when the performance reaches 0.5. The GA slowly converges to the nucleolus. Its best solution at 500 generations is very close to (30,10,20) within the 0.5 performance.

5 Conclusion

The key of this approach has five basic properties to be a successful tool for solving the nucleolus. They are: (1) representation of the imputation of solution space, (2) fitness associated with the lexicographic order, (3) selection process in reproduction, (4) recombination probability vector for fast convergence, and (5) mutation for the alternative path. Hidden behind the basic properties are a variety of parameters and policies such as recombination rate, mutation generation, and replacement policy. All of those settings may affect performance. There is empirical support for the statement that within reasonable ranges the values of such parameters are not critical.

Highlights of the several interesting discoveries are as follow:

1. The highly heuristic approach can be applied to the optimization of multi-objective functions with respect to lexicographic order.

2. With a large population and high mutation rate, random mating will be better than mate-selection heuristics. In the experiment, the mate-selection heuristic performs well, because it selected a smaller population.

3. Recombination is based on a random vector of probabilities to mate genotypes that performed well in the convex space of the core. The crossover of chromosomes of binary strings of bits is not suitable to our application.

4. The search domain forms a unit hypercube of imputations in which the nucleolus of AGs is the optimal point with the highest performance value (equal to zero). Further investigation of the hypercube structure suggests interesting research.

References

Shapley, L. S. and M. Shubik, "The assignment Game I: The Core," International Journal of Game Theory, 1, pp. 111-130, 1972.

Schmeidler, D., "The Nucleolus of a characteristic function game," SIAM Journal on Applied Mathematics, 17, pp. 1163-1170, 1969.

Kuhn, H. W., "The Hungarian Method for the assignment Problem," fNaval Research Logistics Quarterly, 2, pp. 83-97, 1955.

Kohlberg, E., "The nucleolus as a solution of a Minimization Problem," SIAM Journal on Applied Mathematics, 23, pp. 34-39, 1972.

Owen, G., "A note on the Nucleolus," International Journal of Game Theory, 3, pp. 101-103.

Solymosi, T. and T. E. S. Raghavan, "An Algorithm for Finding the Nucleolus of Assignment Games," International Conference on Game Theory at Stony Brook, New York, July 1992.

Maschler, M., B. Peleg, and L. S. Shapley, "Geometric Properties of the Kernel, Nucleolus, and Related Solution concepts," Mathematics of Operations Research, 4, pp. 303-338, 1979.

Maschler, M, J. A. M. Potters, and S. H. Tijs, "The General Nucleolus and the Reduced Game Property," International Journal of Game Theory, 21, pp.85-106, 1992.

Sankaran, J. K., "On Finding the Nucleolus of an N-person Cooperative Game," International Journal of Game Theory, 19, pp. 329-338, 1991.

Holland, J. H., "Adaptation in Natural and Artificial system," University of Michigan Press, 1975.

DeJong, K. A., "Analysis of the Behavior of a Class of Genetic Algorithms," University of Michigan, Ph.D. Thesis, Ann Arbor, MI., 1975.

Brindle, A., "Genetic Algorithms for Function Optimization," University of Alberta, Ph.D. Thesis, 1980.

Bethke, A. D., "Genetic Algorithms as function Optimizers," University of Michigan, Ph.D. Thesis, 1981.

Goldberg, D., "Computer Aid Gas Pipeline Operation Using Genetic Algorithms and Rule Learning," University of Michigan, Ph.D. Thesis, 1983.

Hubert H. Chin
Computer Science Department
New York Institute of Technology
Old Westbury, NY 11568

SOME RECENT ALGORITHMS FOR FINDING THE NUCLEOLUS OF STRUCTURED COOPERATIVE GAMES

T.E.S. Raghavan[1,2]

Abstract *Nucleolus is one of the fundamental solution concepts in cooperative game theory. There has been considerable progress in locating the nucleolus in the last three years. The paper motivates through examples how the recent algorithms work efficiently for certain structured class of coperative games. Though the data of a cooperative game grows exponentially in size with the number of players, assignment games, and balanced connected games, grow only polynomially in size, on the number of players. The algorithm for assignment games is based on an efficient graph theoretic algorithm which counts the longest paths to each vertex and trimming of cycles to quickly arrive at the lexicographic geometric centre. Connected games are solved by the technique of feasible direction, initiated in the assignment case. The sellers market corner of the core for assignment games has its counterpart, the lexmin vertex in balanced connected games. Nucleolus has also been characterized via a set of anxioms based on subgame consistency. This is exploited for standard tree games to arrive at an efficient and intuitively explainable algorithm. Improvements on the pivoting manipulations to locate coalitions with constant excess are possible and the paper initially discusses such an algorithm at the beginning.*

1 Nucleolus via a prolonged simplex

A cooperative TU-game is defined by a finite set $N = \{1, 2, \cdots, n\}$ and a map $v : 2^N \rightarrow \Re$ with $v(\emptyset) = 0$. Here 2^N is the set of all subsets of N. Intuitively, if N denotes the set of players in a game, then for each $S \subset N$, $v(S)$ denotes the worth of coalition S. Thus $v(S)$ measures what coalition S can achieve by their own joint effort. This definition does not say anything about which coalitions would form and how the coalitions would share their joint worth. The main problem of cooperative game theory is to propose reasonable solutions when the grand coalition N forms. There are several solution concepts that address

[1]The author would like to thank Ms Evangelista Fe, Tamas Solymosi and N. Etemadi for some critical discussions in the preparation of this paper

[2]Partially funded by NSF-Grants: DMS-9301052 and INT 9511592, U-S India Binational Workshop on Game Theory and its Applications, Jan 2-6, 1996, Bangalore, India

T. Parthasarathy et al. (eds.),
Game Theoretical Applications to Economics and Operations Research, **207–238.**

this problem, including the core, the stable set, the bargaining set, the kernel and the τ-value. However, the two solutions that stand out are the Shapley value and the nucleolus. Shapley proposed a value for individual participants axiomatically (Shapley [1953]). Schmeidler [1969], attempting to locate a special element of the bargaining set[3], landed on the notion of nucleolus. Unlike the Shapley value, the nucleolus always lies in the core when it exists. It is an element of the kernel and the bargaining set. More recently the nucleolus has also been axiomatized via consistency properties for induced subgames (Sobolev [1975]).

The Shapley value is expressed by a closed form formula involving all the data defining the game. In a pioneering paper, Maschler, Peleg and Shapley [1979] characterized the nucleolus in an iterative fashion. From an algorithmic point of view, this iterative definition is a very powerful tool which makes the nucleolus more amenable for actual computations. The main difficulty in computing the nucleolus lies in storing the data defining the game. Since the data can increase exponentially with the number of players, any iterative method is limited when the number of players is large. Fortunately, there are certain subclasses of games whose characteristic function v is completely determined once we know the worth for coalitions in a set of much smaller size. We call the coalitions in this set *essential*. Inessential games are generally called *additive games*. Assignment games (Shapley and Shubik [1972]) are determined completely by coalitions of size at most two. These coalitions are called essential coalitions. If the number of essential coalitions grows only polynomially with the number of players, it might be possible to compute the nucleolus for such games. We will introduce various efficient algorithmic schemes and compute the nucleolus for some typical examples of such families of games.

Given a game (N, v), the preimputation consists of vectors $x = (x_1, \ldots, x_n)$ such that $x(N) = \sum_{i \in N} x_i = v(N)$. If $x_i \geq v(i)$ for all i, then the preimputation is called an imputation. Let $I(v)$ denote all imputations of a game (N, v). Any $x \in I(v)$ is a potential division of the worth of the grand coalition N. While $x \in I(v)$ only guarantees individuals what they can get on their own, players in a coalition S get $x(S)$ which may or may not give as much as $v(S)$, the worth of coalition S. Any vector $x \in I(v)$ that satisfies $x(S) \geq v(S) \quad \forall S \subset N$ is called a coalitionally rational payoff vector to players. The set $C = \{x \in I(v) : x(S) \geq v(S)\}$, called the *core*, is often empty. In case $C \neq \emptyset$, coalitions may prefer one element of C over another, based on the satisfaction an imputation in C gives. We measure the *satisfaction* of a coalition for the imputation x by $f(x, S) = x(S) - v(S)$. The negative satisfaction is called the *excess*. Thus $e(x, S) = -f(x, S)$.

While the satisfaction for the extreme coalitions $\emptyset$ and N are identically zero in $I(v)$, they could vary quite considerably for all other coalitions. Given any two imputations x and y, one can arrange the 2^n- tuple of satisfactions in an

[3]Private communication by Professor Maschler

increasing order. Rearranging the set of all coalitions as $S_1, S_2, \ldots S_{2^n}$ for x and $T_1, T_2, \ldots T_{2^n}$ for y, let

$$f(x, S_1) \leq f(x, S_2) \leq \cdots \leq f(x, S_{2^n})$$

$$f(y, T_1) \leq f(y, T_2) \leq \cdots \leq f(y, T_{2^n})$$

When x is proposed as the worth of a coalition, coalition S_1 is the least satisfied followed by S_2 and so on. Coalition T_1 is the least satisfied when y is proposed while T_{2^n} is the most satisfied. We say x is better than y if the vector $\theta(x) = (f(x, S_1), f(x, S_2), \cdots f(x, S_{2^n}))$ is lexicographically larger than the vector $\theta(y) = (f(y, T_1), f(y, T_2), \cdots f(y, T_{2^n}).)$ The nucleolus is defined as the set of imputations which are are lexicographically maximal. It is known that the nucleolus consists of just one imputation. (Schmeidler [1969]).

Lexicographic Center

An alternative characterization of the nucleolus called the lexicographic center was proposed by Maschler, Peleg and Shapley [1979]. Virtually all the algorithms studied here depend on this alternative characterization.

Let X be a given subset of imputations. A coalition S is called *settled* on X if the satisfaction $f(x, S)$ is constant for all $x \in X$. Let Δ be the collection of settled coalitions. The coalitions outside Δ are called *unsettled coalitions*. If $\Delta = \{\emptyset, N\}$ and $X = I(v)$, then trivially the coalitions in Δ are settled on X, namely $f(x, S) \equiv 0$ if $S = \emptyset$ or $S = N$. Let $\mathcal{M} = 2^N$. For $r = 0, 1, 2, \cdots, S$, we will iteratively define sets X^r, Δ^r, Σ^r such that $X^0 \supset X^1 \supset X^2 \supset \cdots$, $\Delta^0 \subset \Delta^1 \subset \cdots \subset \mathcal{M}$ and $\Sigma^0 \supset \Sigma^1 \supset \Sigma^2 \supset \cdots$. The termination occurs when $\Delta^r = \mathcal{M}$, $\Sigma^r = \emptyset$.

Iteration $r = 0$. $\Delta^0 = \{\emptyset, N\}$. $\Sigma^0 = M \setminus \Delta^0$. $X^0 = I(v)$. $\Sigma_0 = \{\emptyset, N\}$. Let

$$\begin{aligned}
\alpha^{r+1} &= \max_{x \in X^r} \min_{S \in \Sigma^r} f(x, S) \\
X^{r+1} &= \{x : \min_{S \in \Sigma^r} f(x, S) = \alpha^{r+1}\} \\
S_{r+1} &= \{S : f(x, S) \text{ is constant on } X^{r+1}\} \\
\Delta^{r+1} &= \Delta^r \cup S_{r+1} \\
\Sigma^{r+1} &= \Sigma^r \setminus S_{r+1}.
\end{aligned}$$

The theorem of Maschler, Peleg, and Shapley [1979] asserts that the iteration must terminate with $X^{\rho+1}$ consisting of a single point and $\Sigma^{\rho+1} = \emptyset$. This is the iterative algorithm to locate the nucleolus for all cooperative TU-games (v, N).

If we consider a generic game, the first difficulty involves enumerating the $2^n - 1$ numbers $v(S)$. The algorithm involves solving a series of linear programs. It is possible to reduce to solving one single linear program with $O(n)$ variables and $2^n!$ constraints (Kohlberg [1972]) or $O(2^n)$ variables and $O(4^n)$ constraints (Owen [1974]) or $O(2^n)$ iterations (Sankaran [1991]). The most recent reductions for general games use $n - 1$ linear programs with $O(2^n)$ rows with only

$O(n)$ non-zero entries and $O(2^n)$ columns (Solymosi [1993], Reijnierse [1995]). Also see Dragan [1981].

The following implementation procedure of the algorithm is due to Potters, Reijnierse and Ansing ([1996]).

Potters, Reijnierse and Ansing algorithm:

Initialization: $X = I(v), \Sigma = 2^N \setminus \{\emptyset, N\}$

Step 1: Replace X by $\{y \in X : \max_{x \in \Sigma} e(S, x)$ attains its minimum at $y\}$

Step 2: Delete from Σ *at least* one coalition S such that $e(S, y)$ is constant for all y in X.

Step 3: If X is empty, terminate; else go to step 1.

A description of the above steps and (X, Σ) correspond to a linear inequality system. Its solution (x, y) satisfies

$$\begin{aligned} Ax + By &= d \geq 0 \\ x, y \geq 0, & \quad x \in R^N, y \in R^\Sigma \end{aligned} \tag{1.1}$$

such that

1. $x \in X \Rightarrow$ system (1) has a solution (x, y).
2. If A has q rows then the system has q basic variables that occur in exactly one equation.
3. For each $x \in X, y$ is unique with (x, y) satisfying (1).

Starting with the maximal excess $c = \max_{S \neq N} v(S)$. variables $y_S, S \in \Sigma$ measure the difference between last iteration's maximal excess and the excess of the current imputation x in the polytope defined by inequalities (1).

It is also convenient to fix a player, say player n, and describe the polytope by

$$x \in R_+^N, \qquad y \in R_+^\Sigma$$

$$x(N) = y(N) \tag{1.2}$$

$$y_S - x(S) = c - v(S) \quad \forall S \in \Sigma, \; S \not\ni n \tag{1.3}$$

$$y_S + x(N \setminus S) = c + v(N) - v(S) \quad \forall S \in \Sigma, \; S \ni n \tag{1.4}$$

Given a coalition $S \in \Sigma$, either $S \ni n$ or $S \not\ni n$. Note that the variables y_S and also the variable x_n occur in exactly one equation. Step 1: We replace the current polytope X and unsettled coalitions Σ by a new polytope and a subcollection of unsettled coalitions. We look for $x \in X$ that minimizes the maximal excess or equivalently that finds the highest possible reduction of the maximal excess in Σ. The guaranteed reduction of the excess is given by $\min_{S \in \Sigma} y_S$. This we would like to maximize over the current X. It is achieved by introducing an auxiliary variable t and solving

$$\max \quad t$$

subject to

$$\begin{aligned} x \geq 0, \qquad & t \geq 0 \\ Ax + By &= d \\ y_S &\geq t, \quad \forall S \in \Sigma \end{aligned} \tag{1.5}$$

The inequalities (5) can be replaced by equalities with the introduction of a variable z_S satisfying $z_S + t = y_S, z_S \geq 0, \ t \geq 0$. Thus the linear programming problem can be written as

$$\max \quad t$$

subject to

$$x \geq 0, t \geq 0, z \geq 0$$

$$Ax + Bz + (Be_S)t = d, \quad S \in \Sigma$$

where e_S is the indicator of S.

Since t has to be a basic variable at intermediate steps, we can as well bring the variable t into the basis, by one pivot operation. The row i that contains the t variable after the first pivot step replaces the objective function. Therefore one should never pivot with entries of the i-th row but treat it as the objective function. suppose row i has some negative coordinates, say

$$t + \Sigma_j \pi_{ij} x_j + \Sigma_{S \in \Sigma} q_{iS} z_S = d_i. \tag{1.6}$$

Since the linear program has an optimal solution, pivoting will terminate with row i looking like (6) but with coefficients $\pi_{ij} \geq 0$ and $q_{iS} \geq 0$. If $\pi_{ij} > 0$ then the variable x_j is nonbasic and if $q_{iS} > 0$ then z_S is nonbasic. Thus we can replace our equation (6) with $x_j = 0$ if $\pi_{ij} > 0$ and $z_S = 0$ if $q_{iS} > 0$. Also, we can replace $t = d_j = \hat{t}$ with the current optimal value. We have cut down the excess by $\hat{t}$ and the current highest excess is $c - \hat{t}$. We can delete the equation $t = \hat{t}$ and rename z_S by y_S. The new y_S measures the difference between $c - \hat{t}$ and $e(S, x)$.

Step 2

Delete any row k which is elementary, namely for some coalition $S \in \Sigma$, $\pi_{kj} = 0 \ \forall j \in N, \ \ q_{kS} \neq 0$, but $q_{kT} = 0 \ \ \forall T \in \Sigma, \ T \neq S$. Such a row will always be present after step 1. Elementary row k determines $z_S =$ constant for the particular S over the polytope. We can delete row k and the column corresponding to y_S in the next iteration. We still have a basis after removing the row and column.

Return to step 1 with the new table. The algorithm terminates once all the rows are elementary corresponding to $I(x) = \nu(v)$ where all y_S are ultimately deleted. An elementary equation should have $d_k = 0$ and all coefficients ≥ 0. Such equations can be removed, if we also remove variables with positive coefficients in this equation and from all other equations. If such a variable is an x-variable, say x_i, one should remember that the nucleolus should have zero as the ith coordinate. One can also add the equation $x_i = 0$ again. An equation like $x_3 + x_6 + 2z_S + 3z_T = 0$ becomes $x_3 = 0, x_6 = 0$ and z_S, z_T are deleted from all equations.

We notice that there is at least one elementary row after each linear programming termination. An important observation is that besides deleting the row and column corresponding to elementary equations and constant variables, we have also replaced equations (6) with $x_j = 0$ and $z_S = 0$ when $\pi_{ij} > 0$ or $q_{iS} > 0$. The fact that equation (6) is not trivial guarantees that you find each time at least one new constant z_S variable and that the algorithm works. We thus get more reduction than what we anticipated. Moreover, it can be proved that the variables that are constant on the *affine subspace* defined by the linear equations (without the constraints $x \geq 0, z \geq 0$). can be detected. Thus any elementary equation determining constancy of a variable in the polytope X is determined by some linear combination of equations determining the hyperplane. Thus the variables that are constant on the hyperplane are already present in the current equations describing the hyperplane.

We illustrate the algorithm via an example.

Example 1: Let $N = \{1,2,3\}, v(i) = 0, i = 1,2,3$. Let $v(1,2) = v(1,3) = 3, v(2,3) = 8$ and $v(1,2,3) = 10$. Here, $c = \max_{S \neq \emptyset, N} v(S) = 8$.

The initial system of equations are from equations (2), (3), and (4) with variable x_3 eliminated from all but the first equation (see Table 1) . The new representation in variables t and x's is in Table 2. Since t is ultimately basic in the first iteration, we bring t into the basis. The simplex ratio test of the column d with column t shows the minimum of $(8/1, 8/1, 18/1, 5/1, 15/1, 10/1) = 5/1$. Thus z_{12} leaves the basis when t enters the basis (see Table 3). The row corresponding to t in Table 2 is the row corresponding to the objective function in the simplex algorithm. As long as any coefficient is negative in this row, the column where the negative coefficient lies is a candidate to enter the basis. The current value of t can be increased by bringing the variable x_1 or x_2 into the basis, say x_2 is to enter. Using the column d and column x_2 in Table 3, the ratio test shows z_1 leaves the basis. Continuing this way we arrive at Table 6 and t attains its maximum;row and column of t and columns representing z_{23} and z_1 are deleted.

Iteration 2 (Table 7) begins with all the z variables changed to $y-$ variables. The $z-$ variables are introduced for the new iteration because all the rows are not elementary (Table 8). The new column $t-$ is easily written by simply adding the coefficients corresponding to $z-$ columns. The ratio test for column $d-$ and column $t-$ indicates that z_{13} must leave for t to enter the basis. We get Table 9 by pivoting on the entry column $t-$, row z_{13}. Observe that the entire row corresponding to row t in Table 9 is nonnegative. Deleting columns corresponding to positive entries of row t in Table 9 we get Table 10 in variables y and x. The final matrix in Table 10 has only elementary rows. The algorithm terminates with the nucleolus as $(x_1^*, x_2^*, x_3^*) = (1, 4.5, 4.5)$.

Table 1. Initial data of the game in the variables y's and x's

y_{12}	y_{13}	y_{23}	y_1	y_2	y_3	x_1	x_2	x_3	d	
0	0	0	0	0	0	1	1	1	10	x_3
0	0	0	1	0	0	-1	0	0	8	y_1
0	0	0	0	1	0	0	-1	0	8	y_2
0	0	0	0	0	1	1	1	0	18	y_3
1	0	0	0	0	0	-1	-1	0	5	y_{12}
0	1	0	0	0	0	0	1	0	15	y_{13}
0	0	1	0	0	0	1	0	0	10	y_{23}

Table 2. Initial data in the new variables t and z's.

t	z_{12}	z_{13}	z_{23}	z_1	z_2	z_3	x_1	x_2	x_3	d	
0	0	0	0	0	0	0	1	1	1	10	x_3
1	0	0	0	1	0	0	-1	0	0	8	z_1
1	0	0	0	0	1	0	0	-1	0	8	z_2
1	0	0	0	0	0	1	1	1	0	18	z_3
1	1	0	0	0	0	0	-1	-1	0	5	z_{12}
1	0	1	0	0	0	0	0	1	0	15	z_{13}
1	0	0	1	0	0	0	1	0	0	10	z_{23}

Table 3. Iteration 1; t enters the basis.

t	z_{12}	z_{13}	z_{23}	z_1	z_2	z_3	x_1	x_2	x_3	d	
0	0	0	0	0	0	0	1	1	1	10	x_3
0	-1	0	0	1	0	0	0	1	0	3	z_1
0	-1	0	0	0	1	0	1	0	0	3	z_2
0	-1	0	0	0	0	1	2	2	0	13	z_3
1	1	0	0	0	0	0	-1	-1	0	5	t
0	-1	1	0	0	0	0	1	2	0	10	z_{13}
0	-1	0	1	0	0	0	2	1	0	5	z_{23}

Table 6. t attains its maximum; row and column of t and columns z_{23} and z_1 deleted.

t	z_{12}	z_{13}	z_{23}	z_1	z_2	z_3	x_1	x_2	x_3	d	
0	1	0	-0.5	-0.5	0	0	0	0	1	6	x_3
0	-1	0	0	1	0	0	0	1	0	3	x_2
0	-1	0	-0.5	0.5	1	0	0	0	0	2	z_2
0	1	0	-1	-1	0	1	0	0	0	5	z_3
1	0	0	0.5	0.5	0	0	0	0	0	9	t
0	1	1	-0.5	-1.5	0	0	0	0	0	3	z_{13}
0	0	0	0.5	-0.5	0	0	1	0	0	1	x_1

Table 7. Iteration 2 begins with new data in variables y's and x's.

y_{12}	y_{13}	y_2	y_3	x_1	x_2	x_3	d	
1	0	0	0	0	0	1	6	x_3
-1	0	0	0	0	1	0	3	x_2
-1	0	1	0	0	0	0	2	y_2
1	0	0	1	0	0	0	5	y_3
1	1	0	0	0	0	0	3	y_{13}
0	0	0	0	1	0	0	1	x_1

Table 8. Variables t, z's are introduced.

t	z_{12}	z_{13}	z_2	z_3	x_1	x_2	x_3	d	
1	1	0	0	0	0	0	1	6	x_3
-1	-1	0	0	0	0	1	0	3	x_2
0	-1	0	1	0	0	0	0	2	z_2
2	1	0	0	1	0	0	0	5	z_3
[2]	1	1	0	0	0	0	0	3	z_{13}
0	0	0	0	0	1	0	0	1	x_1

Table 9. Variable t enters; z_{13} leaves the basis.

t	z_{12}	z_{13}	z_2	z_3	x_1	x_2	x_3	d	
0	0.5	-0.5	0	0	0	0	1	4.5	x_3
0	-0.5	0.5	0	0	0	1	0	4.5	x_2
0	-1	0	1	0	0	0	0	2	z_2
0	0	-1	0	1	0	0	0	2	z_3
1	0.5	0.5	0	0	0	0	0	1.5	t
0	0	0	0	0	1	0	0	1	x_1

Table 10. Iteration ends; row- t, columns t, z_{12}, z_{13} are deleted. Rows are elementary and algorithm terminates

y_2	y_3	x_1	x_2	x_3	d	
0	0	0	0	1	4.5	x_3
0	0	0	1	0	4.5	x_2
1	0	0	0	0	2	y_2
0	1	0	0	0	2	y_3
0	0	1	0	0	1	x_1

This prolonged simplex will in general be unmanageable if the number of players is large. One needs methods to store only the relevant data for the effiecient useage of the storage space.

In his thesis Solymosi [1993] shows (Theorem 3.4) that the sequence of dual linear programs in which the maximum increments in satisfactions and settled-unsettled partitions could be determined directly. The computation of an optimal payoff is postponed till the end of the sequence, when the only feasible payoff is the lexicographic center itself. The implementation of this algorithm via the revised simplex method could turn out to be an efficient procedure to locate the nucleolus. This method was initially proposed by Dragan [1981]

2 Assignment Games

Assignment games model two-sided markets (buyers-sellers) where prices competitively balance supply and demand corresponding to elements in the core. The prices corresponding to the nucleolus allocation favor neither buyers nor sellers and hence provide some stability for the market. We start with an example.

Example : There are four houses for sale. The owners approach a real estate agent and indicate that an offer less than 200 thousand dollars on house 1 will be rejected by the owner. For houses 2, 3, and 4, the figures are respectively 240, 300 and 320 thousand dollars. Five potential buyers are interested in buying one house each. The houses are valued differently by the five buyers. Buyer 1 feels that the maximum price for houses 1, 2, 3, and 4 are respectively 260, 280, 290 and 340 thousand dollars. The columns of the following matrix summarizes the initial price ceilings for all the buyers. The agent keeps the anonymity of buyers and sellers till an actual sale is executed. Till then the agent acts as the liason between the two parties.

Buyer's ceiling prices on each house

	Buyer 1	Buyer 2	Buyer 3	Buyer 4	Buyer 5
Seller 1 expects $\geq$ 200	260	270	240	250	290
Seller 2 expects $\geq$ 240	280	270	310	320	270
Seller 3 expects $\geq$ 300	290	310	330	360	340
Seller 4 expects $\geq$ 320	340	340	370	390	400

For example a sale could take place between seller 2 and buyer 5 if they could agree on a price p where $240 \leq p \leq 270$. However the seller will prefer someone like buyer 4 who values the house even more. The agent has an interest in collecting real estate commissions proportional to the net gain from both parties. For example, if seller 2 sells his house to buyer 5 for a price p, the seller's gain is $p-240$ and the buyer's gain is $270-p$. Thus the net gain is $p-240+270-p=30$. When the net gain is high the two sides are keen on the actual transaction. Thus the agent would find it profitable to pair suitable buyers with sellers that will maximize the total gain and hence also his total commission. Suppose buyer

j is willing to give a fixed commission w_j nickels provided the agent finds an acceptable seller. In the same way, suppose seller i is willing to give a commission u_i nickels. The agent expects a 5% commission on the net gains. Thus he will not mind if $u_i + w_j \geq v(i,j)$ where the agent's expectation is $v(i,j)$ nickels for his efforts. In general he expects to collect from any coalition S of buyers and sellers $v(S) = \max_{\sigma} \Sigma_{(ij)\in\sigma, i\in S, j\in S} a_{ij}$ Thus we have a cooperative game with $v(S)$ as the characteristic function. Let $v(N) = \max_{\sigma} \Sigma_{(ij)\in\sigma} v(i,j)$ where σ runs over all matchings of sellers i with buyers j. In case some buyers cannot buy or some sellers cannot sell, we pretend them selling to a dummy buyer called 0 or a dummy seller also called 0. We need dummy sellers and dummy buyers even if all sellers and buyers match. The entries $(i,0), (0,j)$ represent the essential 1-coalitions $\{i\}$, and $\{j\}$. Assignment games were introduced by Shapley and Shubik [1972] who showed that these games have a nonempty core; core imputations are simply the dual optimal solutions to the optimal assignment problem of maximizing the gain for the grand coalition. They also showed that the core has certain lattice structure, namely if $(u', w'), u'', w'')$ are any two core elements for the set of sellers and buyers, then $(u' \vee u'', w' \wedge w'')$ also lies in the core.

Huberman [1980], introduced essential coalitions for balanced games and proved that only essential coalitions need to be considered for the nucleolus. Thus for assignment games, only one and two person mixed coalitions are essential. Thus for these games, the general iterative scheme of Maschler, Peleg and Shapley reduces to considering only singletons and two person (buyer-seller) coalitions. Let $M = \{0, 1, \cdots, m\}$ be sellers and let $N = \{0, 1, 2, \cdots, n\}$ be buyers. To compute the nucleolus we need dummy buyers and dummy sellers even if all the sellers and buyers match. The two person coalitions $(i, 0)$ and $(0, j)$ represent essential 1-coalitions $\{i\}$ and $\{j\}$.

Let $a_{ij} = v(i,j)$ for any seller i, buyer j. Let $a_{0j} = a_{i0} \equiv 0 \ \forall i \in M, j \in N$. Let σ be an optimal matching. For any imputation (u, w), the satisfaction for coalition (i,j), $i \in M, j \in N$ is $f_{ij}(u,w) = u_i + w_j - a_{ij}$.

Let $(\Delta^0, \Sigma^0), (\Delta^1, \Sigma^1), \cdots$, be a sequence of partitions of $M \times N$ and let $X^0 \supset X^1 \supset \cdots$ be subsets of the imputation set defined recursively as follows:

Initially $\Delta^0 = \{(i,j) : (i,j) \in \sigma\}$. $\Sigma^0 = M \times N \setminus \Delta^0$.

$X^0 = \{(u,w) : (u,w) \geq (0,0), f_{ij}(u,w) = 0 \ \forall (i,j) \in \Delta^0, f_{ij}(u,w) \geq \alpha^0 \ \forall (i,j) \in \Sigma^0\}$ where $\alpha^0 = \min_{(i,j)\in\Sigma^0} f_{ij}(u^0, v^0)$ with $u_i^0 = a_{i\sigma(i)} \ \forall i \in M$, $v_j^0 = 0 \ \forall j \in N$.

For $r = 0, 1, 2, \cdots, \rho$ define recursively

1. $\alpha^{r+1} = \max_{(u,w)\in X^r} \min_{(i,j)\in\Sigma^r} f_{ij}(u,w)$

2. $X^{r+1} = \{(u,w) \in X^r : \min_{(i,j)\in\Sigma^r} f_{ij}(u,w) = \alpha^{r+1}\}$

3. $S_{r+1} = \{(i,j) \in \Sigma^r : f_{ij}(u,w) = \text{constant} on X^{r+1}\}$

4. $\Sigma^{r+1} = \Sigma^r \setminus S_{r+1}$, $\Delta^{r+1} \cup S_{r+1}$

where ρ is the last index r for which $\Sigma^r \neq \emptyset$.

Theorem:

In the above recursive scheme, $x^{\rho+1}$ is a unique point and is the nucleolus of the assignment game.

While this recursive scheme has drastically reduced the search procedures to coalitions of size 2 or less, a further substantial reduction occurs when buyers and sellers are put in equivalence classes.

Let (Δ, Σ) be an intermediate iteration with feasible set X. By assumption, $f_{ij}(u,w)$ is constant on X $\forall$ $(i,j) \in \Delta$. Coalitions in Δ are called *settled coalitions*. Coalitions in Σ are called *unsettled coalitions*. We say seller i is *equivalent* to seller k, $(i \sim k)$ provided $(i, \sigma(k)) \in \Delta$. It is not hard to check that $\sim$ is an equivalence relation. (See Lemma 4.1 and Corollary 4.3 in Solymosi and Raghavan [1994]). This equivalence relation is a crucial step in the development of the algorithm to follow.

Let $M_0, M_1, \cdots, M_d$ be the partition of sellers into equivalence classes and let $N_p = \sigma(M_p)$, $p = 0, 1, \cdots, d$.

Theorem:

In the above partition (Δ, Σ) with feasible set X

1. $\Delta = \cup_{p=0}^{d} M_p \times N_p$ and $\Sigma = \cup_{p \neq q} M_p \times N_q$

2. If $(i,j), (k,l) \in M_p \times N_q$, then $f_{ij} - f_{kl}$ is constant on X.

Theorem:

Let $(\Delta, \Sigma, X, \alpha)$ be generic intermediate iterations of the algorithm starting with $\Delta^0, \Sigma^0, X^0, \alpha^0$. Let ν be the nucleolus. Then

$$
\begin{aligned}
X = \{(u,w) : \ f_{ij}(u,w) &= f_{ij}(\nu) \ \forall (i,j) \in \Delta \\
f_{ij}(u,w) &\geq \alpha \ \ \forall (i,j) \in \Sigma\}
\end{aligned}
$$

The set X is a lattice in the sense that (u', w'), $(u'', w'') \in X \Rightarrow (u' \vee u'', w' \wedge w'') \in X$. Thus X admits a unique u-best, w-worst point.

We have all the ingredients to develop an efficient algorithm. We can use the induced equivalence classes to improve on the recursive scheme. We only keep track of the u-best w-worst point of set X^r, in each iteration r. To move to the u-best, w-worst point of the next iteration all we need is the direction of search and step size for updating. This is facilitated by a directed graph associated with each iteration.

Given the settled coalitions Δ^r, Σ^r and X^r, coalitions $(i,j) \in M \times N$ are put in, say, $d+1$ equivalence classes $M_0, M_1, \cdots, M_d$. We will define a graph $G(X,d)$ (later called $G(X_r, \alpha)$) as follows: associate with each equivalence class M_p a vertex p. Associate with each $(i,j) \in M_p \times N_q$ with $p \neq q$ and $(i,j) \in S_{r+1}$ a directed arc from vertex p to vertex q. Thus if $f_{ij}(u, w) = \alpha_{r+1}\ \forall(u, w) \in X^{r+1}$, we have $(i, j) \in S_{r+1}$. In fact we evaluate only at $(\overline{u}^r, \underline{w}^r)$. The graph is called proper as long as there is no arc from any vertex i to 0 and there are no cycles. In the beginning with Δ^0, the graph would consist of just a set of vertices. In each iteration there will be a new arc added to the current graph. Ultimately the graph will be improper. Once it is improper, we collapse any cycle to one of its vertices with all other incoming and outgoing arcs inherited by this vertex. Also we combine any vertex $q \neq 0$ with 0 if $q \to 0$ is an arc. The vertex 0 inherits all the incoming and outgoing arcs of vertex q.

When the graph is proper it amounts to the possibility of strict improvement. (Theorem 5.7, Solymosi and Raghavan [1994]). We can compute $l(p)$, the length of the longest path to vertex p. For each seller i and buyer j we will associate integers s_i, t_j respectively where $s_i = -l(p)$ if $i \in M_p$ and $t_j = l(q)$ if $j \in N_q$ for any given i, j. Thus we get a direction vector $(s, t) = (s_0, s_1, \cdots, s_m; t_0, t_1, \cdots, t_n)$. If $(i, j) \in \Delta$, then by the above theorems $p = q$ and $s_i + t_j = 0$. Choose a step size β such that

$$\beta = \max\{r : \min_{(i,j)\in\Sigma} f_{ij}(\overline{u} + \underline{w}) + r(s, t) = \alpha + r\}.$$

Then $(\overline{u} + \underline{w}) + \beta(s, t)$ is the u-best corner of the next set $X' = X(r, \alpha + \beta)$.

The following is the algorithm:

Iteration r

While $\Sigma \neq \emptyset$, do (1) to (10)

1. Build the graph $G := G(r, \alpha)$.
2. Make G proper if necessary by melting each directed cycle to one of its vertices. If an arc $q \to 0$ exists then delete arc $q \to 0$ and melt vertex q. Let vertex 0 inherit the incoming arcs to q and and out going arcs from q
3. Find direction (s, t) and step size β.
4. Update arcs in the graph $G := G(r, \alpha + \beta)$.
5. Update payoff $(u, w) = (u, w) + \beta(s, t)$.
6. Update satisfactions $f_{ij} = f_{ij} + \beta(s_i + t_j)\ \forall\,(i, j) \in \Sigma$.
7. Update guaranteed satisfaction level $\alpha := \alpha + \beta$.
8. Find coalitions to be settled $\overline{\Sigma} := S_{r+1}$.
9. Update partition $\Sigma := \Sigma \setminus \overline{\Sigma}$, $\Delta := \Delta \cup \overline{\Sigma}$.
10. Set $r := r + 1$.

We will illustrate the algorithm for our example. We can as well take the gain matrix (dividing by 10)

Matrix 1

$$A = \begin{bmatrix} 6 & \boxed{7} & 4 & 5 & 9 \\ 4 & 3 & \boxed{7} & 8 & 3 \\ 0 & 1 & 3 & \boxed{6} & 4 \\ 2 & 2 & 5 & 7 & \boxed{8} \end{bmatrix} = (a_{ij})_{i=0,\dots,4;j=0,\dots,5}.$$

We will initially solve the optimal assignment problem. In this case the optimal assignment is the boxed assignment, namely house 1 sold to buyer 2, house 2 to buyer 3, house 3 to buyer 4, and house 4 to buyer 5. Buyer 1 is left alone. We tie a dummy seller 0 to him. Thus we will also include a dummy seller 0 and dummy buyer 0.

Initially we select the $\overline{u}$-best, $\underline{w}$-worst point which gives all the gains of sales to the sellers. We get $(\overline{u}) = (0,7,7,6,8)^T$ and $\underline{w} = (0,0,0,0,0,0)$. We will include $\overline{u}$ in the leftmost column and $\underline{w}$ at the topmost row by extending the original matrix to a 5×6 matrix (Matrix 2) given below

Matrix 2

$$A = \left[\begin{array}{c||cccccc} \boxed{0} & \boxed{0} & 0 & 0 & 0 & 0 \\ \hline 7 & 6 & \boxed{7} & 4 & 5 & 9 \\ 7 & 4 & 3 & \boxed{7} & 8 & 3 \\ 6 & 0 & 1 & 3 & \boxed{6} & 4 \\ 8 & 2 & 2 & 5 & 7 & \boxed{8} \end{array}\right]$$

$= (a_{ij})_{i=1,\dots 4;\ j=0,\dots 5}$.

Throughout the algorithm these optimally matched pairs share the exact gains they make. (Remember that $u_0 = w_0 = 0$.

Iteration $r = 0$ begins:

$\Delta = \{(0,0),(0,1),(1,2),(2,3),(3,4),(4,5)\}$ Since $(0,1) \in \Delta, w_1 = 0$. The satisfaction matrix $f_{ij} = u_i + w_j - a_{ij}$ is updated. We keep $u_i = u_i + w_0 - a_{i0} = f_{i0}$. Thus the current payoffs u_i, w_j are in the 0-th column and 0-th row as given below with the updated satisfaction matrix (Matrix 3)

Matrix 3

$\boxed{0}$	$\boxed{0}$	0	0	0	0
7	1	$\boxed{0}$	3	2	-2^*
7	3	4	$\boxed{0}$	-1	4
6	6	5	3	$\boxed{0}$	2
8	6	6	3	1	$\boxed{0}$

We build the graph G with $\alpha = \alpha^0 = -2$, namely the initial guaranteed satsifaction. It is for coalition $(1,5)$. This coalition is starred to indicate that it is an active yet unsettled coalition. The boxed entries are the settled coalitions. The rest are unsettled but passive coalitions. Currently the graph G has just 5 vertices $0,1,2,3,4$. Since seller $1 \in M_1$ and buyer $5 \in N_4$(buyer 5 buys from seller 4) $= \sigma(M_4)$ we draw an arc from 1 to 4 and update the length $l(p)$ of the longest path to each vertex p. We get Graph 1. (See the last page) The direction t is determined by the longest path $l(p)$ reaching each vertex p and appears in the bottom row of the matrix below. The direction s is on rightmost column of that matrix. The direction s is determined by $s_i + t_{\sigma(i)} = 0$ where σ is the optimal assignment. The new matrix (Matrix 4) is given below:

Matrix 4

$\boxed{0}$	$\boxed{0}$	0	0	0	0	$s_0 = 0$
7	1	$\boxed{0}$	3	2	-2^*	$s_1 = 0$
7	3	4	$\boxed{0}$	$-1^{♣}$	4	$s_2 = 0$
6	6	5	3	$\boxed{0}$	2	$s_3 = 0$
8	6	6	3	1	$\boxed{0}$	$s_4 = -1$
$t_0 = 0$	$t_1 = 0$	$t_2 = 0$	$t_3 = 0$	$t_4 = 0$	$t_5 = 1$	

The step size β is so chosen that at least one passive coalition becomes active in the next step. The currently passive coalition with a ♣ entry will become an active but unsettled coalition in the next step. The updated satisfaction matrix is computed as follows: For example the updated f_{43} for seller-4 buyer-3 is $3 + \beta(t_3 + s_4) = 3 + \beta(0 + (-1)) \geq -2$ Working out for all unsettled passive coalitions, we get the step size $\beta = 1$. We get the following updated satisfaction matrix (Matrix 5)

Matrix 5

With $\beta = 1$, coalition (2,4)(♣ entry) has become active at the current step. Since $(2,4) \in M_2 \times N_3$ new arc $2 \to 3$ is added to Graph 1 and we get Graph 2.

$\boxed{0}$	$\boxed{0}$	$0^{\clubsuit}$	$0^{\clubsuit}$	0	1	$s_0 = 0$
7	1	$\boxed{0}$	3	2	-1^*	$s_1 = 0$
7	3	4	$\boxed{0}$	-1^*	5	$s_2 = 0$
6	6	5	3	$\boxed{0}$	3	$s_3 = -1$
7	5	5	2	$0^{\clubsuit}$	$\boxed{0}$	$s_4 = -1$
$t_0 = 0$	$t_1 = 0$	$t_2 = 0$	$t_3 = 0$	$t_4 = 1$	$t_5 = 1$	

Matrix 6

With step size $\beta = 1$, ♣ coalitions above , namely (0,2),(0,3) and (4,4) have become active in Matrix 6

$\boxed{0}$	$\boxed{0}$	0^*	0^*	1	2	$s_0 = 0$
7	$1^{\clubsuit}$	$\boxed{0}$	3	3	0^*	$s_1 = -1$
7	3	4	$\boxed{0}$	0^*	6	$s_2 = -1$
5	5	4	2	$\boxed{0}$	3	$s_3 = -3$
6	4	4	$1^{\clubsuit}$	0^*	$\boxed{0}$	$s_4 = -2$
$t_0 = 0$	$t_1 = 0$	$t_2 = 1$	$t_3 = 1$	$t_4 = 3$	$t_5 = 2$	

Once again the last row and column of Matrix 6 above give the new direction vectors t and s determined by $l(p)$, the longest path to each vertex in Graph 3 below. New active coalition $(0,2) \in M_0 \times N_1, (0,3) \in M_0 \times N_2$ and $(4,4) \in M_4 \times N_3$. add new arcs $0 \to 1, 0 \to 2$ and $4 \to 3$ to Graph 2 and we get Graph 3

Matrix 7

With $\beta = \frac{1}{2}$. ♣ Coalitions (1,1), (4,3)above are newly active but an active (2,4) turns passive.

$\boxed{0}$	$\boxed{0}$	$\frac{1}{2}^*$	$\frac{1}{2}^*$	$\frac{5}{2}$	3	$s_0 = 0$
$\frac{13}{2}$	$\frac{1}{2}^*$	$\boxed{0}$	3	4	$\frac{1}{2}^*$	$s_1 = 0$
$\frac{13}{2}$	$\frac{5}{2}$	4	$\boxed{0}$	1	$\frac{13}{2}$	$s_2 = -2$
$\frac{7}{2}$	$\frac{7}{2}$	3	1	$\boxed{0}$	$\frac{5}{2}$	$s_3 = -2$
5	3	$\frac{7}{2}$	$\frac{1}{2}^*$	$\frac{1}{2}^*$	$\boxed{0}$	$s_4 = -1$
$t_0 = 0$	$t_1 = 0$	$t_2 = 0$	$t_3 = 2$	$t_4 = 2$	$t_5 = 1$	

Since the newly emerging active coalitions $(1,1) \in M_1 \times N_0, (4,3) \in M_4 \times N_2$, we add new arcs $1 \to 0, 4 \to 2$. to Graph 3, resulting in the improper Graph 4. We delete arcs $0 \to 1, 1 \to 0$ and delete vertex 1. vertex 0 inherits arcs incoming or outgoing from 1. We thus get Graph 5. This graph is proper. The induced vectors are found using $l(p)$. The associated matrix (Matrix 8) is given by

Matrix 8

Iteration $r = 1$ begins. $\beta = \frac{1}{2}$. (2,4) and (3,3) become newly active. See the improper Graph 6

0	0	$\frac{1}{2}$	$\frac{1}{2}^*$	$\frac{5}{2}$	3	$s_0 = 0$
$\frac{13}{2}$	$\frac{1}{2}$	0	3	4	$\frac{1}{2}^*$	$s_1 = 0$
$\frac{13}{2}$	$\frac{5}{2}$	4	$\boxed{0}$	$1^{\clubsuit}$	$\frac{13}{2}$	$s_2 = -2$
$\frac{7}{2}$	$\frac{7}{2}$	3	$1^{\clubsuit}$	$\boxed{0}$	$\frac{5}{2}$	$s_3 = -2$
5	3	$\frac{7}{2}$	$\frac{1}{2}^*$	$\frac{1}{2}^*$	$\boxed{0}$	$s_4 = -1$
$t_0 = 0$	$t_1 = 0$	$t_2 = 0$	$t_3 = 2$	$t_4 = 2$	$t_5 = 1$	

Collapsing 2 and 3 to one node 2, we get after 2's inheritance of arcs the Graph 7 This is proper. The last row in the next matrix (Matrix 9) is (0, 0, 0, 2, 2, 1) because 0,1,2 are in one equivalence class and 3,4 are in another equivalence class and 5 is in the third equivalence class. We have the updated satisfactions with the reduced set of 3 equivalence classes, one for each vertex. The coalitions $(2,3), (3,3)$, and $(3,4)$ all belong to the same equivalence class and thus the active coalitions among them when tied to settled coalitions of the same equivalence class get settled. The updated satisfactions with settled coalitions are

Matrix 9

With $\beta = \frac{1}{6}$ (2,1) becomes active in the next matrix

0	0	$\frac{1}{2}$	$\frac{3}{2}$	$\frac{7}{2}$	$\frac{7}{2}$	$s_0 = 0$
$\frac{13}{2}$	$\frac{1}{2}$	0	4	5	1^*	$s_1 = 0$
$\frac{11}{2}$	$\frac{3}{2}^{\clubsuit}$	3	$\boxed{0}$	1	6	$s_2 = -2$
$\frac{5}{2}$	$\frac{5}{2}$	2	1	$\boxed{0}$	2	$s_3 = -2$
$\frac{9}{2}$	$\frac{5}{2}$	3	1^*	1^*	$\boxed{0}$	$s_4 = -1$
$t_0 = 0$	$t_1 = 0$	$t_2 = 0$	$t_3 = 2$	$t_4 = 2$	$t_5 = 1$	

The coalition $(0,3)$ which was originally unsettled and active with increased satisfaction become passive. Coalitions $(1,5), (4,3), (4,4)$ are active. The longest arc lengths $l(p)$ for the graph yields vectors s and t. the last row and the

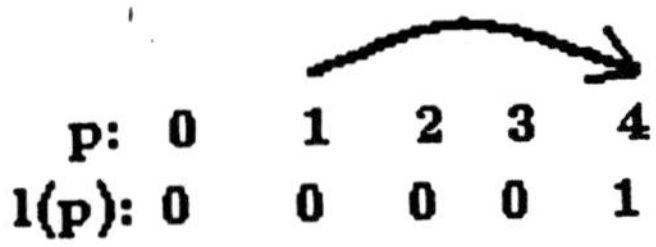

Graph 1

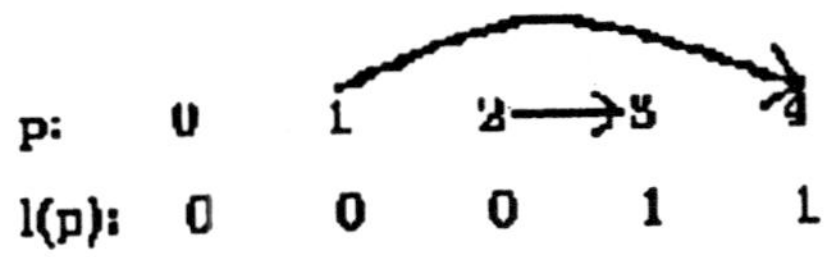

Graph 2

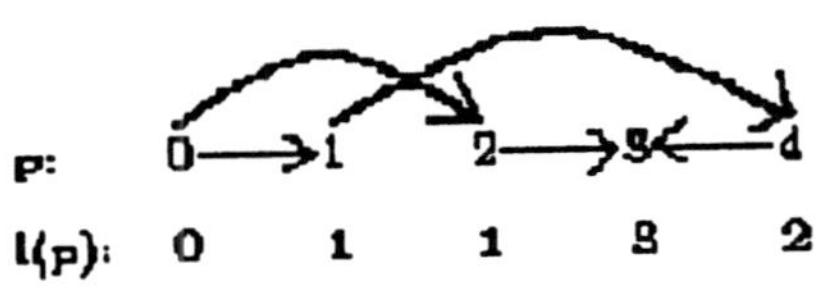

Graph 3

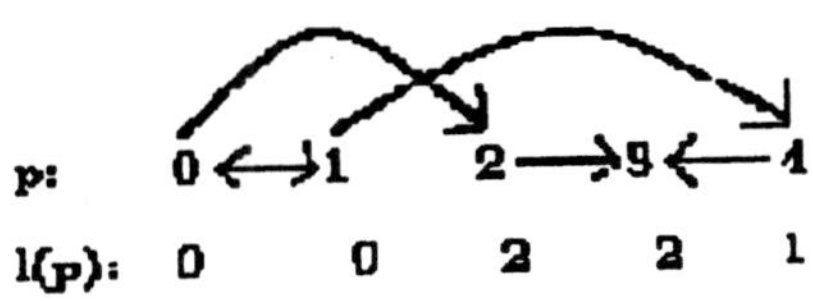

Graph 4

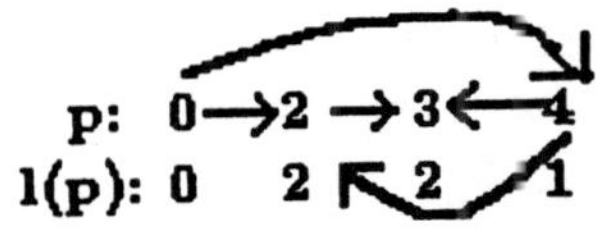

Graph 5

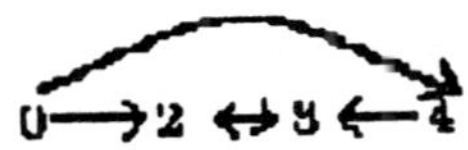

Graph 6

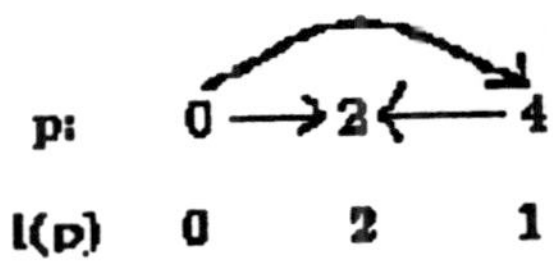

Graph 7

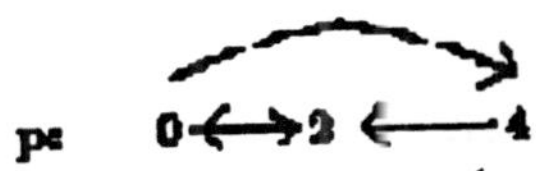

Graph 8

last column in the matrix above. The active coalitions $(1,5),(4,3),(4,4)$ can all be updated with a satisfaction $7/6 = 1 + \beta$ for $\beta = 1/6$ that makes the just passive coalition $(2,1)$ active with updated satisfaction $\frac{3}{2} - \frac{1}{6} = \frac{7}{6}$. Since $(2,1) \in M_2 \times N_0$, a new arc $2 \to 0$ is added to the existing graph. We get the new improper Graph 8. Since it has a full cycle, vertices are melted to vertex 0. The algorithm terminates after updating Matrix 9. The nucleolus is given by the leftmost column and topmost row when Matrix 9 is updated to matrix 10 of satisfactions.

Matrix 10

The cycle $0 \to 4 \to 2 \to 0$ forms. Algorithm stops.

0	0	$\frac{1}{2}$	$\frac{11}{6}$	$\frac{23}{6}$	$\frac{11}{3}$
$\frac{13}{2}$	$\frac{1}{2}$	0	$\frac{13}{3}$	$\frac{16}{3}$	$\frac{7}{6}$
$\frac{31}{6}$	$\frac{7}{6}$	$\frac{8}{3}$	0	1	$\frac{35}{6}$
$\frac{13}{6}$	$\frac{13}{6}$	$\frac{5}{3}$	1	0	$\frac{11}{6}$
$\frac{13}{3}$	$\frac{7}{3}$	$\frac{17}{6}$	$\frac{7}{6}$	$\frac{7}{6}$	0

The nucleolus for sellers 0, 1, 2, 3, 4, and buyers 0, 1, 2, 3, 4, 5 is given by $\left(0, \frac{13}{2}, \frac{31}{6}, \frac{13}{6}, \frac{13}{3}; 0, 0, \frac{1}{2}, \frac{11}{6}, \frac{23}{6}, \frac{11}{3}\right)$. Omitting dummy players, the nucleolus for (real) sellers and (real) buyers is $\left(\frac{13}{2}, \frac{31}{6}, \frac{13}{6}, \frac{13}{3}; 0, \frac{1}{2}, \frac{11}{6}, \frac{23}{6}, \frac{11}{3}\right)$.

3 Balanced connected games

Consider a cooperative game (N, v). A coalition S of the type $S = \{k : i \leq k \leq j\}$ is called an interval coalition. Let $\mathcal{I}$ be the collection of all interval coalitions.

A coalition S is *essential* for (N, v) if $v(S) > \Sigma_{T \in \Gamma} v(T)$ for any Γ which is an arbitrary, non- trivial partition of S. The game is called *connected* iff any essential coalition is an interval coalition. Our notion of connectedness is somewhat narrow. More general formulations are possible. Connected games appear naturally in many situations.

Example:

Consider a repairman leaving his house 0 and visiting customers in the order

$$0 \to 1 \to 2 \to \cdots \to n \to 0.$$

Let c_{ij} be the cost of travel to go from i to j. He would have to collect $\Sigma_0^n c_{ii+1}$ (here, $c_{nn+1} = c_{n0}$) from the customers. Let $S = \{i_1, i_2, \cdots, i_k\}$ be a set of customers with $i_1 < i_2 < \cdots < i_k$. Let $C(S)$ be the cost of travel exclusively to customers S with the restriction that

$$0 \to i_1 \to i_2 \to \cdots \to i_k \to 0.$$

We can associate with this cost a game $v^*(S) = C(N) - C(N-S)$, where players in S contribute the cost in excess of what others outside S (namely $N-S$) have to bear. We can think of $v^*(S)$ as a contribution of coalition S to the total cost. Interestingly v^* is a connected game. We will assume that the core $C(N,v) \neq$.

Analogous to assignment games, starting with $\alpha^0 = 0$ define recursively just using interval coalitions $(\Delta^r, \Sigma^r), X^r, \alpha^{r+1}, S_{r+1}$ and Δ^{r+1}. Let ρ be the last index for which $\Sigma^r \neq$

The set $X^{\rho+1}$ has a unique element, namely the nucleolus.

Example:

A cargo plane carries cargo from home base 0 to cities $1, 2, \cdots, n$ in the order

$$0 \to 1 \to 2 \to \cdots \to n \to 0.$$

The transportation cost $c_{01} + c_{12} + \cdots + c_{n0}$ for this flight schedule is to be paid by cities $1, 2, \cdots, n$. If the cargo service is available also to any set S of cities $i_1 < i_2 < \cdots < i_k$ where the route is

$$0 \to i_1 \to i_2 \to \cdots \to i_k \to 0.,$$

then S will be charged $C(S) = c_{0i_1} + c_{i_1 i_2} + \cdots + c_{i_n i_0}$. Here $c_{\alpha\beta}$ = direct travel cost from city α to city β . The induced game (N, C) is called a routing game (Potters, Curiel and Tijs [1992]). Routing games have the following interesting property.

Theorem: [Derks and Kuijpers]

Let (N, C) be a routing game. The dual game (N, v) where $v(S) = C(N) - C(N \setminus S)$ is a connected game. The two games have the same core and the same nucleolus. The core is nonempty iff $C(N) \leq C(S) + C(N \setminus S)$ for all coalition S.

The following algorithm efficiently finds a core element for a connected game (N, v).

Given $\mathcal{I}$ the collection of interval coalitions, let $\hat{x}$ be the lexmin vertex of the polytope $\hat{C} = \{x : x(S) \geq v(S) \ \forall S \in \mathcal{I}\}$. Since $\hat{x_i} \geq v\{i\} \ \forall\, i$, the polytope is bounded below and its lexmin is well-defined.

The lexmin vertex $\hat{x}$ can be found by solving recursively the n linear programs

$$\begin{aligned}
\hat{x_1} &= \min\{x_1 : x \in \hat{C}\} \\
\hat{x_2} &= \min\{x_2 : x \in \hat{C}, x_1 = \hat{x_1}\} \\
\cdots & \\
\hat{x_n} &= \min\{x_n : x \in \hat{C}, x_i = \hat{x_i}, 1 = 1, 2, \cdots, n-1\}
\end{aligned}$$

The following theorem characterizes connected games with nonempty core.

Theorem: (Derks and Kuipers [1992])

A connected game (N, v) has a nonempty core iff the lexmin payoff $\hat{x}$ is efficient, that is, $\hat{x}[1\ n] = v(N)$.

We will illustrate with an example the recent algorithm of Solymosi, Aarts and Driessen [1995] to locate the nucleolus of balanced connected games in at most $O(n^4)$ in time and $O(n^2)$ in space, where n is the size of the player set. The algorithm adapts the improving direction method that was successfully applied to solving assignment games. The sellers corner for assignment games is replaced by the lexmin corner. Starting with the lexmin vertex of the core, one moves to a shrinking subset of the core in the next iteration , specificaly to its lexmin vertex. Thus the algorithm reduces to determining the lexmin vertex, a new direction and a new stepsize for each iteration.

We need the following notations to describe the algorithm.

For any interval coalition $S = [i\ j]$, we denote the left end i by $i(S)$ and the right end j by $j(S)$. For a collection Γ of coalitions, let $I(\Gamma) = \{i(S) : S \in \Gamma\}$, $J(\Gamma) = \{j(S) : S \in \Gamma\}$,

$$\vee(\Gamma) = \{[i, j] : i \in I(\Gamma),\ j \in J(\Gamma),\ i \le j\},$$

$$\wedge(\Gamma) = \{[i, j] : i \in I(\Gamma),\ j = \min\{k : i \le k, k \in J(\Gamma)\}\}.$$

Among settled coalitions in $\Delta^r, r \ge 1$, some are marginally settled. For example, a coalition $S \in \Sigma^{r-1}$ with constant satisfaction $f(x, S) = \alpha^r\ \forall\ x \in X^r$ is called a marginally settled coalition. A coalition $T \in \Sigma^{r-1}$ with $f(x, T) = \nu_T, (\nu_T > \alpha^r)\ \forall x \in X^r$is called a non-marginally settled coalition for iteration r. We denote by $\Delta^r_=$ all marginally settled coalitions up to iteration r. Thus $\Delta^r_=$ contains marginally settled coalitions of $\Delta^1, \Delta^2, \cdots, \Delta^r$. The same applies to non-marginally settled coalitions $\Delta^r_>$. They include all settled coalitions of past iterations which are non-marginal.

Theorem: For any $0 \le r \le \rho + 1$,

1. Δ^r is the union of interval partitions of N
2. $\Gamma^r := \wedge(\Delta^r_=)$ is an interval partition of N.
3. $\Delta^r := V(\Gamma^r)$
4. $|\Gamma^r|$ = number of sets in the collection $\Gamma^r < |\Gamma^{r+1}|$ if $r \le \rho$.

Theorem:

For any $\alpha^r \le \alpha \le \alpha^{r+1}$, let

$$X^r(\alpha) = \{x : x(N) = v(N), x(S) - v(S) \ge \alpha\ \forall\ S \in \Sigma^r, x(T) - v(T) = \nu_T\ \ \forall T \in \Delta^r\}$$

(Recall that ν_T is the constant level of satisfaction for coalition T for every $x \in X^r$) Let

$$\begin{aligned} v^{r,\alpha}(S) &= v(S) + \alpha \quad \text{if } S \in \Sigma^r \\ &= v(S) + \nu_S \quad \text{if } S \in \Delta^r \\ &= v(S) \quad \text{if } S \in 2^N \setminus \mathcal{I} \end{aligned}$$

Then the game $v^{r,\alpha}$ for $0 \le r \le \rho$ is connected and balanced. Its core is $C(v^{r,\alpha}) = X^r(\alpha)$.

Coalitions in iterations $0 \le r \le \rho$ have a satisfaction level $\alpha^r \le \alpha \le \alpha^{r+1}$. We call $S \in \Sigma^r$ *active* if the lexmin payoff $x^{r,\alpha}$ for $v^{r,\alpha}$ satisfies $x^{r,\alpha}(S) = v^{r,\alpha}(S)$ and *passive* otherwise. Let $A^{r,\alpha}, C^r$ be respectively the set of indicator vectors of active coalitions and marginally settled coalitions in iteration r. Let $H^{r,\alpha}$ be all these vectors arranged reverse alphabetically, namely if $S = \{p_1, p_2, \cdots, p_k,\}$ and $R = \{q_1, q_2, \cdots, q_l\}$ are coalitions then row e^S precedes row e^R if $\Sigma_{i=1}^{r} 2^{p_i} < \Sigma_{j=1}^{l} 2^{q_j}$. Since $x^{r,\alpha}$ is the lexmin vertex of $X^{r,\alpha}$ is also an extreme point at least one interval from $\mathcal{I}_j = \{I : I = [k\ j], k \le j\}$. Thus the matrix $H^{r,\alpha}$ contains a lower triangular basis for $\boldsymbol{R}^n$.

Thus $J(A^{r,\alpha}) \cup J(C^r) = N$. Also $J(A^{r,\alpha}) \ne \emptyset$ for any $0 \le r \le \rho$ and $\alpha^r \le \alpha \le \alpha^{r+1}$.

Theorem:

Given $0 \le r \le \rho$ and $\alpha^r \le \alpha \le \alpha^{r+1}$, the following are equivalent:

1. $\{d : A^{r,\alpha} d \ge 1,\ C^r \cdot d = 0\} = D^{r,\alpha} \ne \emptyset$
2. $\alpha^{r+1} > \alpha$
3. $J(A^{r,\alpha}) \cap J(C^r) = \emptyset$.

Theorem:

If $d \in D^{r,\alpha}$ is the lexmin vertex of $D^{r,\alpha}$, and

$$\beta^{r,\alpha} = \min \frac{\{x^{r,\alpha}(S) - v^{r,\alpha}(S)}{(1 - d(S))}, d(S) \le 0, S \in \Sigma^r\},$$

then for any $0 \le r \le \rho$ and $\alpha^r \le \alpha < \alpha^{r+1}$, the lexmin vertex of $X^r(\alpha + \beta) = x^{r,d} + \beta d\ \forall\ 0 \le \beta \le \beta^{r,\alpha}$.

We are ready to spell out the algorithm.

Algorithm:

Set $r = 0, \alpha = 0, \Gamma = \{N\}, \Sigma = \mathcal{I} \setminus \{N\}$.
Find $x = \text{lexmin}\ \{y : y(N) = v(N), y(S) \ge v(S),\ \forall S \in \Sigma\}$.
Compute $f(S) = x(S) - v(S)\ \forall S \in \Sigma$.
While $|\Gamma| < n$, do
Find $\Theta = \{S : f(S) = \alpha\}$

While $J(\Theta) \cap J(\Gamma) = \emptyset$, do

Find $d = \text{lexmin}\ \{\boldsymbol{y} : \boldsymbol{y} \in \boldsymbol{R}^n,\ \boldsymbol{y}(\boldsymbol{S}) \geq 1\ \forall \boldsymbol{S} \in \Theta,\ \boldsymbol{y}(\boldsymbol{T}) = 0\ \forall\ \boldsymbol{T} \in \Gamma\}$

compute $d(S)\ \forall S \in \Sigma$

Find $\beta = \min\{\frac{(f(S)-\alpha)}{(1-\alpha(S))} : \ S \in \Sigma, d(S) \leq 0\}$

Update $\alpha := \alpha + \beta$

Update $x := x + \beta d$

Update $f(S) := f(S) + \beta d(S)\ \forall\ S \in \Sigma$

Update $\Theta := \{S \in \Sigma : \ f(S) = \alpha\}$.

Return

While $J(\Theta) \cap J(\Gamma) \neq \emptyset$, do

Find $S \in \Theta$ with $j(S) = \min(J(\Theta) \cap J(\Gamma))$

Set $\Pi = \{S\}$.

Repeat

do

Find $T \in \Theta$ with $j(T) = i(S) - 1$

Update $\Pi := \Pi \cup \{T\}$

Set $S := T$

Return

Until $i(S) \in I(\Gamma)$

Update $\Gamma := \wedge(\Gamma \cup \Pi)$, $\Theta := \Theta \setminus \vee\Gamma$

Return

Update $\Sigma := \Sigma \setminus \vee(\Gamma)$, set $r = r + 1$, and return.

Example:

The following is a 4-person balanced connected game where $\overline{ij}$ denotes the coalition $[ij] = \{k : i \leq k \leq j\} \cap \{1, 2, 3, 4\}$. Here v is the characteristic function.

S:	$\overline{11}$	$\overline{22}$	$\overline{12}$	$\overline{33}$	$\overline{23}$	$\overline{13}$	$\overline{44}$	$\overline{34}$	$\overline{24}$	$\overline{14}$
v:	3	0	10	2	9	14	2	14	20	25

The lexmin vertex x is determined as follows: $x_1 = v[1\ 1] = 3 = v\{1\}$. To determine the lexmin value of x_2 we minimize x_2 subject to $x_2 \geq v[22]$, $x_1 + x_2 \geq v[12]$, and $x_1 = 3$, i.e. $x_2 \geq 0, 3 + x_2 \geq 10$ and x_2 a minimum. This gives $x_2 = 7$. We will fill up

$$\min\{x_3 : x_3 \geq 2, x_2 = 7, x_2 + x_3 \geq 9, x_1 = 3, x_1 + x_2 + x_3 \geq 14\}.$$

Thus $x_3 = 4$. Continuing in this way, $x(\overline{ij}) = \Sigma_{k=i}^{j} x_k$ is given by

S:	$\overline{11}$	$\overline{22}$	$\overline{12}$	$\overline{33}$	$\overline{23}$	$\overline{13}$	$\overline{44}$	$\overline{34}$	$\overline{24}$	$\overline{14}$
v:	3	0	10	2	9	14	2	14	20	25
x:	3	7	10	4	11	14	11	15	22	25
f:	0*	7	0*	2	2	0*	9	1	2	$\boxed{0}$

Here, $f(\overline{ij}) = x(\overline{ij}) - v(\overline{ij})$ = satisfaction for coalition $[ij]$ at lexmin x. Let $\alpha = 0$, $\Gamma = \{\overline{14}\}$. The lexmin vertex is $x = (3, 7, 4, 11)$. Initially the collection Γ has just one element, namely the grand coalition.

Iteration 0 begins:

The least satisfied value $f = 0$ are attained for coalitions $\overline{11}, \overline{12}, \overline{13}, \overline{14}$. Among them the unsettled ones belong to Θ and the rest to $\Delta_{=}$. In the intial step the grand coalition $\overline{14}$ has constant satisfaction 0 over the whole core. Thus $\overline{14}$ is a settled coalition. Thus $\Theta = \{\overline{11}, \overline{12}, \overline{13}\}, \Gamma = \{\overline{14}\}, J(\Theta) = \{1, 2, 3, \}. J(\Gamma) = 4$. Since $J(\Theta) \cap J(\Gamma) = \emptyset$, an improvement iteration follows. We indicate by a * the satisfactions of active but unsettled coalitions $\in \Theta$. The direction of motion d is determined by the vector that is lexmin for: $d(S) \geq 1\ \forall S \in \Theta, d(S) = 0\ \forall\ S \in \Gamma$. This is found by finding the lexmin point of

$$\begin{aligned} d_1 &\geq 1 \\ d_1 + d_2 &\geq 1 \\ d_1 + d_2 + d_3 &\geq 1 \\ d_1 + d_2 + d_3 + d_4 &= 0 \end{aligned}$$

The first three inequalities correpond to the current Θ; the last equation corresponds to the current Γ.

We find $d_1 = 1, d_2 = 0, d_3 = 0, d_4 = -1$ as the lexmin point. From now on we keep track of just S, f, and d.

$\alpha = 0$.

S:	$\overline{11}$	$\overline{22}$	$\overline{12}$	$\overline{33}$	$\overline{23}$	$\overline{13}$	$\overline{44}$	$\overline{34}$	$\overline{24}$	$\overline{14}$
f:	0*	7	0*	2	2	0*	9	1	2	$\boxed{0}$
d:	1	0	1	0	0	1	-1	-1	-1	0

Here the d row represents $d(\overline{ij}) = \Sigma_{k=1}^{j} d_k$ for various interval coalitions. To determine step size β, we look at $d(\overline{ij}) \leq 0$ and $\overline{ij}$ unsettled and not active. They are for $\overline{ij} = \{\overline{22}, \overline{33}, \overline{23}, \overline{44}, \overline{34}, \overline{24}\}$.

$$\beta = \min\left(\frac{f - \alpha}{1 - d}\right)(S) = \min\left(\frac{7}{1}, \frac{2}{1}, \frac{2}{1}, \frac{9}{2}, \frac{1}{2}, \frac{2}{2}\right) = 1/2.$$

Thus the step size is 1/2. We get update on satisfaction as $\alpha = \frac{1}{2}$. The update on f is given by $f := f + \beta d$.

S:	$\overline{11}$	$\overline{22}$	$\overline{12}$	$\overline{33}$	$\overline{23}$	$\overline{13}$	$\overline{44}$	$\overline{34}$	$\overline{24}$	$\overline{14}$
f:	$\frac{1}{2}^*$	7	$\frac{1}{2}$	2	2	$\frac{1}{2}^*$	$\frac{17}{2}$	$\frac{1}{2}$	$\frac{3}{2}$	$\boxed{0}$

The coalition $\overline{34}$, which was originally unsettled but inactive becomes active with minimal satisfaction 1/2. Now $\Theta = \{\overline{11}, \overline{12}, \overline{13}, \overline{34}\}$, $\Gamma = \{\overline{14}\}$. Also,

$J(\Theta) = \{1,2,3,4\}$; $J(\Gamma) = \{4\}$. Since $J(\Theta) \cap J(\Gamma) = \{4\}$, the settling step 0 begins. The coalition $S \in \Theta$ with $\min(J(\Theta) \cap J(\Gamma)) = \{4\}$ is $\overline{34}$. Let $\Pi = \{\overline{34}\}$. Its left neighbor in Θ is $\overline{12}$. Thus updated Π is $\{\overline{12}, \overline{34}\}$. Now we have to update Γ by considering

$$New\ \ \Gamma = \wedge\{old\ \ \Gamma \cup new\ \ \Pi\} = \wedge\{\overline{14}, \overline{12}, \overline{34}\} = \{\overline{12}, \overline{34}\}$$

$$New\ \ \Theta = old\ \ \Theta \setminus \vee(new\ \ \Gamma) = \{\overline{11}, \overline{13}\}.$$

Thus iteration $r = 0$ ends by removing $\overline{12}, \overline{34}$ from unsettled intervals and setting $r = 1$. Since $\Gamma = \{\overline{12}, \overline{34}\}, |\Gamma| < 4$ and $r = 1$ another iteration follows with the updated $\alpha, \Gamma, \Sigma, \Theta$.
Iteration $r = 1 : \alpha = 1/2, \Theta = \{\overline{11}, \overline{13}\},\ \Gamma = \{\overline{12}, \overline{34}\}$.

S:	$\overline{11}$	$\overline{22}$	$\overline{12}$	$\overline{33}$	$\overline{23}$	$\overline{13}$	$\overline{44}$	$\overline{34}$	$\overline{24}$	$\overline{14}$
f:	$\frac{1}{2}^*$	7	$\boxed{\frac{1}{2}}$	2	2	$\frac{1}{2}^*$	$\frac{17}{2}$	$\boxed{\frac{1}{2}}$	$\frac{3}{2}$	$\boxed{0}$
d:	1	-1	0	1	0	1	-1	0	-1	0

We box the satisfaction of settled interval coalitions.

The current active coalitions are $\overline{11}, \overline{13}$. Since $J(\Theta) = \{1,3\}$ and $J(\Gamma) = \{2,4\}$, $J(\Theta) \cap J(\Gamma) = \emptyset$. Thus improving step 1 begins.

The d's for this have to be the lexmin vertex of the system

$$\begin{aligned} d_1 &\geq 1 \\ d_1 + d_2 &= 0 \\ d_1 + d_2 + d_3 &\geq 1 \\ d_3 + d_4 &= 0 \\ d_1 + d_2 + d_3 + d_4 &= 0 \end{aligned}$$

. In the above table we have $d(S)$ corresponding to the above d for passive coalitions S.

$$\beta = \min\left(\frac{7-\frac{1}{2}}{2}, \frac{2-\frac{1}{2}}{1}, \frac{2-\frac{1}{2}}{2}, \frac{\frac{17}{2}-\frac{1}{2}}{2}, \frac{\frac{3}{2}-\frac{1}{2}}{2}\right) = \frac{1}{2}$$

The next best with updated f:

S:	$\overline{11}$	$\overline{22}$	$\overline{12}$	$\overline{33}$	$\overline{23}$	$\overline{13}$	$\overline{44}$	$\overline{34}$	$\overline{24}$	$\overline{14}$
f:	1^*	$\frac{13}{2}$	$\boxed{\frac{1}{2}}$	$\frac{5}{2}$	2	1^*	8	$\boxed{\frac{1}{2}}$	1^*	$\boxed{0}$

Notice $\Theta = \{11, 13, 24\}, \Gamma = \{12, 34\}$. Since $J(\Theta) \cap J(\Gamma) \neq \emptyset$, a settling step starts.

$\Pi = \{\overline{11}, \overline{24}\}$

$$\begin{aligned} New\ \Gamma &= \wedge\{\Pi \cup old\Gamma\} \\ &= \wedge\{\overline{11}, \overline{24}, \overline{12}, \overline{34}\} \\ &= \{\overline{11}, \overline{24}, \overline{34}\} \\ \vee(New\ \Gamma) &= \{\overline{11}, \overline{14}, \overline{24}, \overline{34}\} \end{aligned}$$

$$New\ \Theta = old\ \Theta \setminus \{\overline{11}, \overline{14}, \overline{24}, \overline{34}\} = \{\overline{13}\}.$$

Thus the updated table is given by

S:	$\overline{11}$	$\overline{22}$	$\overline{12}$	$\overline{33}$	$\overline{23}$	$\overline{13}$	$\overline{44}$	$\overline{34}$	$\overline{24}$	$\overline{14}$
f:	$\boxed{1}$	$\frac{13}{2}$	$\boxed{\frac{1}{2}}$	$\frac{5}{2}$	2	1	8	$\boxed{\frac{1}{2}}$	$\boxed{1}$	$\boxed{0}$
d:	0	0	0	1	1	1	-1	0	0	0

The lexmin d must satisfy

$$\begin{aligned} d_1 &= 0 \\ d_1 + d_2 &= 0 \\ d_1 + d_2 + d_3 &\geq 1 \\ d_3 + d_4 &= 0 \\ d_2 + d_3 + d_4 &= 0 \\ d_1 + d_2 + d_3 + d_4 &= 0 \end{aligned}$$

. Thus $d = (0, 0, 1, -1)$ is the lexmin solution whose d vector for coalitions is given above.
$\beta = \frac{7}{2}$ giving the updated table:

S:	$\overline{11}$	$\overline{22}$	$\overline{12}$	$\overline{33}$	$\overline{23}$	$\overline{13}$	$\overline{44}$	$\overline{34}$	$\overline{24}$	$\overline{14}$
f:	$\boxed{1}$	$\frac{13}{2}$	$\frac{1}{2}$	6	$\frac{11}{2}$	$\frac{9}{2}^*$	$\frac{9}{2}^*$	$\frac{1}{2}$	$\boxed{1}$	$\boxed{0}$

$\Theta = \{\overline{13}, \overline{44}\}$. With $\Gamma = \{11, 24, 34\}$ we find $J(\Theta) \cap J(\Gamma) = \{4\}$. Thus $\Pi = \{13, 44\}$. The new $\Gamma = \wedge\{11\quad 24\quad 34\} \cup \{13, 44\} = \{11, 24, 34, 44\}$. Since $|\Gamma| = 4$, the algorithm terminates. The nucleolus is given by $f(S) + v(S)$ for $S = \{\overline{11}, \overline{22}, \overline{33}, \overline{44}\}$, *i.e.*$(4, \frac{13}{2}, 8, \frac{13}{2})$.

4 Standard Tree Enterprise:

Consider a tree with given root. Let V be the set of vertices and let E be the set of edges with exactly one edge emanating from the root. Let $N = \{1, 2, \cdots, n\}$ be a set of n players also called residents where each player occupies exactly one vertex other than the root. Let each vertex be occupied by at least one player. If each edge e has a nonnegative cost $a(e)$ associated with it, we call the game a *standard tree game*.

Example: Imagine a city center to which residents from different villages commute daily through freeways. Often, residents belonging to a cluster of villages use a fixed exit/entry ramp of a fixed freeway. Thus we can identify any exit/entry ramp as a vertex of a tree, the city center as the initial vertex, and any cluster of villages whose residents use the same ramp , namely the vilages can be identified as the players residing at the given vertex. A freeway will represent a path from the intial vertex to some terminal vertices. If two distinct clusters of villages are closer to two ramps of the same freeway, then residents of one village from one cluster can go to another village of another cluster of the same freeway, avoiding the city. However, any two distinct clusters of villages not on the same freeway is assumed to be reachable only by passing through the city center. We can associate a vertex for a cluster of such nearby villages whose residents use a fixed exit/entry ramp. We can call such a cluster of villages neighboring players. Given any set of villages S, let $C(S)$ be the minimal cost to maintain the portion of the freeways that connects the main city to all the villages in S. We immediately have an induced cost game $C(V, E, a, N)$. Here vertices of the graph correspond to the exit ramp used by any given cluster of village residents. The edges correspond to the road between two adjacent exits of the same freeway. The cost $a(e)$ correspond to the maintainance cost for the road between the two adjacent exits.

Cost allocation problems on trees were first considered by Bird [1976]. Megiddo [1978] was the first one to compute the nucleolus of a standard tree game in $O(n^3)$ steps where n is the number of vertices of the tree. later it was improved by Galil [1979] to $O(n \log n)$.

The theoretical properties on the nucleolus for tree games were first studied by Granot and Huberman [1982]. Finally the theorems culminated in developing an efficient algorithm by Granot, Maschler, Owen and Zhu [1996] to locate the nucleolus for standard tree games like the one above.

There are two key approaches to understanding the nucleolus. One uses the iterative geometric center approach that was useful in locating the nucleolus for assignment games (Solymosi and Raghavan [1994], balanced connected games (Solymosi, Aarts and Driessen [1994] and also tree enterprise (Megiddo [1978]). Yet another powerful idea is to exploit the consistency property for reduced games (See Sobolev [1975], Driessen [1991], Maschler [1992]). This property can be fruitfully used to derive certain necessary equations. For tree enterprises by the reduction to two person games type of 2-person games, each player at the end of one edge yields necessary conditions for the nucleolus that turn out to be sufficient too.

The following are the essential ingredients to develop an efficient algorithm.

Theorem: (Granot-Huberman [1982]). Standard tree games satisfy the inequality
$C(T \cup i) - C(T) \leq C(S \cup i) - C(S)$ for $i \notin T$ and $S \subset T$.

Such games are called convex (See, Owen [1994]]) and they admit nonempty core. Since $\Sigma_{i \in S} C(i) - C(S) \leq \Sigma_{i \in T} C(i) - C(T)$ for any $S \supset T$, it satisfies the *zero- monotonicity* condition. The prekernel for a game consists of all $n-$tuples x satisfying $x(N) = \Sigma_{i \in N} x_i = C(N)$ and $\min\{C(S) - x(S) : S \ni k, s \not\ni l\} = \min\{C(S) - x(S) : S \ni l, S \not\ni k\}$.

Theorem: (Maschler-Peleg-Shapley [1972]) A standard tree game is zero-monotonic and has a unique element in its prekernel that coincides with the nucleolus.

Let i, j be players occupying end vertices v_i, v_j of an edge e where v_i is the immediate neighbor of v_j on the path from v_j to the root. Every edge splits the tree into two parts. One contains the root and is a rooted tree. The other consists of branches via v_j. if the edge (v_i, v_j) is added to them we get a subtree B_{ij} of the standard game tree (V, E, a, N) whose root is v_i where all paths in B_{ij} pass through v_j. We will also denote by B_{il} any branch of B_{ij} where v_l is located somewhere after v_j, adjacent to v_i. Let T_{ij} be the rooted subtree with the same root v_i as the original tree and containing vertices and arcs not in B_{ij}. For any preimputation x, let $x(v) =$ the sum of all x_i where players i reside in vertex v. If $S \subset V$, $x(S) = \Sigma_{v \in S} x(v)$.

Theorem: (Granot, et al. [1996])

Let x be the nucleolus of a standard tree game. Then for every pair of adjacent players i, j and players p, q residing at the same vertex

$$
\begin{aligned}
x_i = a(T_{ij}) - x(T_{ij}) \quad &\text{if} \quad x_j \geq a(T_{ij}) - x(T_{ij}) \\
x_i = x_j \quad &\text{if} \quad x_j \leq a(T_{ij}) - x(T_{ij}) \\
x_p = x_q \quad & \\
x(N) = C(N). \quad &
\end{aligned}
$$

Theorem:

Let (V, E, a, N) be a standard tree game. Then the system of equations

$$
\begin{aligned}
x_i = x(B_{ij}) - a(B_{ij}) - x(v^i) \quad &\forall \text{ adjacent pairs } (i,j) = \text{ edge } e_{ij} \\
x_p = x_q \quad &\text{when } p, q \text{ reside at the same vertex.} \\
x(N) = C(N) \quad &
\end{aligned}
$$

has a unique solution.

In the above description, the unique solution is called the proto- nucleolus. In case $x_i \leq x_j$ for all edges $[v_i, v_j]$ where vertex v_i precedes vertex v_j on the path from the root to vertex v_j, the proto-nucleolus will satisfy the inequalities of the previous theorem and hence will be the nucleolus. Any edge $[v_i, v_j]$) with $x_i > x_j$ for residents at v_i, v_j in the proto-nucleolus is called a bad edge. The algorithm involves fine tuning the proto-nucleolus by identifying the bad edges and removing them in a proper order.

We use an example due to Granot et al. to describe the algorithm to compute the proto-nucleolus. The following Tree describes the game. The costs are given for all edges.

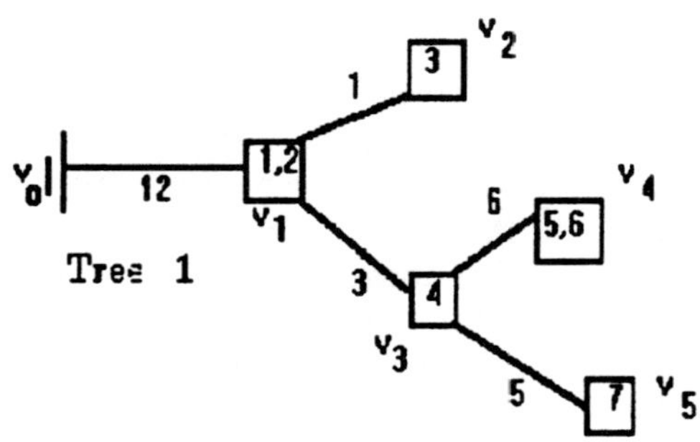

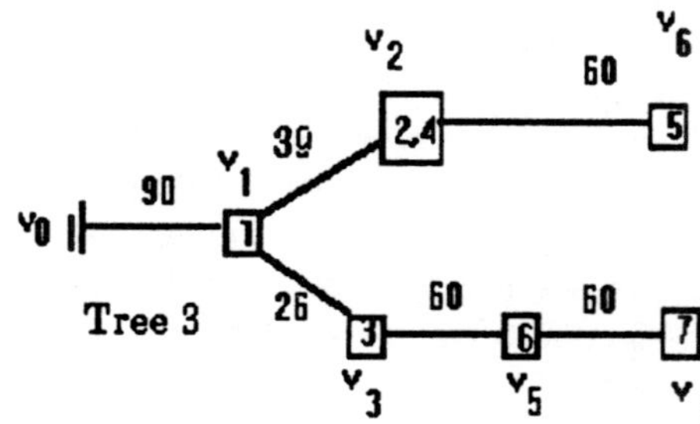

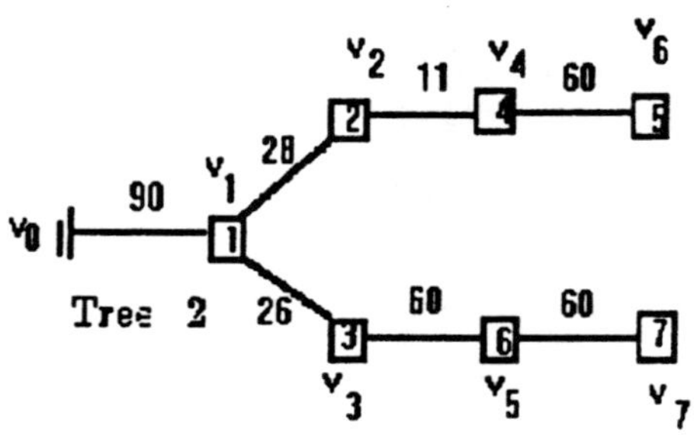

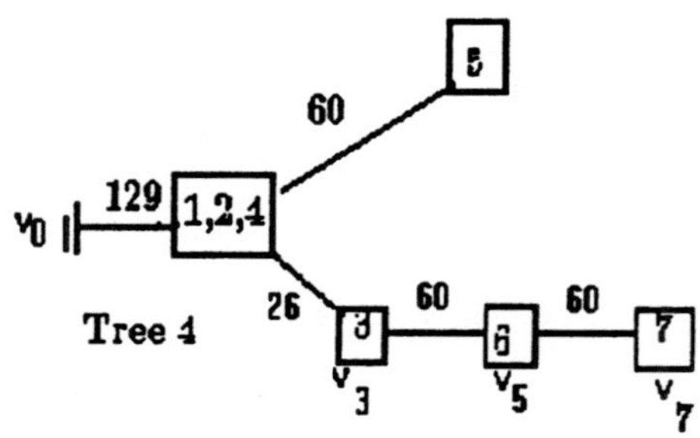

Consider player 4 at v_3. Certainly players 1, 2, and 3 cannot be charged for the edge cost 3 for the edge $e = [v_1 \;\; v_3]$. they never use. Edge $[v_3 \;\; v_4]$ is not the responsibility of players 1, 2, 3, or 7. However, $[v_0 \;\; v_1]$ is used by all and every one is responsible. For example players 1 and 2 at v_1 may feel that if each edge has a users' union to maintain the edge then players 1 and 2 would like to distribute the 12-unit cost of $[v_0 \;\; v_1]$ equally among the residents of v_1 the edge users union $[v_1 \;\; v_2]$ and edge users union $[v_1 \;\; v_3]$. While the edge user of $[v_1 \;\; v_2]$ is just player 3, users of $[v_1 \;\; v_3]$ are 4, 5, 6, and 7. Exclusive neighborhoods like 3 have greater cost burdens on intermediate edges. Thus $x_1 = x_2 = \frac{12}{4} = 3$. Now player 3 gets $3 + 1 = 4$ for his share of using $[v_0 \;\; v_1]$ and $[v_1 \;\; v_2]$. The edge users union $[v_1 \;\; v_3]$ adds this burden of 3 units as an an additional tax to $[v_1 \;\; v_3]$. Repeating this procedure, starting with v_1, player 4 at v_3 and edge users union $[v_3 \;\; v_4]$ and $[v_3 \;\; v_5]$ share the sum of the edge cost with tax equally. Thus while player 4 pays $\dfrac{3+3}{3} = 2$, player 7 pays $2 + 5 = 7$ and players 5 and 6 pay $\frac{1}{2}(6 + 2) = 4$ each. The proto-nucleolus is $(3, 3, 4, 2, 4, 4, 7)$.

it seems to be fair that players farther away from the root pay more than those closer to the root. Since player 4 pays only 2 which is less than player 1 staying closer to the root, $[v_1 \;\; v_3]$ is a bad edge. In general there may be many bad edges. The order of elimination of bad edges is important. We should choose among bad edges , the one for which

$$h_j = \frac{\text{cost (edge}[v_i \;\; v_j])}{\#\text{residents at } v_i + \text{edges } at \ v_i - 1}$$

is the least. (If more than one bad edge has this least value, we can choose them in any order) Lump the players at v_j with those at v_i and add the edge cost to the previous edge on the unique path to the root. We compute the proto-nucleolus of the new tree all over and continue till no more bad edges are found. The proto-nucleolus in that case coincides with the nucleolus. The algorithm has the following steps.

Algorithm:

Step 1: Number all vertices of the tree according to depth first search. Two vertices of the same depth are numbered in any order.

Step 2: Initialize by charging each resident i of the vertex v_1 the amount

$$x_{iv_1} = \frac{\text{cost}([v_0 \;\; v_1])}{\#\text{ residents } at \ v_1 + \#\text{ edges leaving } v_1}$$

Proceed to v_2; otherwise terminate.

Suppose the vertex v_h has just been processed and say v_k folows v_h (i.e. $[v_h v_k]$ is an edge). Process v_k by charging its residents

$$\frac{x_h + \text{cost}[v_h \ \ v_k]}{\# \text{ residents at } v_k + \# \text{ edges leaving } v_k}$$

When v_k is an end point of the tree, backtrack to find a vertex v_l that has been processed, but not its successor v_t. Proceed to charge residents of v_t the amount

$$\frac{x_l + \text{cost}[v_l \ \ v_t]}{\# \text{ residents at } v_l + \# \text{ edges leaving } v_l}$$

If no such vertex v_l exists, go to step 3.

Step 3: Look for bad edges where residents at a vertex pay more than some followers. When bad edges $[v_i \ v_j]$ are found, eliminate the one with the least value for

$$h_j = \frac{\text{cost}[v_i \ \ v_j]}{\# \text{ residents at } v_i + \text{edges leaving } v_l - 1}$$

Merge the residents at v_j with residents at v_i and add the edge cost $[v_i \ v_j]$ to the previous edge on the path from the root to v_i. Go to Step 1

In case no bad edges exist, terminate. The x_i's will be the nucleolus for the standard tree games. The example in Granot, et. al. [1996] will be used to compute the nucleolus. It is described by (Tree 2).

$x_1 = \frac{90}{3} = 30, \quad x_2 = \frac{30+28}{1+1} = 29, \quad x_4 = \frac{29+11}{1+1} = 20, \quad x_5 = 20 + 60 = 80,$
$x_3 = \frac{30+26}{2} = 28, \quad x_6 = \frac{28+60}{2} \quad x_7 = 107.$

We will calculate h_j's. For $[v_1 \ v_2]$, $h_2 = 28, h_3 = 26, h_4 = 11, \ldots$ We eliminate $[v_2 \ \ v_4]$ first by combining the nodes 2, 4 and by adding the two costs to the previous edge $[v_1 \ \ v_2]$. We get the Tree 3. For lack of notation, let h_S be the edge cost for the last edge to reach coalition S via the unique path from the root v_0. Thus $h_{24} = \frac{39}{2+1-1} = 19.5$. Since the rest of the h's are the same as before, h_{24} is still the least. Again shrinking the edge e joining the vertices of 1 and 2, 4 with the vertex of 1, we get Tree 4. Let us calculate the proto-nucleolus again.

$x_1 = x_2 = x_4 = \dfrac{129}{3+2} = 25.8, \quad x_5 = 85.8, \quad x_3 = \dfrac{51.8}{2} = 25.9, \quad x_6 = 42.95, \quad x_7 = 102.95.$

The costs to players further along the path maintain the inequalities such as $x_1 \leq x_5, \ \ x_3 \leq x_7$ and so on. Thus the proto-nucleolus is the nucleolus given by $(25.8, 25.8, 25.9, 25.8, 85.8, 42.95, 102.95)$

The nucleolus computations are much simpler for any tree with a unique path containing all edges. Here we eliminate bad edges in any order. By the time the last edge is processed we reach the nucleolus (Littlechild and Owen [1977]).

Acknowledgement: The author would like to thank Professor J.A. M. Potters and an anonymous referee for many constructive suggestions.

References

Bird, C. [1976] On cost allocation for a spanning tree: a game theoretic approach.*Networks*: 6, 335-350.

Curiel, I., Pederzoli, G. and S. Tijs. [1988] Sequencing Games. *European Journal of Operational Research*, 40: 344-351.

Curiel, I., Potters, J., Rajendra Prasad, V., Tijs, S. and B.Veltman, [1994] Sequencing and cooperation. *Operations Research*, 42: 366-368.

Derks, J and J. Kuipers. [1992] On the core and nucleolus of routing games. Tech Rept. University of Limburg, Maastrict.

Dragan, I. [1981] A procedure for finding the nucleolus of a cooperative n person game. *Zeitschrift für Operations Research*, 25: 119-131.

Driessen, T. [1991] A survey of consistency properties in cooperative game theory. *SIAM Review*, 33: 43-59.

Galil, Z. [1980] Application of efficient mergeable heaps for optimization problem on trees*Acta Informatica*,13:53-58.

Granot, D. and F. Granot. [1992] On some network flow games. Mathematics of Operations Research, 17: 792-841.

Granot, D. and G. Huberman [1984] On the core and nucleolus of minimum cost spanning tree games,*On the core and nucleolus of minimum cost spanning tree games, mathematical Programming*, 29:323-347.

On some spanning network games. Working paper, The University of British Columbia, Vancouver, British Columbia, Canada.

Granot, D., Maschler, M., Owen, G. and W. Zhu. [1996] The kernel/nucleolus of a standard tree game. *International J. Game Theory*, 25:219-244.

Huberman, G. [1980] The nucleolus and the essential coalitions. *Analysis and Optimization of Systems*, Springer, Berlin. 416-422.

Kohlberg, E. [1972] The nucleolus as solution of a minimization problem. *SIAM Journal of Applied Mathematics*, 23: 34-39.

Kuhn, H. [1955] The Hungarian Method for assignment problem,*Naval Reserach Logistic Quarterly*, 2:83-97.

Kuipers, J. [1994] *Combinatorial methods in cooperative game theory.* Ph.D. Thesis, Maastricht.

Littlechild, S.C. and G. Owen. [1977] A further note on the nucleolus of the 'airport game.'*International J. Game Theory*, 5:91-95.

Maschler, M. [1992] The bargaining set, kernel, and nucleolus.In: Aumann, R.J. and S. Hart (eds).*Handbook of Game Theory*. Vol. I.Elsevier science Publ.. BV Amsterdam, North Holland. 591-667.

Maschler, M., Peleg, B. and Shapley, L. [1972] The kernel and the bargaining set for convex games. *International J. Game Theory*, 1: 73-93.

Maschler, M., Peleg, B. and Shapley, L. [1979] Geometric properties of the kernel, nucleolus, and related solution concepts. *Mathematics of Operations Research* 4: 303-338.

Megiddo, N. [1978] Computational complexity of the game theory approach to cost allocation for a tree, *Mathematics of Operations Research*, 3:189-196.

Noltmier, H. [1975] An algorithm for the determination of the longest distances in a graph.*Mathematical Programming*, 9: 350-357.

Owen, G. [1974] A note on the nucleolus. *International Journal of Game Theory*, 3: 101-103.

Potters, J., Reijnierse, J. and Ansing, M. [1996] Computing the nucleolus by solving a prolonged simplex algorithm. *Mathematics of Operations Research*,21:757-768.

Sankaran, J. [1991] On finding the nucleolus of an n-person cooperative game. *International Journal of Game Theory*, 19: 329-338.

Schmeidler, D. [1969] The nucleolus of a characteristic function game. *SIAM Journal of Applied Mathematics*, 17: 1163-1170.

Shapley, L. [1953] A value for n-person games. *Contributions to the theory of games II*, (Eds. H. Kuhn and A.W. Tucker). Princeton University Press, Princeton, New Jersey, 307-317.

Shapley, L. and Shubik, M. [1972] The assignment game I: the core. *International Journal of Game Theory*, 1: 111-130.

Sobolev, A. [1975] A characterization of optimality principles in cooperative games by functional equations (Russian). *Mathematical Methods in the Social Sciences*, 6: 94-151.

Solymosi, T. [1993] *On computing the nucleolus of cooperative games*. Ph.D. Thesis, University of Illinois at Chicago.

Solymosi, T. and Raghavan, T. [1994] An algorithm for finding the nucleolus of assignment games. *International Journal of Game Theory*, 23: 119-143.

Solymosi, T, Aarts, H and T. Driessen. [1994] On computing the nucleolus of a balanced connected game. Tech Rept. University of Twente.

T.E.S.Raghavan
Department of Mathematics & Computer Science
University of Illinois at Chicago
Chicago, U.S.A

THE CHARACTERISATION OF THE UNIFORM REALLOCATION RULE WITHOUT SIDE PAYMENTS

Bettina Klaus [1]

Abstract: We consider the problem of reallocating the total endowment of an infinitely divisible commodity among agents with single-peaked preferences and study several properties of reallocation rules such as individual rationality, endowment monotonicity, no-envy, and bilateral consistency. Our main result is the proof that individual rationality and endowment monotonicity imply Pareto optimality. This result is used to provide two characterizations of the uniform reallocation rule. The first characterization states that the uniform reallocation rule is the unique reallocation rule satisfying individual rationality, endowment monotonicity, and no-envy. In the second characterization, no-envy is replaced by bilateral consistency.

1 Introduction

1.1 Allocation and Reallocation Problems

We study the problem of reallocating the individual endowments of agents with single-peaked preferences. Such re*allocation* problems may occur when we are concerned with *allocation* problems where preferences might change over time. Consider the following example. A task or a certain amount of work has to be divided among a group of workers. If we assume that they are rewarded proportionally to their shares, then preferences over individual shares are single-peaked: each worker has an optimal share below and above which his welfare is decreasing. Suppose now that we solved the *allocation* problem. If after a certain period of time preferences have changed, then it might be the case that the *allocation* can be improved upon by re*allocation*. In other settings individual endowments are directly

[1] I thank Ton Storcken and the referee for helpful comments. I am particularly grateful to William Thomson for detailed suggestions for improvements.

T. Parthasarathy et al. (eds.),
Game Theoretical Applications to Economics and Operations Research, **239–255.**

given, *e.g.*, in fixed-price exchange economies (see Benassy (1982)), or can be interpreted as natural claims or priorities, *e.g.*, investments, as described by Barberà, Jackson, and Neme (1997).

In this paper we are concerned with the axiomatic analysis of *reallocation* rules. We show that if a rule satisfies "individual rationality" and "endowment monotonicity", then it is Pareto optimal. Using this result, we obtain two characterizations of the so-called uniform *reallocation* rule. Since we build on the existing "axiomatic literature" on *allocation* and *reallocation* rules, we first give a short overview of the relevant articles.

1.2 A Short Review of the Literature

Benassy (1982) introduced the uniform *reallocation* rule in the slightly different setting of rationing, and noted that it is strategy-proof. However, under special assumptions[2] *reallocation* problems reduce to *allocation* problems.

A wide literature is concerned with the axiomatic analysis of *allocation* rules. For *allocation* problems with single-peaked preferences, the *allocation* rule featured preeminently is the "uniform *allocation* rule". Sprumont (1991) started the axiomatic analysis of this class of problems and gave the first characterizations of the uniform *allocation* rule. Since then, a variety of axiomatic studies, which also led to this rule, have been published. Without claiming completeness, we refer the reader to Ching (1992, 1994), Dagan (1995), de Frutos and Massó (1994), Otten, Peters, and Volij (1996), Sönmez (1994), and Thomson (1994a,b, 1995).

It is only recently that the axiomatic study of (*reallocation*) rules began. Here is a short description of the state-of-the-art.

Barberà, Jackson, and Neme (1997) characterize the class of Pareto optimal, strategy-proof, and "replacement monotonic" rules, a class that contains the uniform rule. Klaus, Peters, and Storcken (1995a) show that the uniform rule is the unique rule satisfying Pareto optimality, strategy-proofness, an "equal-treatment" condition, and in addition a "reversibility" condition. Reversibility requires a symmetric treatment of each problem and its "reversed image", the problem in which endowments and peaks are interchanged: demand in the former equals supply in the latter. Klaus, Peters, and Storcken (1995c) study some variations of the model, *e.g.*, they

[2]For instance, when individual endowments are ignored in the reallocation. Other examples are fixed-price exchange economies, where agents on the short side of the market, *e.g.*, suppliers in case of excess demand, receive their preferred consumptions and their "excess" is allocated among the remaining agents.

allow for debts and consider different preference domains. Thomson (1996) suggests an extension of the model to reallocation situations where an additional amount of the commodity, an "obligation" to or from the outside world, has to be allocated. The compatibility of, and the trade-offs between, a variety of properties are explored. Furthermore, characterizations of the uniform rule and of its extended version are established. Among the properties are: monotonicity with respect to individual endowments, monotonicity with respect to the obligation, endowment strategy-proofness, (preference) strategy-proofness, population monotonicity, and consistency. Population monotonicity is also considered by Moreno (1995), who characterizes the uniform rule in terms of Pareto optimality, no-envy, and population monotonicity.[3] Finally, Klaus, Peters, and Storcken (1995b) focus on properties describing the effect of population and endowment variations on the reallocations, *e.g.*, population monotonicity, bilateral consistency, endowment monotonicity, and endowment strategy-proofness. They too establish several characterizations of the uniform rule.

1.3 The Results of this Study

In most of the papers summarized above, Pareto optimality is a basic condition imposed on a rule. Here, Pareto optimality is not initially imposed, but our main result is that this property is implied by individual rationality and endowment monotonicity (Theorem 1): individual rationality states that after the reallocation agents are not worse-off than at their individual endowments, and endowment monotonicity requires that certain changes in the endowments, which permit a Pareto improvement over the reallocation, make no agent worse-off. Next, we show that for two-agent problems, individual rationality and endowment monotonicity characterize the uniform rule (Theorem 2). For more than two agents, Theorems 1 and 2, together with two former characterizations due to Klaus, Peters, and Storcken (1995b), are used to obtain two new characterizations of the uniform rule.

The first characterization (Corollary 1) states that the uniform rule is the unique *reallocation* rule that satisfies individual rationality, endowment monotonicity, and no-envy: no agent prefers another agent's allotment change to his own allotment change. The second characterization can be seen as an extension of the following result for *allocation* rules due to

[3]This characterization is presented independently in Klaus, Peters, and Storcken (1995b). However, the proofs of the characterization are different.

Sönmez (1994): the uniform *allocation* rule is the only rule that is individual rational from equal division, one-sided resource-monotonic, and consistent. The properties of *reallocation* rules corresponding to these "allocation properties", individual rationality, endowment monotonicity, and (bilateral) consistency, characterize the uniform *reallocation* rule. Consistency for *reallocation* rules can be described as follows. Suppose a group of agents leaves with the amounts assigned to them by the *reallocation* rule. By doing so, they might create a positive or negative "leftover". Distributing this leftover as equally as possible among the remaining agents, defines the so-called "reduced problem". Then, by consistency, applying the *reallocation* rule to the reduced problem yields the same allotments for the remaining agents as in the original problem. Bilateral consistency only requires consistency for situations where all but two agents leave with their allotments.

Because most of the properties that are central in this study are introduced in Klaus, Peters, and Storcken (1995b), we refer to this paper for details concerning their motivation and their relation to the corresponding "allocation properties". The results presented here (see also Table 1) contribute to the understanding of the trade-offs between several other properties presented in Klaus, Peters, and Storcken (1995b), Table 3.

The paper proceeds as follows. After introducing the model in Section 2, we prove in Section 3 that individual rationality and endowment monotonicity imply Pareto optimality (Theorem 1). In Section 4, we introduce the uniform *reallocation* rule and characterize it by individual rationality and endowment monotonicity for two-agent problems (Theorem 2). We proceed with the characterizations for arbitrary problems (Corollaries 1 and 2). Finally, we discuss the independence of the properties in the characterizations.

2 The Model

There is an infinite population of potential agents, indexed by the positive integers $\mathbb{N}$. Each agent i is described in terms of an *individual endowment* $e_i \in \mathbb{R}_+$ of an infinitely divisible commodity, and a continuous and single-peaked preference relation R_i defined over the non-negative reals $\mathbb{R}_+$. Single-peakedness of R_i means that there exists a point $p(R_i) \in \mathbb{R}_+$, the *peak of agent* i, with the following property: for all $\alpha, \beta \in \mathbb{R}_+$, with $p(R_i) \leq \alpha < \beta$ or $p(R_i) \geq \alpha > \beta$ we have $\alpha P_i \beta$.[4] We denote the set of all

[4] By P_i we denote the asymmetric part of R_i. As usual, $\alpha R_i \beta$ is interpreted as "α is weakly preferred to β", and $\alpha P_i \beta$ as "α is strictly preferred to β". The symmetric part

continuous and single-peaked preferences by $\mathcal{R}$. For a finite set of agents $N \subset I\!N$, $\mathcal{R}^N$ denotes the set of *profiles* $R = \langle R_i \rangle_{i \in N}$ of all such preferences.

A *reallocation problem with single-peaked preferences*, or in short a *problem*, is a triple $\langle N, e, R \rangle$, where $N \subset I\!N$ is a non-empty and finite set of agents, $e \in I\!R_+^N$ is a vector of individual endowments, and $R \in \mathcal{R}^N$ is a profile of continuous and single-peaked preferences.

Let $\langle N, e, R \rangle$ be a problem. We call agent i a *demander* if his endowment is strictly less than his peak: he "demands" $p(R_i) - e_i$ units of the commodity. We denote this *demand* by $d_i(N, e, R)$ and the set of demanders by $D(N, e, R)$. We call agent i a *supplier* if his endowment is strictly greater than his peak: he wants to "supply" $e_i - p(R_i)$ units of the commodity. We denote this *supply* by $s_i(N, e, R)$ and the set of suppliers by $S(N, e, R)$. We call agent i a *non-trader* if his endowment is equal to his peak: he favors no trade.[5] Let $d(N, e, R) := \sum_{i \in D(N,e,R)} d_i(N, e, R)$ denote *total demand* and $s(N, e, R) := \sum_{i \in S(N,e,R)} s_i(N, e, R)$ *total supply*. Furthermore, let $z(N, e, R) := d(N, e, R) - s(N, e, R)$ denote *excess demand*. The latter may be positive, zero, or negative. If it is positive, then $\langle N, e, R \rangle$ is a problem with *excess demand*. If it is zero, then $\langle N, e, R \rangle$ is *balanced*. If it is negative, then $\langle N, e, R \rangle$ is a problem with *excess supply*.

A vector $x = \langle x_i \rangle_{i \in N} \in I\!R_+^N$ is called *feasible* for $\langle N, e, R \rangle$, or is a *reallocation*, if $\sum_{i \in N} x_i = \sum_{i \in N} e_i$. A *reallocation rule* φ, or in short a *rule*, assigns to every problem $\langle N, e, R \rangle$ a reallocation $\varphi(N, e, R)$.

Let $\langle N, e, R \rangle$ be a problem and $i \in N$. Then $\varphi_i(N, e, R)$ denotes the *allotment* of agent i and $\Delta\varphi_i(N, e, R) := \varphi_i(N, e, R) - e_i$ his *allotment change*. Agent i is *non-satiated at* $\varphi(N, e, R)$ if $p(R_i) \neq \varphi_i(N, e, R)$.

3 Pareto optimality, Individual Rationality, and Endowment Monotonicity

In this section we show that *individual rationality* and *endowment monotonicity* together imply *Pareto optimality*.

First, we define *Pareto optimality*: reallocations assigned by the rule

of R_i is denoted by I_i: $\alpha I_i \beta$ means that agent i is indifferent between α and β.

[5] Our notion of demanders, suppliers, and non-traders is somewhat different from what it is in "classical exchange economies". There, the types of the agents not only depend on their own characteristics but also on prices. However, for economies with individual endowments and single-peaked preferences derived from fixed-price exchange economies, the notions coincide.

cannot be changed in such a way that no agent is worse-off and some agents are better-off.

Pareto optimality A rule φ is *Pareto optimal* if for all $\langle N, e, R\rangle$ there is no reallocation $x \in I\!R_+^N$, such that for all $i \in N$,

$$x_i R_i \varphi_i(N, e, R),$$

and for some $j \in N$,

$$x_j P_j \varphi_j(N, e, R).$$

By the single-peakedness of preferences it is easy to see that a rule is *Pareto optimal* if and only if it is *same-sided*, *i.e.*, for all problems $\langle N, e, R\rangle$, either $\varphi_i(N, e, R) \leq p(R_i)$ for all $i \in N$, or $\varphi_i(N, e, R) \geq p(R_i)$ for all $i \in N$. In Sprumont (1991) same-sidedness is used as definition of *Pareto optimality*.

Next, we introduce the standard notion of *individual rationality*: after reallocation, no individual is worse-off than at his individual endowment.

Individual rationality A rule φ is *individual rational* if for all $\langle N, e, R\rangle$ and all $i \in N$,

$$\varphi_i(N, e, R) R_i e_i.$$

The last property we introduce in this section is *endowment monotonicity*, which is concerned with changes of the reallocation in response to certain endowment variations. This property can be seen as an extension of the one-sided resource-monotonicity condition for *allocation* problems introduced by Thomson (1994b). If in case of excess demand some of the individual endowments decrease, then no agent is made better-off in the reallocation that follows this change. Similarly, if in case of excess supply some of the individual endowments increase, then no agent is made better-off in the reallocation that follows this change.

Let x and y be two vectors in $I\!R^N$. Then $x \geq y$ means that $x_i \geq y_i$ for all $i \in N$.

Endowment monotonicity A rule φ is *endowment monotonic* if for all $\langle N, e, R\rangle$ and $\langle N, e', R\rangle$ with $e' \leq e$,

if $z(N, e, R) \geq 0$, then $\varphi_i(N, e, R) R_i \varphi_i(N, e', R)$ for all $i \in N$, and
if $z(N, e', R) \leq 0$, then $\varphi_i(N, e', R) R_i \varphi_i(N, e, R)$ for all $i \in N$.

Theorem 1 *Let φ be an individual rational and endowment monotonic rule. Then, φ is Pareto optimal.*

Proof Let φ satisfy *individual rationality* and *endowment monotonicity*, and suppose by contradiction that φ is not *Pareto optimal*. Then, a problem $\langle N, e, R\rangle$ exists such that the reallocation $\varphi(N, e, R)$ is not same-sided. Without loss of generality, suppose that $\langle N, e, R\rangle$ is a problem with excess demand or which is balanced (the excess supply case is handled similarly).

In the remainder of the proof, we only vary the endowment vector of the problem $\langle N, e, R\rangle$ and leave the set of agents N and the profile R unchanged. Therefore, a problem $\langle N, \bar{e}, R\rangle$ is now denoted by $\bar{e}$.

The proof proceeds as follows. In (i), we show that in e the rule φ assigns to all suppliers and all non-traders their peaks.[6] In (ii), we define a sequence of problems $\{e^q\}_{q\in I\!N}$ such that for any $q \in I\!N$ the corresponding allotment $\varphi(e^q)$ is not same-sided. Finally, in (iii), for q large enough we derive a contradiction to feasibility.

(i) Suppose by contradiction that for some agent $j \notin D(e)$,

$$\varphi_j(e) \neq p(R_j).$$

By *individual rationality*,

$$p(R_j)P_j\varphi_j(e)R_je_j.$$

Now, let $e' \in I\!R^N_+$ be such that $e'_i = e_i$ for all $i \neq j$ and $e'_j = p(R_j)$. Then, in e', agent i is a non-trader. Hence, by *individual rationality* in e',

$$\varphi_j(e') = p(R_j). \tag{1}$$

Note that $e' \leq e$ and $z(e) \geq 0$. Then, by *endowment monotonicity* applied to e' and e,

$$p(R_j)P_j\varphi_j(e)R_j\varphi_j(e').$$

Thus,

$$\varphi_j(e') \neq p(R_j).$$

This contradicts (1). Hence,

$$\varphi_j(e) = p(R_j) \text{ for all } j \notin D(e). \tag{2}$$

[6]This statement holds for arbitrarily chosen problems with excess demand or which are balanced. Similarly, for problems with excess supply, all demanders and non-traders receive their peaks. See also Thomson (1996), Lemma 4.

(ii) Since $\varphi(e)$ is not same-sided, by (i), there exists a demander $k \in D(e)$ such that

$$\varphi_k(e) > p(R_k) > e_k.$$

By *individual rationality* in e, there exists a point $a^1 \in I\!R_+$ such that

$$p(R_k) > a^1 \geq e_k \text{ and } \varphi_k(e) I_k a^1. \tag{3}$$

Let $d^1 := p(R_k) - a^1$.

In e, by feasibility and *individual rationality* for all demanders, there exists at least one supplier l. Now, let $e^1 \in I\!R_+^N$ be such that $e_i^1 = e_i$ for $i \neq l$ and $e_l^1 = e_l - \min\{e_l - p(R_l), d^1\}$. Note that $D(e) = D(e^1)$. Also, note that $e^1 \leq e$ and $z(e) \geq 0$. Then, by *endowment monotonicity* applied to e^1 and e,

$$\varphi_i(e) R_i \varphi_i(e^1) \text{ for all } i \in N. \tag{4}$$

Hence, by (3) and (4) it follows that for agent k, either

$$\varphi_k(e^1) \geq \varphi_k(e) > p(R_k), \tag{5}$$

or

$$p(R_k) > a^1 \geq \varphi_k(e^1) \geq e_k. \tag{6}$$

If (5), then

$$\varphi_k(e^1) > p(R_k). \tag{7}$$

If (6), then the allotment of agent k moves to the other side of his peak so that by construction $\varphi_k(e) > \varphi_k(e^1) + \min\{e_l - p(R_l), d^1\}$.

By (2), applied to both e and e^1, and by feasibility,

$$\begin{aligned}\sum_{i \in D(e)} \varphi_i(e) &= \sum_{i \in D(e)} \varphi_i(e^1) + \min\left\{e_l - p(R_l), d^1\right\} \\ &= \sum_{i \in D(e),\ i \neq k} \varphi_i(e^1) + \varphi_k(e^1) + \min\left\{e_l - p(R_l), d^1\right\}.\end{aligned}$$

Since $\varphi_k(e) > \varphi_k(e^1) + \min\{e_l - p(R_l), d^1\}$, in e^1 there exists a demander m such that $\varphi_m(e^1) > \varphi_m(e)$. Hence, by (4),

$$\varphi_m(e^1) > p(R_m). \tag{8}$$

Since either (7) or (8) hold, it follows that $\varphi(e^1)$ is not *Pareto optimal*. Now, we repeat the argument, proceeding from e^1 instead of e.

At each step $n \in I\!N$, the allotment $\varphi(e^n)$ is not same-sided and there exists $i \in D(e^n)$ such that $\varphi_i(e^n) > p(R_i) > a^n \geq e_i^n$ and $\varphi_i(e^n) I_i a^n$. Hence,

$$d^{n+1} = p(R_i) - a^n > 0. \tag{9}$$

In step $n+1$ we lower the endowment of some $j \in S(e^n)$ by $\min\{e_j^n - p(R_j), d^{n+1}\}$. This procedure yields

$$\begin{array}{ll} D(e) = D(e^q) & \text{for all } q \in I\!N \text{ and} \\ \{\varphi(e^q)\}_{q \in I\!N}, & \text{an infinite sequence of reallocations} \\ & \text{which are not } \textit{Pareto optimal}. \end{array} \tag{10}$$

(iii) Since the set of demanders $D := D(e) = D(e^q)$ is finite, by (10) it follows that at least one demander must be several times on the "wrong side" of his peak. Consider such an $i \in D$. Let $s > n$ be such that $\varphi_i(e^s) > p(R_i)$ and $\varphi_i(e^n) > p(R_i)$. Note that $e^s \leq e^n$ and $z(e^n) \geq 0$. Then, by *endowment monotonicity* applied to e^s and e^n, $\varphi_i(e^s) \geq \varphi_i(e^n)$. Hence, $a^s \leq a^n$ and therefore

$$d^{s+1} \geq d^{n+1} \text{ for } s > n. \tag{11}$$

Because the number of demanders is finite, (9) and (11) imply that the sequence $\{d^q\}_{q \in I\!N}$ is bounded from below by a term $d > 0$ of the sequence. So, at each step, the endowment of a supplier is lowered by at least d without this turning him into a demander. Repeating this procedure, all suppliers are made non-traders after a finite number t, $t \in I\!N$, of steps. For problem e^t, it follows that

$$\varphi_j(e^t) = p(R_j) = e_j^t \text{ for } j \notin D(e^t).$$

Then, by feasibility and *individual rationality* in e^t,

$$\varphi_i(e^t) = e_i \text{ for } i \in D(e^t).$$

Note that $\varphi(e^t)$ is a term of the sequence $\{\varphi(e^q)\}_{q \in I\!N}$ which is same-sided and therefore *Pareto optimal*. This contradicts (10). □

4 The Uniform Reallocation Rule

In this section, we concentrate on the uniform *reallocation* rule, or uniform rule for short. This rule satisfies many appealing properties (see for example Klaus, Peters, and Storcken (1995a,b,c), Moreno (1995), and Thomson (1996)).

First, we define the uniform rule and prove that for two-agent problems the uniform rule is the only rule that is *individual rational* and *endowment monotonic*.

Next, by applying Theorem 1 and using two characterizations of the uniform rule established by Klaus, Peters, and Storcken (1995b), two new characterizations of this rule are obtained. The first characterization states that the uniform rule is the unique rule satisfying *individual rationality*, *endowment monotonicity*, and *no-envy*. In the second characterization *no-envy* is replaced by a consistency property. We conclude with a discussion of the independence of the properties. A table illustrates the trade-offs between the properties.

A particular rule that satisfies *Pareto optimality*, *individual rationality*, and *endowment monotonicity* is the *uniform* (*reallocation*) *rule* U^r, *e.g.*, introduced in Klaus, Peters, and Storcken (1995a). For all $\langle N, e, R\rangle$,

$$U_i^r(N,e,R) := \begin{cases} \min\{p(R_i), e_i + \lambda\} & \text{if } z(N,e,R) > 0 \quad \text{(excess demand)} \\ p(R_i) & \text{if } z(N,e,R) = 0 \quad \text{(balancedness)} \\ \max\{p(R_i), e_i - \lambda\} & \text{if } z(N,e,R) < 0 \quad \text{(excess supply)} \end{cases}$$

for every $i \in N$, where $\lambda \geq 0$ solves $\sum_{j \in N} U_j^r(N,e,R) = \sum_{j \in N} e_j$.

The uniform rule works as follows. In case of excess demand, all suppliers and non-traders get their peaks. The total supply is distributed so that either a demander gets his peak, or his allotment change is maximal. In a balanced problem, all agents get their peaks. In case of excess supply, the reallocation is dual to the excess demand case: all demanders and non-traders get their peaks. The total demand is distributed so that either a supplier gets his peak, or his allotment change is minimal.

For the next theorem see also Thomson (1996), Theorem 2.

Theorem 2 *For two-agent problems the uniform rule is the unique rule satisfying individual rationality and endowment monotonicity.*

Proof It is easy to show that the uniform rule satisfies the properties named in the theorem. Conversely, let φ be a rule satisfying *individual rationality* and *endowment monotonicity*. Let $\langle N, e, R\rangle$ be a two-agent problem with $N = \{i, j\}$.

Case 1 $U^r(N,e,R) = e$.
Then, by *individual rationality*, $\varphi(N,e,R) = e = U^r(N,e,R)$.

Case 2 $U^r(N,e,R) \neq e$.

This can only occur if agents are of "opposite" types, *i.e.*, N consists of one demander and one supplier. Suppose that $z(N,e,R) \geq 0$ and, without loss of generality, let $i \in S(N,e,R)$. Then, similarly as in the proof of Theorem 1, part (i), it follows that $\varphi_i(N,e,R) = p(R_i) = U_i^r(N,e,R)$. Thus, $\varphi_j(N,e,R) = e_j + (e_i - p(R_i)) = U_j^r(N,e,R)$. For $z(N,e,R) < 0$, $\varphi(N,e,R) = U^r(N,e,R)$ follows similarly. □

Remark 1 Theorem 2 only applies to the two-agent case. As already established in the proof, a "real reallocation" only takes place when agents of opposite types exist. In this case one of the agents receives his peak; suppose, without loss of generality, that this agent is the supplier. Then, his total supply has to be allocated to the demander. For more than two agents, again the agents of one group receive their peaks, *e.g.*, the suppliers and the non-traders, and the total supply of this group is distributed among the demanders. This can be done in different ways. All rules that are same-sided and distribute the total supply "monotonically" among the demanders are *individually rational* and *endowment monotonic* for problems with excess demand. A similar statement holds for excess supply. Examples of such rules are the uniform rule, but also the hierarchical rule (see Example 2) and the proportional rule (see for instance Klaus, Peters, and Storcken (1995a), Example 2).

For the first characterization of the uniform rule, we introduce a notion of no-envy in terms of allotment changes: no agent strictly prefers the allotment change of another agent to his own allotment change.

No-envy A rule φ is *envy-free*, or satisfies *no-envy*, if for all $\langle N,e,R \rangle$ and all $i,j \in N$ with $e_i + \Delta\varphi_j(N,e,R) \geq 0$,

$$\varphi_i(N,e,R) R_i (e_i + \Delta\varphi_j(N,e,R)).$$

So, i envies j if i prefers j's allotment change, added to his endowment, to his own allotment—provided the former is feasible. The uniform rule is *envy-free*. For instance, in case of excess demand, only demanders can be non-satiated and, if so, they obtain the same, maximal, allotment change.

The well-known property of no-envy is introduced by Foley (1967) for resource allocation problems. No-envy for *allocation* problems with single-peaked preferences is one of the conditions that led Sprumont (1991) to one of his characterizations of the uniform *allocation* rule. Ching (1992) gives a simple and elegant proof of this result. The concept of no-envy in terms of net allotment changes is formulated by Schmeidler and Vind (1972) in the more general context of exchange economies. Klaus, Peters, and Storcken (1995b) introduce a stronger notion of no-envy[7] for *reallocation* problems.

Corollary 1 *The uniform rule is the unique rule satisfying individual rationality, endowment monotonicity, and no-envy.*

In the proof of Corollary 1, we use the following characterization of the uniform rule.

Theorem 3 *The uniform rule is the unique rule satisfying Pareto optimality, endowment monotonicity, and no-envy.*

For a proof of Theorem 3 we refer to Klaus, Peters, and Storcken (1995b), Theorem 4.4. There, a characterization of the uniform rule in terms of *Pareto optimality*, *endowment monotonicity*, and the stronger no-envy condition mentioned above, is provided (Theorem 4.4). However, the proof of Theorem 4.4 remains valid if we impose *no-envy*.

Proof of Corollary 1 It is easy to prove that the uniform rule satisfies the properties named in the theorem (see also Klaus, Peters, and Storcken (1995b)). Conversely, let φ be a rule satisfying *individual rationality*, *endowment monotonicity*, and *no-envy*. By Theorem 1, φ is *Pareto optimal*. Then, by Theorem 3, $\varphi = U^r$. □

In the second characterization, we replace the *no-envy* condition by a consistency property. Consistency essentially describes the stability of the reallocation in situations where agents leave with their allotments: by leaving, these agents might create a positive or negative "leftover". Consider the reduced problem where this leftover is divided as equally as possible[8] among the remaining agents. Then, consistency requires that the agents' allotments in this reduced problem be equal to their allotments in the original problem. For exchange economies, this notion of consistency can be found in Thomson (1992).

[7] At an envy-free reallocation as defined in Klaus, Peters, and Storcken (1995b) an agent, in addition to the no-envy notion as introduced here, does not envy the feasible part of another agent's allotment change.

[8] As equal as possible with respect to the domain restrictions.

In the sequel, we consider the weaker notion of bilateral consistency which applies to situations where all but two agents leave with their allotments.

Bilateral consistency A rule φ is *bilaterally consistent* if for all $\langle N, e, R\rangle$ and all $i, j \in N$, $i \neq j$,

$$\varphi_i\left(\{i,j\}, e(i,j), R_{|\{i,j\}}\right) = \varphi_i(N, e, R).$$

Here, the *reduced problem* $\langle \{i,j\}, e(i,j), R_{|\{i,j\}}\rangle$ is defined as follows. By $R_{|\{i,j\}} = \langle R_i, R_j\rangle$ we denote the restriction of R to $\{i,j\}$. To define the adjusted endowment vector $e(i,j) \in I\!R_+^{\{i,j\}}$ assume, without loss of generality, that $\Delta\varphi_i(N,e,R) \leq \Delta\varphi_j(N,e,R)$. Then, dividing the leftover $\sum_{k\in N\setminus\{i,j\}}(e_k - \varphi_k(N,e,R))$ as equally as possible among i and j yields the adjusted endowments

$$\begin{aligned} e(i,j)_j &:= \max\left\{0, e_j + \frac{1}{2}(\Delta\varphi_i(N,e,R) + \Delta\varphi_j(N,e,R))\right\} \geq 0 \text{ and} \\ e(i,j)_i &:= e_i + (\Delta\varphi_i(N,e,R) + \Delta\varphi_j(N,e,R)) - (e(i,j)_j - e_j) \geq 0. \end{aligned}$$

So, endowment adjustments are as close as possible to the mean allotment changes of i and j.[9]

It is straightforward to prove that the uniform rule satisfies *bilateral consistency*.[10] *Bilateral consistency* for *reallocation* problems as introduced above can be seen as an extension of the bilateral consistency condition for *allocation* problems introduced by Thomson (1994a). By the construction of the reduced problem an equity condition is embedded in our notion of *bilateral consistency*.

Corollary 2 *The uniform rule is the unique rule satisfying individual rationality, endowment monotonicity, and bilateral consistency.*

In the proof of Corollary 2, we use the following characterization of the uniform rule (Klaus, Peters, and Storcken (1995b), Theorem 5.2) for reallocation problems where the set of potential agents might is a finite subset of $I\!N$.

[9] By just applying mean allotment changes, negative endowments, which are not admissible in this model, might occur.

[10] The uniform rule also satisfies consistency which we have not formalized here.

Theorem 4 *Let the set of potential agents contain at least three agents. Then, the uniform rule is the unique rule satisfying* Pareto optimality, individual rationality, *and* bilateral consistency.

For a proof of Theorem 4 we refer to Klaus, Peters, and Storcken (1995b), proof of Theorem 5.2.

Proof of Corollary 2 As already mentioned, the uniform rule satisfies the properties named in the theorem (see also Klaus, Peters, and Storcken (1995b)). Conversely, let φ be a rule satisfying *individual rationality*, *endowment monotonicity*, and *bilateral consistency*. By Theorem 1, φ is *Pareto optimal*. Then, by Theorems 2 and 3 we obtain that for all problems $\langle N, e, R\rangle$ with $|N| \geq 2$, $\varphi(N, e, R) = U^r(N, e, R)$. By feasibility, it follows that for all problems $\langle N, e, R\rangle$ with $|N| = 1$, $\varphi(N, e, R) = U^r(N, e, R)$. Hence, $\varphi = U^r$. □

In order to discuss the logical independence of the properties in the characterizations, consider the following examples.

Example 1 The following *no-trade rule* φ^0 satisfies *individual rationality*, *no-envy*, and *bilateral consistency*, but neither *endowment monotonicity* nor *Pareto optimality*. For all $\langle N, e, R\rangle$,

$$\varphi^0(N, e, R) := e.$$

Example 2 The following *hierarchical rule* φ^h satisfies *Pareto optimality*, *individual rationality*, and *endowment monotonicity*, but neither *no-envy* nor *bilateral consistency*.[11] In case of excess demand, the rule satiates all suppliers. Then, the demander with the lowest index is "served" as well as possible, *i.e.*, he receives his peak if that is possible and the total supply otherwise. If there is something left, the demander with the second lowest number is served, *etc.* In case of excess demand ($z(N, e, R) \geq 0$),

$$\varphi_i^h(N, e, R) := \begin{cases} p(R_i) & \text{if } i \notin D(N, e, R) \text{ and} \\ \min\{p(R_i), e_i + s(N, e, R) - \sum_{j \in D(N,e,R),\ j<i} \Delta\varphi_j^h(N, e, R)\} & \text{otherwise.} \end{cases}$$

In case of excess supply ($z(N, e, R) \leq 0$), $\varphi^h(N, e, R)$ is defined similarly.

[11] The hierarchical rule is often called serial, or lexicographic, dictatorship.

Example 3 The following rule $\bar{\varphi}$ satisfies *Pareto optimality*, *endowment monotonicity*, and *bilateral consistency*, but neither *individual rationality* nor *no-envy*. It coincides with the uniform rule in case of excess demand and balancedness. In case of excess supply the total endowment $E := \sum_{i \in N} e_i$ is divided so that all agents have the same supply with respect to their peaks. In case of excess demand $(z(N, e, R) \geq 0)$,

$$\bar{\varphi}(N, e, R) := U^r(N, e, R).$$

In case of excess supply $(z(N, e, R) \leq 0)$,

$$\bar{\varphi}_i(N, e, R) := p(R_i) + \lambda,$$

for every $i \in N$, where $\lambda \geq 0$ solves $\sum_{j \in N} \bar{\varphi}_j(N, e, R) = E$.

As the examples show, the properties of Corollary 2 and Theorem 3 are independent. For Corollary 1 and Theorem 2 we were neither able to prove nor able to disprove the independence of the properties. To be more precise, it is an open question whether *Pareto optimality*, or *individual rationality* respectively, is independent of the other properties.

Table 1

	Cor. 1	Th. 3	Cor. 2	Th. 4
Pareto optimality	×		×	
individual rationality		×	×	×
endowment monotonicity	×	×		×
no-envy	×	×		
bilateral consistency			×	×

× The condition in question (row) is part of the corresponding characterization (column).

References

Barberà, S., M.O. Jackson, and A. Neme (1997): "Strategy-Proof Allotment Rules", *Games and Economic Behavior*, 18, 1–21.

Benassy, J.P. (1982): *The Economics of Market Disequilibrium*, San Diego: Academic Press.

Ching, S. (1992): "A Simple Characterization of the Uniform Rule", *Economics Letters,* 40, 57–60.

Ching, S. (1994): "An Alternative Characterization of the Uniform Rule", *Social Choice and Welfare,* 11, 131–136.

Dagan, N. (1995): "A Note on Thomson's Characterization of the Uniform Rule", *Journal of Economic Theory*, 69, 255–261.

Foley, D. (1967): "Resource Allocation and the Public Sector", *Yale Economic Essays,* 7, 45–98.

de Frutos, M.A., and J. Massó (1994): "More on the Uniform Rule: Equality and Consistency", Working Paper.

Klaus, B., H. Peters, and T. Storcken (1995a): "Strategy-Proof Division with Single-Peaked Preferences and Individual Endowments", forthcoming in *Social Choice and Welfare.*

Klaus, B., H. Peters, and T. Storcken (1995b): "Reallocation of an Infinitely Divisible Good", forthcoming in *Economic Theory.*

Klaus, B., H. Peters, and T. Storcken (1995c): "Strategy-Proof Reallocation of an Infinitely Divisible Good", forthcoming in *Charlemagne and the Liberal Arts: 1200 Years of Civilization and Science in Europe, Volume 2: Mathematical Arts*, eds. P.L. Butzer, H.Th. Jongen, W. Oberschelp, Aachen: Thouet Verlag.

Moreno, B. (1995): "The Uniform Rule in Economies with Single-Peaked Preferences, Endowments, and Population Monotonicity", Working Paper, University of Alicante.

Otten, G.-J., H. Peters, and O. Volij (1996): "Two Characterizations of the Uniform Rule for Division Problems with Single-Peaked Preferences", *Economic Theory,* 7, 291–306.

Schmeidler, D., and K. Vind (1972): "Fair Net Trades", *Econometrica,* 40, 635–642.

Sönmez, T. (1994): "Consistency, Monotonicity, and the Uniform Rule", *Economics Letters*, 46, 229–235.

Sprumont, Y. (1991): "The Division Problem with Single-Peaked Preferences: A Characterization of the Uniform Allocation Rule", *Econometrica,* 59, 509–519.

Thomson, W. (1992): "Consistency in Exchange Economies", Working Paper, Economics Department, University of Rochester.

Thomson, W. (1994a): "Consistent Solutions to the Problem of Fair Division when Preferences are Single-Peaked", *Journal of Economic Theory,* 63, 219–245.

Thomson, W. (1994b): "Resource-Monotonic Solutions to the Problem of Fair Division when Preferences are Single-Peaked", *Social Choice and Welfare,* 11, 205–223.

Thomson, W. (1995): "Population-Monotonic Solutions to the Problem of Fair Division when Preferences are Single-Peaked", *Economic Theory,* 5, 229–246.

Thomson, W. (1996): "Endowment-Monotonicity in Economics with Single-Peaked Preferences", Working Paper, Economics Department, University of Rochester.

Bettina Klaus
Department of Quantitative Economics
Maastricht University
P.O. Box 616
6200 MD Maastricht
The Netherlands

TWO LEVEL NEGOTIATIONS IN BARGAINING OVER WATER

Alan Richards and Nirvikar Singh

Abstract : *The paper analyzes the impact of a two-level game for water allocations. For a model with two domestic groups and two countries, and with both domestic and international negotiations, Nash bargaining theory is used to derive several propositions on the consequences of different bargaining rules for water allocations. The effect on international negotiations of the ability to commit to having domestic negotiations is examined. The importance of the nature and timing of complementary investments, and whether they are included in negotiations, in affecting the efficiency of the negotiated outcome is also explored.*

1 Introduction

The problem of increasingly acute water scracity plagues several regions of the developing world. Water demands are rapidly increasing, thanks to population growth, urbanization, and agricultural policies which usually subsidize water users. Nature meanwhile ensures that supplies in arid zones are inevitably limited. Compounding the problem is the fact that many countries or regions depend for water on rivers or aquifers which are shared with other nations/regions. Achieving agreement among water-scarce, often hostile nations has proved very difficult. Even in federal states such as India, inter-state conflict over water rights poses difficult issues.

Managing water scarcity will require two fundamental changes : domestic economic policies will have to be reformed to rationalize water use (e.g. World Bank, 1993); and transnational (or trans-state) agreements on water sharing must be forged. Each of these problems is difficult enough by itself; however they are intimately interrelated. In fact, each national government is engaged in a 'two-level game' (e.g. Putnam, 1988) : it is simultaneously trying to reform its domestic water regime while also negotiating with its neighbours over how to share the river's resources. But a move in one game will typically have implications for the outcome of the other. International water negotiators are looking over their shoulders at domestic political conflicts over economic reform, while advocates and opponents of economic reform monitor international developments for its domestic implications.

There are several examples of such two-level games over water. The Colorado river has been the subject of negotiations between the United States and Mexico, as well as among the riparian states in the U.S. [1] Sharing the waters of the Ganges river has been a point of contention between India and Bangladesh for several decades, while both countries have had to devise schemes for apportioning their own shares among different user groups and/or regions.[2] The Indus and its tributaries became the focus of intense bargaining between India and Pakistan after India's partition in 1947. Simultaneously, India had to allocate its share of water from this basin to several different states.[3] Finally, Israel, Jordan, and the Palestinian

[1] See Friedkin (1987) and Ramana (1992), pp.68-69.

[2] See Crow (1995), Chaudry and Siddigi (1987), National Water Development Agency (1992).

[3] See Barrett (1994) and Dhillon (1983).

National Authority, are negotiating over sharing the Jordan and Yarmouk rivers, as well as mountain aquifers, while Israel and Jordan face issues of reallocating water between urban and rural users internally.[4]

In the above examples, there are often more than two players at the subnational level. As a first approximation, we focus on only two groups within a country. One may think of of this as 'urban' versus 'agricultural' interests; in fact, the interests of these groups are often in conflict. For example, the city of Amman faces increasingly serious water shortages while farmers in the (irrigated) Jordan Valley enjoy substantial subsidies to grow water-intensive crops such as bananas, in which Jordan has no comparative advantage. Alternatively, we may think of groups as political entities, e.g., the states of Punjab and Haryana in India.

In this paper, we begin a theoretical study of this process of the interaction of domestic economic reform (viewed here as domestic negotiation over water allocations) and international water negotiation using the theory of cooperative bargaining. The structure of the paper is as follows. In section 2, we construct a specific model of the two-level bargaining process, capturing the key feature of two-levels of negotiation in a framework of Nash cooperative bargaining. Although one might object that this neglects the highly conflictual nature of actual water negotiations, the Nash and multiperson extension of the Nash solutions approximate the equilibria of, respectively, the Rubinstein (1982) noncooperative bargaining game, and a multiperson generalization of that noncooperative game, when those sequential games have high enough discount factors.[5]. We assume two countries and two groups within each country or state; that state actors are benevolent; and that the initial allocation of water is sub-optimal for either of two reasons : (1) formerly water was not scarce and hence institutions have arisen which now no longer serve to allocate water efficiently, or (2) the initial bargaining conditions have changed. It follows that transfers are necessary to achieve Pareto efficiency. We examine several alternative (simultaneous and sequential) cooperative bargaining structures, and derive some equivalence results and welfare comparisons. Taken together, our results provide a demonstration of why there is often disagreement not only over water, but even over how to negotiate about water. The form of negotiations (the institutional arrangements) have very different consequences for different actors. A key feature in our framework is that being able to commit to domestic negotiations can improve the bargaining position at the international level.[6]

In section 3, we discuss the generalization of the model, to include the possibility of complementary, productive investments, and of noncooperative actions. Water by itself rarely gives utility; it must be stored, moved, channeled, pumped, and piped to be useful. Investments which enhance the utility of water may be local, national, or international. Here we focus on the impact of national investment on water negotiations. We offer three

[4]See Fisher (1995), Just, Netanyahu and Horowitz (1996).

[5]See Harsanyi (1977), pp. 110-112 for a basic discussion of the relationship between cooperative and noncooperative games, the Binmore, Rubinstein and Wolinsky (1986) and Krishna and Serrano (1995) for analyses of the connection between the axiomatic approach and the sequential strategic approach to bargaining. The latter two paper show that the Nash and multiperson extension of the Nash solutions approximate the equilibria of, respectively, the Rubinstein (1982) noncooperative bargaining game, and a multiperson generalization of that noncooperative game, when those sequential games have high enough discount factors. Thus, we can justify our approach as being a shortcut, approximating the outcomes of noncooperative games, even when the assumption of cooperative behaviour is questionable

[6]Of course, this is not the only possible consequence of domestic politics of international bargaining. Most obviously, strong domestic farm lobbies may make it more difficult for a national government to make concessions. The argument that 'our hands are tied is often quite useful in negotiations (e.g. Schelling, 1960). Alternatively, an unfinished international water negotiation may retard domestic reform, as the negotiating parties fear that any domestic water-savings achieved reform will reduce their ability to claim a larger share of international water. One abstacle to reform of domestic Jordanian water policy before the peace treaty with Israel was precisely the fear by some Jordanian officials that greater efficiency in Jordanian domestic water use would be seized upon by Israeli negotiators, who would argue that Jordan 'needed less water'.

additional propositions in this extension of our basic two interest groups/two country model. First, the optimal allocation of water in one country will depend on domestic investment in the other country, even in the absence of direct externalities, as long as such investment affects the marginal utility of water. Second, if domestic investment affects the marginal utility of water, then, even in the absence of direct externalities, negotiating only over water, and not over domestic investments leads to an inefficient outcome if domestic investments can be precommitted. Finally, whether or not domestic investments can be precommitted before internal negotiations, if they take place after international negotiations, then, in the absence of direct externalties, the outcome in terms of water allocation and domestic investments is efficient. Section 4 is a brief conclusion.

2 Cooperative Bargaining Structures

The structure of our model allows us to focus on a key issue, the nature of the interaction between international and domestic bargaining. We use a cooperative bargaining framework, assuming that parties to an agreement can commit to it. We first restrict attention to the allocation of water only, ignoring for now the other dimensions of domestic and international negotiations. [7] In section 3, we introduce the important issue of investment. We first present the model and notation, followed by analysis.

Model and Notation

There are two nations or countries, denoted by A and B, and two groups in each country. To reduce multiple subscripts, we number the groups sequentially, from 1 to 4. Groups 1 and 2 are in country A, while groups 3 and 4 are in country B. As suggested, these may be thought of as 'city dweller' and irrigated land farmers'. There are two goods consumed by each group : water, denoted w, and numeraire good, denoted y. There is no collective decision problem within any group, so each group has a well-specified utility function. We assume for tractability that all utility functions are quasi-linear, so that utility is transferable, and all Pareto frontiers are straight lines or hyperplanes. This consiberably simplifies analysis. We will attempt to relax this assumption in future work. Initial allocations of the goods are assumed to be available to each group, and are denoted by bars over the corresponding letters, e.g., $\bar{w}$. Hence, with this notation, the initial utility of group i is given by

(1) $u_i(\bar{w}_i, \bar{y}_i) = v_i(\bar{w}_i) + \bar{y}_i$

We assume that the functions $v_i(w_i)$ are strictly concave and differentiable. Within the countries, the total quantities of water available are $\bar{w}^A, \bar{w}^B$, in A and B respectively.

The essential problem faced by the bargainers is that the initial allocation of water is suboptimal, both across groups within a country and across countries. This can be explained by changing circumstances. Water may have been relatively abundant in the past, and all parties were able to get as much as they wanted at a zero price. As populations and economic values change, however, the allocation determined by historical circumstances and traditional institutional mechanisms may no longer be optimal. In addition, the climate for international negotiations may also fluctuate : it may be more propitious now than in the past.

We denote the optimal amounts of water by asterisks, e.g., w_i* for group i. The conditions for the optimal allocation of water are given by the requirement that the marginal utility of water be equated for each group, i.e. $v'(w_i*) = v'(w_j*)$. Clearly, such a solution, tranferring water from one group to another, will not feasible without some compensatory transfers. Thus we have that final utilities are given by

[7] As a referee has pointed out, this means that some of our anlaysis is not specific to water issues only: the bargaining could be over any resource. We hope this generality can be viewed as a positive feature.

(2) $u_i(w_i*, y_i*) = v_i(w_i*) + y_i * \equiv v_i(w_i*) + \bar{y}_i + t_i*$

The identity above defines the optimal transfer for group i as the difference between the final and initial amounts of the numeraire good. Unlike the optimal amounts of water, however, the transfers are not uniquely determined. Instead, they depend on the bargaining game. In particular, since we are going to use the Nash bargaining solution, the transfers will depend on the threat points of the parties engaged in bargaining. Note that where it is unambiguous, we will abbreviate expression such as $u_i(w_i*, y_i*)$ to v_i*.

It is instructive to provide an alternative institutional determination of the transfers, for comparison purposes. Suppose that the result of the negotiations is an agreement to recognize initial allocations of water as representing *de facto* property rights, with an international market for water also being created. Let the price of water be p_w. Then, as long as the price is set appropriately, there will be market clearing at the optimal allocation of water. To see this, consider the problem of i, which is now to maximize.

(3) $v_i(w_i) + \bar{y}_i + p_w(\bar{w}_i - w_i)$

The solution to this is to set the marginal utility of water equal to the price of water. Hence, all marginal utilities are equalized. Furthermore, there will be an excess supply or demand for water unless the price equals the shadow price of the constraint in the problem.

(4) $\max \sum u_i(w_i, y_i)$ subject to $\sum w_i = \bar{w}$

But with that price (which is the marginal utility of water for any of the groups at the optimum), the market solution will correspond to the optimal solution. The tranfers are then the result of market transactions, with

(5) $t_i = p_w(\bar{w}_i - w_i)$

Returning to the bargaining case, at the optimum allocation of water, the transfers are not uniquely determined unless one specifies the bargaining game. Any cooperative bargaining game will depend on recognizing the set of Pareto optimal points in utility space. The equation of the utility possibility frontier is given by

(6) $\sum u_i(w_i*, y_i*) = \sum v_i(w_i*) + \sum \bar{y}_i \equiv H$

The identity above defines the maximum total utility available. Note that in writting the middle expression, we use the fact that the sum of the numeraire good is unaffected by the transfers, or the sum of the transfers is zero. A final piece of initial notation is given by

(7) $d_i = u_i(\bar{w}_i, \bar{y}_i) = v_i(\bar{w}_i) + \bar{y}_i$

Here the notation captures the fact that the initial utility level is potentially the disagreement payoff or threat point for the water bargaining game. We will introduce further notation as necessary, as we turn to analyzing several possible cases of bargaining.

Domestic Negotiations Only

To begin with, consider the case where only domestic negotiations and reallocation of water are possible. We will focus on country A, with the analysis for country B being parallel. Let w_1^A, w_2^A denote the optimal allocation of water in country A, i.e., the solution to

(8) $\max v_1(w_1) + v_2(w_2)$ subject to $w_1 + w_2 = \bar{w}^A$

Furthermore, let H^A be the maximized utility that results from this optimal allocation within country A. Using the Nash bargaining solution, the problem in country A is

(9) $\max (u_1 - d_1)(u_2 - d_2)$ subject to $u_1 + u_2 = H^A$

The solution is easily seen to be

(10) $u_1^D = \frac{1}{2}(H^A + d_1 - d_2)$, $u_2^D = \frac{1}{2}(H^A + d_2 - d_1)$

Focusing on group 1, the utility from the Nash bargaining solution can be rewritten more explicitly as

(11) $u_1^D = \frac{1}{2}[(v_1(w_1^A) + v_2(w_2^A) + v_1(\bar{w}_1)) - v_2(\bar{w}_2)] + \bar{y}_1$

Furthermore, the associated transfer is

(12) $t_1^D = \frac{1}{2}[v_2^A - \bar{v}_2) - (v_1^A - \bar{v}_1)]$
where we use the abbreviated notation.

It is instructive to compare the Nash bargaining solution with the market solution, which, for group 1, is given by $t_1^{DM} = v_1'(w_1^A)(\bar{w}_1 - w_1^A)$.[8] Comparing this with (12), it is clear that the two approaches, bargaining and markets, give different answers in general. [9]

International Negotiations Only

Suppose that both countries agree that domestic negotiations will be effectively conducted at the international level, as a way of reaching an overall agreement in one step. This can be achieved by bringing all four groups simultaneously to the bargaining table. It may seem unusual for domestic groups to participate in international negotiations, but it is certainly feasible. With four groups, one can use Harsanyi's (1963) generalization of the Nash bargaining solution. [10] The problem is now to solve

(13) $\max \prod_{i=1}^4 (u_i - d_i)$ subject to $\sum_{i=1}^4 u_i = H$

Note that the disagreement payoffs here incorporate the assumption that separate domestic negotiations will not occur. The solution to the maximization is given by

(14) $u_i^I = \frac{1}{4}H + \frac{3}{4}d_i - \frac{1}{4}\sum_{j \neq i} d_j$

This implies a transfer for group of

(15) $t_i^I = \frac{1}{4}\sum_{j \neq i}[v_j^* - \bar{v}_j) - (v_i^* - \bar{v}_i)]$

How does this outcome compare with the case where only domestic negotiations are possible ? Clearly, greater efficiency is achieved by allowing reallocation of water across the two countries. However, the Nash bargaining solution above may not represent a Pareto improvement over the case of separate domestic bargaining. This is because the presence of new parties at the bargaining table affects the relative impact of the threat points. We may illustrate this with a simple example. Suppose that the utility functions for all four groups are identical, and given by $\log w_i$, while the initial allocations of water are 2,4,6 and 8 for the four groups, respectively. It is possible to show that in this case, $u_i^I > u_i^D$ for each group. However, if the initial allocations of water are 2,6,6,6, then group 2 is worse off in the all party international negotiation. Essentially, in domestic negotiations, this group is in a strong position relative to group 1 in country A, through the impacts of the initial allocations on the threat points. This position is diluted by the presence of the other two groups in the bargaining. To see this more generally, consider group 1, for example. ¿From (10) and (14), we obtain

$$u_1^D - u_1^I = \frac{1}{2}(h^A - d_1 - d_2) - \frac{1}{4}(H - d_1 - d_2 - d_3 - d_4),$$

which is the difference in the gains from agreement in the two cases, and the first term can outweigh the second. We summarize the implications of this comparison in the following statement.

Proposition 1 : In our two country model with two groups in each country, it is possible that a group may prefer to have only domestic negotiations over all-party international bargaining without the option of separate domestic negotiations.

---

[8] Similar expressions can be derived for the second group in country A. In fact, since the transfers sum to zero, group 2's transfer is just the negative of group 1's. The utility expressions are symmetric. The same results can be derived for group 3 and 4 in the country B.

[9] Note that the market solution depends linearly on the initial allocation, while the Nash bargaining solution depends on it nonlinearly, through the threat point of the two groups. Hence, it is not possible to make a general comparison of the two outcomes in terms of the magnitudes of the transfers.

[10] Harsanyi (1977), Ch. 10, provides a complete treatment.

All-Party International Negotiations with Domestic Negotiations as Fallback

An alternative possibility is that if international negotiations among all four groups fail, domestic bargaining can still take place. In this case, the participants in the international round should correctly assume that the domestic negotiations will succeed. The bargaining outcome is now the solution to

(16) $\max \prod_{i=1}^{4} (u_i - u_i^D)$ subject to $\sum_{i=1}^{4} u_i = H$

where the u_i^D 's were derived in the analysis of domestic negotiations (equations (10) and (11). The solution is given by

(17) $u_i^{ID} = \frac{1}{4}H + \frac{3}{4}u_i^D - \frac{1}{4}\sum_{j \neq i}^{4} u_j^D$

We will analyze this outcome further after we have presented some alternative possiblities for the conduct of international negotiations. In particular, it is realistic to consider situations where only nations bargain with each other, and within-country bargaining, among domestic groups, takes place separately.

International Negotiations: If Successful, Followed by Domestic Negotiations

Suppose that international negotiations take place between the two countries, where each national government has a purely utilitarian objective. This objective seems plausible in the absence of any bias towards one group or another. Clearly, other objectives are possible : the main point to note is that national governments are assumed to be benevolent. We assume here that domestic negotiations will take place only if the international negotiations are successful. This may seem restrictive, but it provides for a useful comparison with other cases. The assumption implies that the international negotiations take the initial allocation as defining the threat point.

With these assumptions, the Nash bargaining game for the international negotiations is given by

(18) $\max(u_1 + u_2 - d_1 - d_2)(u_3 + u_4 - d_3 - d_4)$ subject to $\sum_{i=1}^{4} u_i = H$

The solution for country A is given by

(19) $u_A \equiv \frac{1}{2}(H + d_1 + d_2 - d_3 - d_4)$,

with a similar expression for country B.

The domestic negotiations then take the outcome of international negotiations as a starting point for dividing the allocation and transfers within each country. We assume that, *if the domestic negotiations break down, the gains from the international agreement can not be implemented.* This implies that the threat points in the domestic negotiations are the d_i's. For example, for country A, the Nash bargaining solution is given by

(20) $\max(u_1 + d_1)(u_2 - d_2)$ subject to $u_1 + u_2 = u_A$.

Analogously to the case of domestic negotiations alone, this can be solved to give, for groups 1 and 2,

(21) $u_1 = \frac{1}{2}(u_A + d_1 - d_2), u_2 = \frac{1}{2}(u_A + d_2 - d_1)$

Now a substitution for u_A shows that the utility of group i is $\frac{1}{4}H + \frac{3}{4}d_i - \frac{1}{4}\sum_{j \neq i} d_j$

But this is the same as in the case four-way international negotations, i.e. equation (14). We state this as our second result.

Proposition 2 : In our two country model with two groups in each country, four-way (group level) international bargaining gives the same outcome as international bargaining at the national level, where in each domestic negotiations *do not occur* if the international round of negotiations fails. In the case of national level international bargaining, domestic negotiations follows its successful conclusion.

This equivalence result between two level and one level negotiations is useful, but the assumption that domestic negotiations only occur if the international round is successful may not be appropriate in all cases. Thus we turn to an alternative formulation of combined international and domestic negotiations.

International Bargaining : Always Followed by Domestic Bargaining

Now we consider the case where the international negotiations assume that if they fail to reach an agreement, domestic negotiations will take place, as in the very first case we analyzed. Thus, the threat points for the groups are given by u_i^D. We also assume that if international bargaining fails, negotiations at that level will not be reopened. In equilibrium, however, our assumption imply that this will not occur: the international bargaining will achieve the cooperative outcome. This could arise, for example, if there were a limited political window of opportunity which may pass: Jordan could get an agreement with in Israeli government headed by Rabin, but not by Shamir. [11] In this case, the Nash bargaining solution for the international negotiations is given by

(22) $\max(u_1 + u_2 - u_1^D - u_2^D)(u_3 + u_4 - u_3^D - u_4^D)$ subject to $\sum_{i=1}^4 u_i = H$.

The solution for country A is given by

(23) $\hat{u}_A \equiv \frac{1}{2}(H + H^A - H^B)$,

with a similar expression for country B. Domestic negotiations proceed based on this expression for the total utility, and again assuming that the international agreement can not be implemented without domestic agreement on how to internally divide the gains (so that the $d_i's$ are the threat points), yielding

(24) $u_1^{ID} = \frac{1}{2}(\hat{u}_A + d_1 - d_2)$, $u_2^{ID} = \frac{1}{2}(\hat{u}_A + d_2 - d_1)$

We use the same notation as for the second four-way bargaining case : this is justified in Proposition 4 below. Substituting for ($\hat{u}_A$ yields an alternative expression for these utilities. Now

(25) $\hat{u}_A - H^A = \frac{1}{2}(H + H^A - H^B) - H^A = \frac{1}{2}(H - H^A - H^B)$

The last expression is clearly positive, as long as there are gains to reallocation of water between the two countries. Hence we have the following result (which contrasts with case 'I', in which some domestic group could prefer only domestic negotiations).

Proposition 3 : In our two country model with two groups in each country, international bargaining which assumes that domestic negotiations will still occur if international agreement is not reached, followed by domestic negotiations, is preferred by all groups to domestic negotiations alone.

In Proposition 2, we demonstrated an equivalence result for particular cases of simultaneous and two-level negotiations, where domestic negotiations did not occur if the international bargaining failed to reach agreement. We have a similar equivalence result for all-party and two-level negotiations when, in each case, failure of international negotiations is assumed to be followed by successful domestic negotiations.

Proposition 4 : In our two country model with two groups in each country, four-way international bargaining gives the same outcome as national level international bargaining, where in each case domestic negotiations are correctly assumed to occur if the international round of negotiations fails. In the case of national-level international bargaining domestic negotiations also follow its successful conclusion.

Proof : The solution for the four way bargaining in this case is, as given earlier,

(17) $u_i^{ID} = \frac{1}{4}H + \frac{3}{4}u_i^D - \frac{1}{4}\sum_{j\neq i}^4 u_j^D$

Substituting for the u_i^D's from (10) this reduces to, for example,

(26) $u_1^{ID} = \frac{1}{4}(H + H^A - H^B) + \frac{1}{2}(d_1 - d_2)$

with similar expressions for the other three groups.

Alternatively, with two level negotiations, from (23) and (24),

[11] Of course this example is merely meant to be suggestive, since there have been many other factors at work in the Middle East.

(27) $u_1^{ID} = \frac{1}{2}(\hat{u}_A + d_1 - d_2)$ and $\hat{u}_A \equiv \frac{1}{2}(H + H^A - H^B)$
A simple substitution establishes the equivalence. [12]

Finally, we can also compare the two cases, with and without domestic negotiation following failure of the international negotiaiton. Consider group 1 for illustrative purposes. We have, after some simplification,
(28) $u_1^{ID} - u_1^I = \frac{1}{4}(H^A - H^B) + \frac{1}{2}(d_1 - d_2) - \frac{3}{4}d_1 + \frac{1}{4}(d_2 + d_3 + d_4)$
Hence we have
(29) $u_1^{ID} > u_1^I \Leftrightarrow H^A - (d_1 + d_2) > H^B - (d_3 + d_4)$
Since this last expressions just depends on country-level differences, it is easy to see that it will hold for group 2 as well. For groups 3 and 4 in country B, the inequality is reversed. Hence we have the following result.

Proposition 5 : Consider countries preferences between two negotiating arrangements. In the first, international bargainers assume that if they fail to reach agreement, domestic negotiations will proceed in both countries . In the second, international bargainers assume that domestic negotiations will not take place if an international agreement can not be reached. The country which gains more from a domestic agreement prefers the first arrangement, whereas the other country will always have the opposite preference (i.e. prefers the second arrangement).

The intuition for the result is clear. The country that gains more from a domestic agreement improves its threat point more than the other country, and therefore gains in having negotiations which assume that a domestic agreement will always occur. This result also illustrates why there is sometimes difficulty in agreeing to how international negotiations are to be conducted.

Domestic Bargaining Followed by International Bargaining

There are several possibilities to consider in this case. Assume that the domestic bargaining anticipates that the international bargaining will succeed if the domestic bargaining does. [13] Different assumptions are possible with respect to how the situation will be perceived if domestic bargaining fails: these will affect the threat point of the domestic bargaining game.

First, consider the case where it is expected that, if domestic bargaining fails, there will be no international negotiations. Then the threat points for the domestic groups in internal negotiations are the d_i's. Hence, the domestic negotiations anticipating the international agreement yield a bargaining game identical to the one where they are subsequent to the international agreement. If the latter is based on the expectation that domestic negotiations will still follow the failure of international bargaining, the solution will be identical to the previous case, labeled with the superscript ID.

Now consider the expectation that international negotiations will still go ahead, even if the domestic negotiations fail to reach an agreement. Again, if this international bargaining

[12] Note that in general, two-level and all-party negotiations may not be equivalent, even for given assumptions about domestic negotiations, and which groups are to be allowed to participate at the international level may matter. We consider briefly the generalization of Propositions 2 and 4. First of all, it should be pointed out that these are not simply consequences of the consistency axiom, which is a property of the multiperson Nasb bargaining solution. (See Krishan and Serrano, 1995 and Lensberg, 1988.) In fact, this is demonstrated by observing that, even for the quasilinear utility case, the result does not apply if there are two group in one country and only one in the other (this is done by simple calculations, omitted here). However, for the symmetric case, with two group in each country, we can establish the following, for general utility functions. Suppose that the threat point in all-groups, international bargaining are denoted by x_i. Then the equivalence result such as in the two propositions holds if and only if $x_1 - x_2 = d_1 - d_2$ and $x_3 - x_4 = d_3 - d_4$. the intution here is that equivalence holds if the relative bargaining strengths of domestic groups vis-a-vis each other are unaffected at the different levels of negotiation.

[13] If international negotiations were expected to fail, or to not occur at all, this would reduce to the case of only domestic bargaining, which we considered first.

is of the kind that assumes domestic agreement will be reached whatever the international negotiation outcome (because internal negotiations can successfully reopen), we will get the 'ID' solution. If, on the other hand, the international bargaining that will take place in the absence of a prior domestic agreement assumes that domestic bargaining will fail or not take place if the international bargaining fails, we will get the second case outcome, indicated by superscript I.

Other Symmetric Cases

Other symmetric cases (in terms of expectations) reduce to one of the previous three. For example, suppose that international negotiations come first, but there is the recognized possibility of reopening them if they succeed but the domestic negotiations fail, or if they fail. We argue that this reduces to the previous case, where domestic negotiations come first. There, we showed that the outcome would be one of the cases 'D', 'I' or 'ID'. Similarly, any chain of possibilities for reopening negotiations will reduce to one of these three cases. The salient issue in working out the outcome of each possible game is what is expected to happen if there is no immediate agreement, since this determines the threat point for the current negotiations. This is what matters for the equilibrium, rather than the exact sequence of the two levels of bargaining.

Assymetric Cases

Now suppose that international negotiations take place, but that, if they fail, domestic negotiations will take place only in country B. [14]. The Nash product in the international bargaining is now

(30) $(u_1 + u_2 - d_1 - d_2)(u_3 + u_4 - u_3^D - u_4^D)$

Solving the two levels of the bargaining game yields, for group 1 in country A,

(31) $u_1^{RI} = \frac{1}{4}(H - H^B) + \frac{3}{4}d_1 - \frac{1}{4}d_2,$

where the superscript 'RI' is meant to suggest that country A is more rigid or restricted in its position. It is a simple matter to check that

(32) $u_1^{RI} < \min\,(u_1^I, u_1^{ID})$

A similar result will obviously hold for group 2 in country A. Thus, being rigid in this way (failing to hold domestic negotiations when international negotiations fail) unambiguously hurts country A. For the groups in country B, we have

(33) $u_i^{RI} > \min\,(u_i^I, u_i^{ID}),\ i = 3, 4.$

If the roles of countries A and B are switched, we denote that case by 'IR' and the analogues of (32) and (33) will hold. These inequalities are used in discussing endogenous commitements to negotiate.

Commitments to Negotiate

The asymmetric cases allow us to tackle the question of how national governments will behave if they can make commitments regarding domestic negotiations. Up to now, we have assumed that whether domestic negotiations take place after the failure of international negotiations was exogenously given. Making it an endogenous choice creates a 2x2 noncooperative game that precedes the two levels of cooperative bargaining. Since the case indexed 'I' is equivalent to the siutation where both countries are rigid, and 'ID' is equivalent to the case where neither one is, equations (32) and (33) and their analogues when country B is rigid provide a simple solution : *it is a dominant strategy for each national government to commit to having domestic negotiations, whatever the possible outcome of the international netotiations.* [15] The intuition is simple : doing so improves the threat point of the subse-

[14] We are grateful to a referee for suggesting this possibility, and the issue of commitments to negotiate discussed next

[15] Thus, even if one country has this choice, it will exercise it in the manner indicated. This might be the case if the other country's politics are more fractious or rigid, so that only successful international cooperation

quent international bargaining game for each country. Note that one country will be better off, the other worse off as a result of this commitment, compared to the case where neither commits : this follows from Proposition 5.

3 Investment and Noncooperative Behaviour

In the model constructed in the previous section, there is essentially only one decision at stake : how much will suppliers of water be compensated ? This may be determined by cooperative bargaining [16], or by a market-type mechanism. We have suggested that bargaining is closer to actual practice than are market mechanisms. The allocation of water itself in this framework is basically determined by the conditions for Pareto optimality. The nature of the allocation mechanism used, in particular the structure of the two-level negotiations, can have an impact on the final transfers of numeraire (money) made, through its effect on threat points, but the allocation of water itself is invariant to the choice of the allocation mechanism, provided it is efficient. The simple structure we used, particularly quasi-linear utilities, ensured this property, and allowed us to focus on exploring the different possible structures for international and domestic bargaining.

We now introduce a significant complication. We recognize that the productivity or utility of a given quantity of water will very likely depend on the level of complementary investments. These may be dams, irrigation projects, or even more general complementary investments in agriculture or industry. [17] In this section, we will focus on the implications of such investment, through its timing and its productivity effects, for the conduct and outcome of two-level water negotiations. We will initially take the most simple case, and assume no direct effects of investments on water availability as such, but will discuss relaxing this after we have analyzed the simpler case.

Hence, we still assume that the total availability of water is $\bar{w}$, with each group having a *de facto* initial allocation of $\bar{w}_i$. We now assume, however, that the utility function of group i, evaluated at the initial allocation, is given by

(34) $$u_i(\bar{w}_i, k^j \bar{y}_i) = v_i(\bar{w}_i, k^j) + \bar{y}_i.$$

Here, k^j is the investment made by country j. Thus, we are keeping matters, simple by assuming that there are no direct externality effects of investment, in addition to assuming that water availability is unaffected by such investments. At this stage, we also assume that investments are made only at the national level : domestic groups do not control them, nor can they make group-specific investments. This assumption will need to be changed, of course, if the groups are regions in a federal system, and we will turn to that subsequently. Note also that the utility function is defined gross of the costs of investment. These are given by the strictly convex, twice differentiable functin $c_j(k^j)$. We will discuss later how these costs may be shared between the two groups in a country.

We use the following notation : $\partial v_{iw}, \partial v_{ik}$ denote respectively, the partial derivative of v_i with respect to w, and the partial derivative of v_i with respect to k. We assume that all these derivatives are positive. In the remainder of this section, it is convenient to suppress the $\bar{y}_i$'s by assuming that they are zero, or that utility is scaled to be measured from the initial numeraire allocation as origin. This reduces inessential notation.

---

will spur domestic agreement.

[16] Recall also that the cooperative bargaining approach may be viewed as obtaining an approximation to some noncooperative bargaining procedure, involving offers and counter offers.

[17] For example, the introduction of HYVs of seeds in India - the Green Revolution - increased the importance of irrigation and the regular availability of water in general.

Optimal Allocation of Water and Investments

First suppose that the investment levels are arbitrarily given. The first order conditions determining the Pareto optimal allocation of water (maximizing total utility) are given by

(35) $\partial v_{iw}(w_i, k^j) = \lambda$

where λ is the multiplier associated with the aggregate resource constraint. As long as the utility functions are not additively separable in w and k, these equations imply that the optimal allocation of water depends on the investments in both countries. Hence, even though there are no direct externalities as a result of the investment, the conditional optimum of water allocation involves a linkage of both countries. What country A does with its investment will affect the optimal amount of water that country B should receive. This effect operates through the aggregate resource constraint. Clearly this will be important in negotiations, and will be an important point in our subsequent analysis.

For example, suppose that we have, for group i in country j,

(36) $v_i = w_i^a (k^j)^{1-a}$

In this case, we can easily solve explicitly for the conditional optimum allocations of water. For country A and group i, i = 1,2, we have [18].

(37) $w_i = \frac{k^a \bar{w}}{2(k^A + k^B)}$

Note that the optimal amount of water for a group in country A decreases with the level of investment by country B, but increases with its own investment. These are consequences of the complementarity of water and investment in the utility functions. We state the point illustrated by this example more formally in the following.

Proposition 6 : In our two country model with two groups in each country, the optimal allocation of water in one country will depend on domestic investment in the other country, even in the absence of direct externalities, as long as domestic investment affects the marginal utility of water.

In general, let the conditional optimum amounts of water be denoted by $w_i^*(k^A, k^B)$. Furtermore, let $v_i^*(k^A, k^B) = v_i(w_i^*(k^A, k^B), k^j)$. Thus the v_i*'s are the utilities assuming that whatever the decisions on investments, the allocation of water will be optimal conditional on those decisions. Finally, for country A, let

(38) $V_A^*(k^A, k^B) = v_1^*(k^A, k^B) + v_2^*(k^A, k^B)$,

with a similar expression for country B. These are the gross utilities, before the costs of investment are subtracted off. It will be convenient to work with these functions.

We can now simply note that the optimal choice of investments (maximizing the sum of net welfare in the two countries) is given by the following first order conditions :

(39) $\partial V_{Ak^A}^*(k^A, k^B) + \partial V_{Bk^A}^*(k^A, k^B) = c_A'(k^A)$

$$\partial V_{Ak^B}^*(k^A, k^B) + \partial V_{Bk^B}^*(k^A, k^B) = c_B'(k^B)$$

These equations determine the optimal investments, and hence the optimal allocation of water is determined by the functions w_i^*.

Now suppose that both investments and the allocation of water are the subject of international negotiations. The above solution will determine the Pareto frontier. The outcome of the negotiations will include a joint agreement on the allocation of water between the countries, as well as a joint agreement on the levels of domestic investments in the two countries. This part of the outcome will be invariant to the specific form of the negotiations, as

[18] With the utlity function in (36), (35) becomes $av_i/w_i = \lambda$. Hence $w_i = (a/\lambda)^{1/(1-a)} k^j$. Adding up across groups and countries gives $2(a/\lambda)^{1/(1-a)}(k^A + k^B) = \bar{w}$. Therefore $2(a/\lambda)^{1/(1-a)} = \bar{w}/k^A + k^B)$. Substituting this in the expression for w_i given (37). Note that (37) does not depend on the preference parameter 'a' because only one good is being reallocated: the rate of substitution between w_i and k^1 does not matter

long as the cooperation on both dimensions is possible. However, the specific form of the negotiations will affect the money transfers that accompany the agreement, as we saw in the previous section. Suppose that, for brevity, we restrict attention to case 'ID' from that section. The threat point for the international negotiations is the outcome of successful domestic negotiations in the absence of an international agreement. Since we are assuming that the domestic investments have not been precommitted, but are included in the international negotiations, they are potentially different for the threat situation. Clearly, we can work out the domestic investments for this case, of only domestic negotiations, in two steps as before. The first step is to obtain the optimal domestic reallocation of water given an arbitrary investment level (and now this no longer depends on investment in the other country, since there are no direct externalities). This reduces the problem to the investment dimension, and the national government can choose the optimal level of investment.

At this stage, we need to address the issue of how costs of the national investment are to be allocated to the two groups. If we were to think of this being done through a market or market-like mechanism, where groups would pay user charges based on operating and capacity costs, we could derive the appropriate allocation of the joint costs of the investment through a standard exercise, to ensure optimal usage of capacity. Here, we make no distinction between capacity and its utilization, and we shall simply assume that the national government decides on a particular split of the costs, say 50:50. Note that if this is outside the control of the domestic groups, it does not affect their behaviour, and the split is irrelevant in the aggregate since utilities of groups are merely added up to obtain the national level objective function.

Noncooperative Investments

While domestic investments such as dams may plausibly be the subject of international negotiation - and have been in cases where there are direct externalities such as effects on water availability in another country [19] - it is less likely that countries are willing or able to negotiate broadly at the international level over general investments that affect the utility or productivity of water in the domestic economy. Hence, we turn to an examination of the consequences of noncooperative investment behaviour.

We introduce another piece of notation to make our expressions more compact. Let $V_A^A(k^A)$ denote the total utility in country A (gross of investment costs) in the absence of an international agreement on water. Note that in this case, there is no externality. There is a similar expression for country B. This amount is relevant for the threat point of the international bargaining. Each country is forward-looking in its investment decisions, which are now assumed to be precommitted before the international bargaining takes place. The international negotiations is now only over water. Following the standard solution for the relevant Nash bargaining game, the outcome of the international bargaining for country A is

(40) $$\tfrac{1}{2}[V_A^*(k^A,k^B) + V_B^*(k^A,k^B) - c_A(k^A) - c_B(k^B) + (V_A^A(k^A) - c_A(k^A)) - (V_B^B(k^B) - c_B(k^B))]$$

The first four terms together give the Pareto optimal aggregate utility, conditional on the domestic investments, while the next two pairs are the disagreement payoffs or threat points for A and B respectively. Because we are now assuming that the investments are precommitted, they are the same with or without an international agreement.

Each country now chooses its investment to maximize its own utility, anticipating the effect on the international bargaining outcome. For country A, the solution satisfies the first order condition

[19] Here the Farakka Dam on the Ganges in India, and its effects on Bangladesh, as well as the proposed Unity Dam on the Yarmouk for Syria, Jordan and Israel, come to mine.

(41) $\frac{1}{2}[\partial V^*_{Ak^A}(k^A, k^B) + \partial V^*_{Bk^A}(k^A, k^B) + \partial V^A_{Ak^A}] = c'_A(k^A)$
For country B, a similar condition is obtained. Clearly, these are different from the case where the domestic investment are chosen cooperatively.

Note also that the outcome is different from the simple noncooperative case, where each country myopically chooses investment to maximizes its utility, taking the other country's investment as given. For example, if country A were to choose its domestic investment noncooperatively but not recognizing the strategic implications for future international negotiations, its choice would satisfy [20]

(42) $\partial V^*_{Ak^A}(k^A, k^B) = c'_A(k^A)$.

This also is not optimal, but in a different way. A comparison of these cases is useful. Suppose that $V^*_B(k^A, k^B)$ did not depend on investment in country A. Then the ordinary noncooperative solution, (42), would coincide with the cooperative solution, given in equation (39). However, the strategic noncooperative solution would still be different from the cooperative solution. [21] We summarize the above in the following.

Proposition 7 : In our two country model with two groups in each country, if domestic investment affects the marginal utility of water, then, even in the absence of direct externalities, negotiating only over water, and not over domestic investments leads to an inefficient outcome if domestic investments can be precommitted.

It is easy to introduce complications into the above model of investments that are complementary to water use. These include allowing for investments to directly affect water availability, or create direct externality effects. It is also easy to allow for asymmetries in the effects of such investments : this would be the case for an upstream versus a downstream country. Clearly, our reduced form expressions above can encompass such cases. A different extension is to allow group-specific investments. This would be plausible if the groups are regional, and there is a federal system in the country in question. Similar considerations will clearly apply at the level of domestic groups : if they can they will choose their investments strategically, to affect the outcome of subsequent negotiations, both domestic and international.

Investment may be though of still more broadly. It is commonly argued that international water negotiations are often 'linked' to other negotiations, such as arms control and security. The treatment of 'investment' here is entirely general. One could think of it as the necessary steps to be taken in arms control, for example, rather than as building a dam. The point is the need for difficult-to-reverse steps which affect overall utility, changing threat points and affecting the credibility of commitments in a repeated game. We leave more detailed examination of this case to future work.

A different sort of modification is the following. Return to the ase of national-level investments. Suppose that international negotiations take place before national level investments are determined. Then, if all agreements over water are binding, the choice of domestic investments will be made given allocations of water. Since the allocation of water will be different at the threat points, the domestic investments in that case will also be different than in the case of international agreement. Now, forward-looking negotiators will recognize the effects of their agreement on investment as well. For example, for country A, the objective function will have the form

(43) $v_1(w_1, k^A(w_1, w_2)) + v_2(w_2, k^A(w_1, w_2)) - c_A(k^A(w_1, w_2))$.

The negotiation at the international level can be thought of as proceeding as follows.

[20] Note that the left hand side of this expression is the partial derivative of (38), which does incorporate the allocation of water.

[21] It is useful to point out that, while our model is quite different in scope and implications, as well as specific formulation, from that of Grossman and Hart (1986), the central insight is similar.

The two countries can work out the full optimum in terms of water allocations and domestic investments. The international agreement, which can only be over water by assumption, implements the optimal allocation of water corresponding to the optimal domestic investments. Now if each country can decide its level of investment before the domestic negotiations take place, since there are no externalities, it will independently choose the optimal level of domestic investment. This is because the possibility of affecting water allocations through investments, which was the focus of the previous analysis, and the source of inefficiency there, does not arise in this case.

Now suppose that domestic negotiations also take place before the domestic investment decisions are made. This is plausible in that investment decisions may take time, while domestic negotiations can follow quickly on an international agreement. We have treated the international agreement as including a domestic allocation of water for each group, but it may actually only be binding at the level of international transfers of water and money (in opposite directions). This did not matter earlier, since investment was fixed or precommited in all those cases. Now, however, the possibility arises that, given the overall international agreement, the domestic groups will negotiate over the internal distribution of water, taking account of the fact that the internal distribution will affect domestic investment decisions. However, if the national government always responds by maximizing total net utility, it is easy to see by the envelope theorem that the condition for the optimal allocation of water is unaffected by this possibility.

The discussion in the preceding two paragraphs can be summarized in the following result, formally proved in the Appendix.

Proposition 8 : In our two country model with two groups in each country, whether or not domestic investments can be precommitted before internal negotiations, if they take place after international negotiations, then, in the absence of direct externalities, the outcome in terms of water allocation and domestic investments is fully efficient.

This result says that in our simple framework the timing of domestic investments matters only relative to the timing of international negotiations, not relative to the domestic negotiations, as long as the latter come after the international negotiations. It applies only to the case of national level investments. It is easy to extrapolate, however, to the case of group (region) specific investments, where the implication would be that their timing relative to domestic negotiations will affect the efficiency of the outcome.

4 Concluding Remarks

In this paper, we have begun the task of understanding the outcome of two level (international and domestic) negotiations for the allocation of a good, using the framework of the Nash bargaining solution, and the case of water. This can be viewed as obtaining an approximation to a particular, but plausible, noncooperative bargaining procedure. The use of the Nash bargaining solution distinguishes our analysis from those of Lax and Sebenius (1991), Iida (1993), and Mo (1995). Mayer (1992) uses the Nash bargaining approach, but less formally, and to ask a different set of questions. The contributions of our analysis are two-fold. First, we compare different simultaneous and two-level bargaining situations, showing when and why some might yield equivalent outcomes. Second, we highlight the importance of the nature and timing of complementary investments, and whether they are included in negotiations, in affecting the efficiency of the negotiated outcome. While the equivalence results are sensitive to our particular formulation, the lessons of the analysis of investment are not restricted to any functional form.

We have not explored specifically the possibility of investments that affect the other nation's threat point, for example through the possibility of military action. This is an interesting issue, closely related to the issue of lack of well-defined initial property rights, but one that requires a separate, detailed analysis. Here, we simply note that we have throughout assumed that the bargaining takes place from well-defined initial positions, reflecting legal or *de facto* property rights. Thus, we focus on the mutual benefits of trade, and how to achieve them optimally. The issue of conflict over initial property rights is less tractable, and must be tackled as an inherently noncooperative game, since gains for one party mean losses for another.

The main insights of our analysis do not rely on the presence of uncertainty or lack of information. However, it will interesting infuture work to extend our analysis to allow for both. In addition to extending our work to allow for uncertainty and lack of information, we will also examine the effect of having self-interested governments that also respond to political pressure, as in recent work by Grossman and Helpman (1995a, 1995b). This will introduce a significant additional dimension to the problem.

Appendix : Proof of Proposition 8

Let $V_j(w_j, k^j)$ be the gross utility of country j, assuming water is allocated optimally within each country. The optimum involves choosing w_A, w_B, k^A, k^B to maximize

(A1) $V_A(w_A, k^A) + V_B(w_B, k^B) - c_A(k^A) - c_B(k^B)$ subject to $w_A + w_B = \bar{w}$

The first order condition are

$$\text{(A2)}\quad \begin{aligned} \partial V_{Aw_A} &= \partial V_{Bw_B} \\ \partial V_{Ak^A} &= c'_A(k^A) \\ \partial V_{Bk^B} &= c'_B(k^B) \end{aligned}$$

The case where domestic negotiations follow domestic investment, which follows the international water agreement, is straightforward. The Nash bargaining solution in country A (country B is similar) requires maximizing (with respect to w_1, w_2)

(A3) $(u_1(w_1, k^A) - \frac{1}{2}c_A(k^A) - d_1)(u_2(w_2, k^A) - \frac{1}{2}c_A(k^A) - d_2)$ subject to $u_1 + u_2 = u_A*$, where u_A* is the (net) national level welfare from the international water agreement, assuming an optimal domestic water allocation, but conditional on investment.

The Nash bargaiing solution for group 1 (group 2 is similar) is

(A4) $u_1 = \frac{1}{2}[u_A(w_A*, k^A) - c(k^A) + d_1 - d_2]$.

Maximizing (A4) with respect to k^A gives the required first order condition for optimality.

Now, consider the case where domestic bargaining is before domestic investment, but again after the international water agreement. Now, potentially we have $k^A(w_1, w_2)$, instead of a fixed k^A in (A3), including in the constraint. The Nash bargaining solution is still as in (A4), but group 1, for example gets

(A5) $\frac{1}{2}[v_1(w_1, k^A(w_1, w_2)) + v_2(w_2, k^A(w_1, w_2)) - c_A(k^A, w_1, w_2))]$.

Here the potential dependence of investment on the domestic bargaining outcome is made explicit. Maximizing this subject to $w_1 + w_2 = w_A$ yields the first order condition

$\partial v_{1w_1} + (\partial v_{1k^A} + \partial v_{2k^A} - c'_A)\partial k^A_{w_1} = \lambda$.

However, the term in parentheses is zero, since the national government will choose $k^A(w_1, w_2)$ to maximize national welfare. Therefore the outcome will be the same as if k^A were given to the domestic groups.

This completes the proof of the proposition.

5 Acknowledgement :

This is a revised version of a paper presented at the International Game Theory Conference, Bangalore, India, January 2-6, 1996. We are grateful to the organizers of the conference, and for the helpful comments of participants, particularly Debraj Ray, Kalyan Chatterjee and Satya Das. We are most indebted to an anonymous referee for incisive comments that helped us to substantially improve the paper. Financial support was received from the University of California, Institute of Global Conflict and Cooperation; Indian Statistical Institute and the University of California, Santa Cruz, Academic Senate and Division of Social Sciences. Nirvikar Singh also acknowledges the hospitality of the Centre for Development Economics, Delhi School of Economics, where he was a Senior Visiting Fellow during December 1995 and January 1996. We received helpful research assistance from Hui Miao. Remaining errors are our responsibility alone.

References :

1. Barrett, Scott, (1994), "Conflict and Cooperation in Managing International Water Resources", Policy Research Working Paper No. 1303, The World Bank, May.

2. Binmore, Kenneth, Ariel Rubinstein, and Asher Wolinsky, (1986), "The Nash Bargaining Solution in Economic Modelling", *Rand Journal of Economics*, 17, 2, Summer, 176-188.

3. Chaudry, M., and M.H. Siddigi, (1987), "Toward a National Water Plan in Bangladesh", Chapter 34 in *Water Resources Policy for Asia*, ed., Mohammed Ali George Radosevich, Akbar Ali Khan, Boston, A.A. Balkema.

4. Crow, Ben, with Alan Lindquist and David Wilson (1995), *Sharing the Ganges : The Politics and Technology of River Development*, Sage Publications : New Delhi.

5. Dhillon, P.S. (1983), *A Tale of Two Rivers*, Chandigarh, Dhillon Publishers.

6. Fisher, Franklin M., (1995), "The Economics of Water Dispute Resolution, Project Evaluation and Management : An Application to the Middle East", *Water Resources Development*, 11,. 4, 377-389.

7. Friedkin, Joseph F., (1987), "International Water Treaties : United States and Mexico", Chapter 25 in *Water Resources Policy for Asia*, ed., Mohammed Ali George Radosevich, Akbar Ali Khan, Boston, A.A. Balkema.

8. Grossman, Gene and Elhanan Helpman (1995a), "Trade Wars and Trade Talks", *Journal of Political Economy*, 103, 4, August, 675-708.

9. Grossman Gene and Elhanan Helpman (1995b), "The Politics of Free-Trade Agreements", *American Economic Review*, 85, September, 667-690.

10. Grossman, S., and O. Hart, (1986), "The Costs and Benefits of Ownership : A Theory of Vertical and Lateral Integration", *Journal of Political Economy*, 94, 4, August, 691-719.

11. Harsanyi, J.C., (1963), "A Simplified Bargaining for the n-Person Cooperative Game", *International Economic Review*, 4, 194-220.

12. Harsanyi, J.C., (1977), *Rational Behaviour and Bargaining Equilibrium in Games and Social Situations*, Cambridge University Press.

13. Iida, Keisuke, (1993), "When and How Do Domestic Constraints Matter ? Two Level Games with Uncertainty", *Journal of Conflict Resolution*, 37, 3, September, 403-426.

14. Just, Richard E., Sinaia Netanyahu, and John K. Horowitz, (1996), "Water Pricing and Water Allocation in Israel", processed, University of Maryland, Department of Agricultural Economics.

15. Krishna, Vijay, and Roberto Serrano, (1996), "Multilateral Bargaining", *Review of Economic Studies*, 63, 1, 61-80.

16. Lax, David A., and James K. Sebenius, (1991), "Negotiating Through an Agent", *Journal of Conflict Resolution*, 35, 3, Septemebr, 474-493.

17. Lensberg, T., (1988), "Stability and the Nash Solution", *Journal of Economic Theory*, 45, 330-341.

18. Mayer, Frederick W., (1992), "Managing Domestic Differences in International Negotiations : The Strategic Use of Internal Side-Payments", *International Organization*, 46, 4, Autumn, 793-818.

19. Mo, Jongryn, (1995), "Domestic Institutions and International Bargaining : The Role of Agent Veto in Two-Level Games", *American Political Science Review*, 89, December 914-924.

20. National Water Development Agency, (1992), *National Perspectives for Water Resources Development*, July, New Delhi : NWDA.

21. Putnam, Robert D., (1988), "Diplomacy and Domestic Politics : The Logic of Two-Level Games", *International Organization*, 42, 427-460.

22. Ramana, M.V.V., (1992), *Inter-State River Water Disputes in India, Madras* : Orient Longman.

23. Rubinstein, Ariel, (1982), "Perfect Equilibrium in a Bargaining Model", Econometrica, 50, 1, 97-109.

24. Schelling, Thomas, (1960), *The Strategy of Confict*, Oxford and London : Oxford University Press.

25. World Bank, (1993), *Water Resources Management : A World Bank Policy Paper.* Washington, D.C. : The World Bank.

Alan Richards
Department of Economics
University of California
Santa Cruz, CA 95064
USA

Nirvikar Singh
Department of Economics
University of California
Santa Cruz, CA 95064
USA

PRICE RULE AND VOLATILITY IN AUCTIONS WITH RESALE MARKETS

Ahmet Alkan

Abstract: *This paper offers a model of sealed bid auctions with resale. The policy question whether the seller would fare better under the multiprice rule(where winners pat their actual bids) or the uniprice rule (where winners pay the highest losing bid) has been on the agenda since the 60's and seen a recent revival. While theory has mostly recommended the uniprice rule, mainly on the argument that it would generate higher revenue, practice has predominantly stay with the multiprice rule. The results here recommend the multiprice rule. They say that, while expected revenue (equivalently, the price level) is an invariant throughout, the range of bids is narrower, (hence the price level less volatile) under the multiprice then under the uniprice rule, the more so the greater the resale component of participation. For auctions of treasury depth shares of public firms to be privatized, or other items, where significant proportions of the issues flow to secondary markets, these results thus support the multiprice rule on grounds of price stability next to revenue equivalence.*

1 Introduction

Auctions are events whereby a seller's item gets priced. A variety of auction formats exist. In particular, an auction may receive open bids or sealed bids and the rule that specifies what a winner pays may be various. The present paper is an exploration on whether the *multiprice* (each winner pays his own bid) or the *uniprice* (each winner pays the highest losing bid) rule is more advantageous to the seller in sealed bid auctions. Pertinent examples are auctions of treasury debt, shares of public firms to be privatized, and pollution rights.

The payment rule issue has been on the agenda ever since Milton Friedman (1960) advocated that the U.S. Treasury should switch from the multiprice to the uniprice rule. The Treasury did in fact experiment so in some auctions in the 70s but switched back not long after. Throughout the controversy that has

T. Parthasarathy et al. (eds.),
Game Theoretical Applications to Economics and Operations Research, 275–290.

practise has mostly stayed with multipricing. Recently, the matter has seen a revival, and following a review of its rules (see Joint Report 1992), the U.S. Treasury has chosen to experiment once again and is currently implementing the uniprice rule in the auctions of two - year and five - year notes. This move was preceded by a strong call from Chari and Weber (1992) for a switch to the uniprice rule. More recently, in their survey of related theory and empirical work, Bikhchandani and Huang (1993) have also expressed support for the uniprice rule although in more cautious terms. The prominent contention behind all this support has been that the average payment made in auction (the *auction price*, or equivalently, seller's revenue) is likely to be higher under the uniprice rule than under the multiprice rule.

A salient aspect of auctions, most prominent in the case of treasury bills, is that participants enter for profit-making in the resale markets that follow an auction. To the extent that the resale price is influenced by what bidding has occurred in an auction, participants would naturally consider and set their behavior in accordance. It is notable that few auction models have attempted any explicit treatment of this aspect. (Bikhchandani and Huang (1989), quoted below, is an exception.) In the present paper, I build and analyze a model which incorporates the resale motive and which allows scrutiny at a dimension largely ignored in previous analyses, namely the volatility of the auction price. The findings are in favor of the multiprice rule.

For background, I will briefly review the arguments put forward on why the uniprice rule would generate higher expected revenue:

(i) When buyers' valuations of the items for sale (or estimates thereof) are positively correlated, auction theory predicts that participants *shade* their bids more in the multiprice auction than in the uniprice auction, to such an extent that the auction price is on average lower in the former than in the latter (Milgrom and Weber (1982)). One way to express this phenomenon is that *winner's curse* is more severe in the multiprice than in the uniprice auction and (sophisticated) bidders take this into account. (Winner's curse is owing to the fact that winners will be those bidders who have the highest valuations/estimates, who will then have to downgrade their valuations in view of the correlation, and so who would have overpaid had they bid naively not foreseeing all this.) Expected revenue is the same when and only when no correlation exists among buyers' valuations/estimates.

(ii) Due to higher severity of the winners' curse effect, which in general diminishes with reduced uncertainty, information gathering has a higher return in the multiprice auction. The uniprice auction necessitates and gives rise to less information acquisition by comparison. In equilibrium ex post, the cost of this incremental information is a loss of revenue to the seller (Chari and Weber (1992).)

(iii) Secondary market buyers are less informed than auction participants. They infer value by considering what bids have been made in the auction. Auction participants take this into account and signal value to future customers.

This signalling effect pulls bids higher, the more so under the uniprice rule (Bikhchandani and Huang (1989)).

(iv) Because bid preparation is simpler hence less costly in the uniprice auction, participation will be broader and so revenue will be higher. So has been put forward by Friedman (1960). Chari and Weber (1992) also subscribe to this point of view while Bikhchandani and Huang (1993) express disbelief.

The effects cited in (iii) and (iv) aside, the upshot of points (i) and (ii) above is that the resale price will exceed the average auction price by a greater margin in the multiprice auction than in the uniprice one. The differential between the two margins is to be attributed to the higher level of information acquisition activity that needs to be conducted under the multiprice rule than under the uniprice rule. Chari and Weber (1992) are of the opinion that the associated incremental cost has " dubious social value" and is "wasteful". Following change of payment rule, they forecast, the return on related investments will over time accrue to the seller.

The policy thrust of the findings in this paper is that the incremental information that needs to be acquired in the multiprice auction, at a social cost, may well have a social value, namely that of curbing the potential volatility of the auction price, which risk-averse buyers and seller would naturally be mindful about.

The model I employ is simply born by the addition of two features to the standard auction model which together intend to capture the effect of the resale motive: I hold that (i) participants resell an exogenously specified portion of their purchase in the secondary market and that (ii) resale occurs at the auction price materialized. (I defer further comments on these features to the concluding section.) I also employ the benchmark assumption that buyers' valuations (on the portion of their purchase which they keep) are uncorrelated. I then show that the ensuing games have unique equilibria at which, the higher the degree "resale orientedness", the narrower is the range of bids (shrunk ultimately to a singleton) under the multiprice rule, whereas bidding remains unaffected under the uniprice rule. Furthermore, this beneficial narrowing of uncertainty regarding the auction price comes with no loss in expected revenue: the expected auction price is invariant under the degree of resale orientedness as well as the payment rule.

I formally state the model in the next section. Section 3 contains the analysis and results. Section 4 is devoted, for illustration of the results, to two-unit auctions in which case the equilibrium (second order differential) equation turns out to be the familiar Gauss hypergeometric equation and bidding strategies are obtainable in explicit form. (Single unit auctions form the known classical case.) Section 5 contains some comments on the assumptions and features of the model.

2 The Model

There are m *identical* items to be auctioned and $n + 1$ buyer participants where $n \geq m$. Each participant puts in a sealed bid for one item. The m highest bidders each *pay* according to a prespecified rule and receive an item each. (Some tie-breaking rule is applied in case of ties, not necessary to be specified here, as ties turn out to be zero probability events.) Buyers maximize expected *gain*, i.e., "value" less payment. The auction is called *multiprice* or *uniprice* depending on whether payment equals one's own bid or the $(m + 1)$st highest bid, i.e., the highest losing bid.

In the so-called *independent private values* model, the value of an item to a buyer is a number known by the buyer not by the others, called his *personal value*, and drawn independently from a common distribution. The model I introduce here is essentially the same in every respect except that buyers' valuations are "semi-endogenous" in the following way: Let us call the average payment made in an auction the *auction price*. I stipulate that the value of an item for an auction participant u with personal value $V(u)$ is $\delta\, p + (1 - \delta)V(u)$ where δ is some prespecified "weight" ($\delta \in \Delta = [0, 1)$) and p is the auction price to materialize in that auction. I suggest δ to be interpreted as the "degree" of resale orientedness of the auction. As one parable, for example, consider that there is a post-auction market expected to clear at the auction price, and that winners sell a portion δ of their purchase in this market while keeping the remainder for personal use.

Formally, I let $\mathbf{I} = [0, 1]$ be the set of all potential buyers u facing an auction, each equally likely to be one of the participants and ordered such that personal values $V(u)$ are *increasing* in u with $V(0) = 0$, $V(1) = 1$. An auction game then is described by a quintuple $\Gamma(m,\ n, V,\ \delta,\ R)$ where, in addition to what has already been specified, the last variable R is either M or U depending on whether the payment rule is multiprice or uniprice. A (pure) strategy of a buyer in such a game is a *bid function* b defined on $\mathbf{I}$ and *employed* by him in the sense that $b(u)$ is the bid he would make if his identity is u. It will be assumed that the specification of $\Gamma(m,\ n, V,\ \delta,\ R)$ is common knowledge among all buyers.

One thus has, for each m, n, and distribution V of personal values, a double family of auction games indexed by $\delta \in \Delta$. The query I will pursue is on how bidding behaviour in general and the auction price in particular are affected by the size of δ, comparatively, in the multiprice and uniprice auctions. To motivate this query, let us briefly consider the two extreme degrees $\delta = 0$ and $\delta = 1$. Study of the former case, which originated more than three decades ago (Vickrey (1960)), has uncovered how a bidder u shades his value $V(u)$ under the multiprice rule while bidding exactly $V(u)$ under the uniprice rule and bears, in particular, the celebrated *revenue equivalence theorem*: the auction price is the same in expectation whether the payment rule is multiprice or uniprice. At the other extreme $\delta = 1$, on the other hand, indeterminacy reigns. The auction game that obtains at this limit, which our model does not allow, has a continuum of Nash equilibria. It is easily seen in fact, then, that any real constant c describes an equilibrium, as bidding c is clearly a best response to all others' bidding c, whence the auction price equals c, whether the payment rule is multiprice or uniprice. Thus, while it is determinate that all buyers identically gain zero in any equilibrium, the auction price is indeterminate in fact entirely arbitrary, in either auction, when $\delta = 1$. I will refer to the games that obtain at $\delta = 0$ and $\delta = 1$ as the classical and the *purely speculative* games respectively.

3 Analysis and results

I will throughout this section consider m, n, V fixed and let $\Gamma(\delta, R) = \Gamma(m, n, V, \delta, R)$ be any auction game as described in Section 2. All results stated below are well known to hold for $m = 1$ when the model reduces to the classical case. I will further be assuming that $m \geq 2$.

3.1 Existence and uniqueness of equilibrium

3.1.1 The uniprice auction: Dominant strategy equilibrium

Proposition[1] It is a dominant strategy for any buyer $u \in I$ in a uniprice auction $\Gamma(\delta, U)$ to bid his personal value $V(u)$.

This result says that, whatever δ is and whatever the personal values of all other participants may actually be, it is impossible for any buyer to achieve a higher profit than what he would achieve by bidding his personal value. The proof is straightforward and the same as in the classical case.

3.1.2 The multiprice auction:Existence and uniqueness of symmetric equilibrium

Following standard methodology, I will restrict attention to the analysis of symmetric Nash equilibria, i.e., those equilibria where every buyer employs the same bid function. Formally, a bid function b constitutes a *symmetric equilibrium* for $\Gamma(\delta, M)$ if every buyer $u \in I$ maximizes his expected gain by bidding $b(u)$ given that every other buyer employs b.

The following is the first main result of the paper.

Theorem 1 There exists for every multiprice auction $\Gamma(\delta, M)$ a unique *increasing* bid function which constitutes a symmetric equilibrium.

Assumption The proof I give below utilizes the assumption that V is differentiable and in (Lemma 2) that $(1-u)V$ is strictly concave, i.e., $(1-u)V' - V$ is decreasing in $u \in I$.

Notation Call F_{mn} the cumulative probability distribution of the mth highest of n independent draws from the uniform distribution on the unit interval and call f_{mn} the density function of F_{mn}.

Towards proving Theorem 1, take any buyer $u \in I$ in the auction $\Gamma(\delta, M)$ and let x be his bid. Suppose b is an increasing bid function employed by all the other n buyers (the *opposition*). Buyer u wins an item if and only if $z \leq x$ where z is the mth highest bid among the opposition. Thus, u wins if and only if $b(y) \leq x$, i.e., $y \leq b^{-1}(x)$. Note that if u wins then the $m-1$ other winners are each uniformly distributed with density $1/(1-y)$ on the subinterval $[y, 1]$. So the expected auction price, conditional on u winning with a bid x and y being the mth largest buyer in the opposition, is given by

$$(x/m) + (m-1)/m \int_y^1 b(w)/(1-y)\, dw.$$

The expected gain of u then having made a bid x is

$$G(x) = \int_0^{b^{-1}(x)} ((1-\delta)V(u) + \delta(x/m + (m-1)/m \int_y^1 b(w)/(1-y))dw)$$
$$-x) f_{mn}(y) dy.$$

Differentiating $G(x)$ with respect to x one gets,

$$G'(x) = f_{mn}(b^{-1}(x))((1-\delta)V(u)$$
$$-(1-\delta/m)x)/b'(b^{-1}(x)) - (1-\delta/m)F_{mn}(b^{-1}(x)) -$$
$$\delta(m-1)f_{mn}(b^{-1}(x))B(b^{-1}(x))/m(1-b^{-1}(x))b'(b^{-1}(x))$$

where it is defined for $y \in I$ that

$$(1) \quad B(y) = -\int_y^1 b(w)\, dw.$$

Note that $b = B'$.

Suppose b constitutes a symmetric equilibrium. Then $x = b(u)$. Now, setting $u = b^{-1}(x)$, defining the new parameter

$$(2) \quad k = (\delta m - \delta)/(m - \delta),$$

and rearranging the optimality condition $G'(x) = 0$, one obtains the second order differential equation

$$(3) \quad F_{mn}B'' + f_{mn}B' + (kf_{mn}/(1-u))B = (1-k)f_{mn}V$$

in $u \in \mathbf{I}$. One has from (1) the condition

$$(4) \quad B(1) = 0.$$

Note that the parameter k increases in δ and takes the unit interval onto itself (for every $m \geq 2$). I will refer to the differential equation (3) as $\mathbf{B}(k)$.

To recapitulate, if b is a bid function which is increasing on $\mathbf{I}$ and which constitutes a symmetric equilibrium for the auction $\Gamma(\delta, M)$, then B defined by (1) is a solution of $\mathbf{B}(k)$ that fulfills condition (4).

Proof of Theorem 1 In view of the preceding paragraph, proof of Theorem 1 follows from Lemmas 1 and 2 stated below.

Lemma 1 There exists a unique solution B of the differential equation $\mathbf{B}(k)$ which is defined on $\mathbf{I}$ and which satisfies (4).

We shall make use of expression

(5) $f_{mn}(u) = (n!/(n-m)!(m-1)!)\, u^{n-m}(1-u)^{m-1}$

and the identity

(6) $(n+1)\,F_{mn} = f_{1(n+1)} + \cdots + f_{m(n+1)}.$

Applying (5) and (6) to (3) and cancelling $(n!/(n-m)!(m-1)!)u^{n-m}$ on both sides, one gets that $\mathbf{B}(k)$ is equivalent to

(7) $u\Phi_{mn}B'' + (1-u)^{m-1}B' + k(1-u)^{m-2}B = (1-k)(1-u)^{m-1}V,$

where

$$\begin{aligned}\Phi_{mn} &= F_{mn}/\left(n!/(n-m)!\,(m-1)!u^{n-m+1}\right)\\ &= (n-m)!\,(m-1)!\,\textstyle\sum_0^{m-1} 1/((n-j)!j!)\,u^{m-1-j}\,(1-u)^j\end{aligned}$$

is positive for all $u \in \mathbf{I}$. It is readily checked that $\mathbf{B}(k)$ has a singularity at $u=0$ and is regular elsewhere on $\mathbf{I}$.

Proof of Lemma 1 (Sketch; see for instance Coddington and Levinson (1995)) The general solution B of $\mathbf{B}(k)$ is of the form

$$B = c_1B_1 + c_2B_2 + B^*$$

where B^* is any particular solution, c_1, c_2 are arbitrary constants, and B_1, B_2 are two independent solutions of the homogeneous equation of $\mathbf{E}(k)$. It is straightforward to compute that 0 and $-(n-m)$ are the two solutions of the indicial equation of $\mathbf{B}(k)$. Hence B_1, B_2 have the form

$$B_1(u) = \textstyle\sum_0^\infty a_j\, u^j,$$

$B_2(u) = d\ B_1(u) \ln u + u^{-(n-m)} \left(\sum_0^\infty d_j u^j\right).$

B defined at $U = 0$ implies $c_2 = 0$. Condition (4) determines c_1 uniquely as asserted.

Lemma 2 If B is a solution of $\mathbf{B}(k)$ as in Lemma 1, then B' is increasing on $\mathbf{I}$.

Proof Let B be solution of $\mathbf{B}(k)$ as in Lemma 1. Then

$$(8)\ u\Phi_{mn}B''' + (\Phi_{mn} + u\Phi'_{mn} + (1-u)^{m-1})B'' - \\ (m-1-k)(1-u)^{m-2}B' \ -k(m-2)(1-u)^{m-3}B \\ = (1-k)(1-u)^{m-2}((1-u)V' - (m-1)V)$$

which one obtains upon differentiating (7). The proof is in two steps.

Step 1 B' is increasing on an interval $[0, u]$ for some $u \in (0, 1)$.

Suppose the contrary, i.e., B' is nonincreasing on $[0, u]$ for some $u \in (0, 1)$, and let u^* be the maximum of all such u if a maximum exists. Then

$$(9)\quad B''(0) \le 0.$$

$$(10)\quad B''(u^*) = 0,\ B'''(u^*) \ge 0.$$

It follows from (8), (9) and (10) that

$$(11)\ (m-1-k)(B'(0) - B'(u^*)) \le (m-2)k(B(u^*)/(1-u^*) - B(0)) \\ +(1-k)((1-u^*)V'(u^*) - (m-1)V(u^*) - (V'(0) - (m-1)V(0)).$$

From (7), (9), and (10), on the other hand,

$$k(B(u^*)/(1-u^*) - B(0)) \le B'(0) - B'(u^*) + (1-k)(V(u^*) - V(0)),$$

using which in (11) gives

$$(B'(0) - B'(u^*)) \le (1-u^*)V'(u^*) - V(u^*) - (V'(0) - V(0)).$$

By assumption, the right hand side above is negative, which says $B'(0) < B'(u^*)$. Upon this contradiction, therefore, it can only be that u^* does not exist, i.e., that B' in nonincreasing throughout $\mathbf{I} = [0, 1]$. Now evaluating (8) at $u = 0$,

$(m-1-k)B'(0) = (\Phi_{mn}(0)+1)B''(0) - k(m-2)B(0) -$
$(1-k)(V'(0)-(m-1)V(0))$.

From (9) and the facts that $V(0)=0$, $V'(0)>0$, then

$$B'(0) < -k(m-2)/(m-1-k)B(0) \leq -B(0) = \int_0^1 B'(w)\,dw \leq B'(0),$$

where the last inequality follows from B' being nonincreasing on $\mathbf{I}$. Step 1 follows from this contradiction.

Step 2 B' is increasing for all $u \in I$.

Suppose the contrary. Then, in view of Step 1 and the fact that $B''(1)=0$ (from (7)), there exist $u, u \in (0,1]$ such that $u < u'$, $B'(u) \geq B'(u')$, and

(12) $B''(u) = 0,\ B'''(u) \leq 0,$

(13) $B''(u') = 0,\ B'''(u') \geq 0.$

From (7), (12) and (13),

(14) $k(B(u)/(1-u) - B(u')/(1-u'))$
$= B'(u') - B'(u) + (1-k)(V(u') - V(u)).$

Using (8), (12), (13) and (14), one gets as in Step 1 that

$$B'(u) - B'(u') \leq (1-u')V'(u') - V(u') - ((1-u)(V'(u) - V(u)) < 0$$

and reaches the contradiction $B'(u') > B'(u)$.

3.2 Independence of expected revenue

I now turn to the second main result of the paper, stated in the form of Theorem 2 below, which says that the expected auction price is the same for all degrees $\delta \in \Delta$ and under either the multiprice or the uniprice rule.

Let us consider first the multiprice auction, let B be the unique solution of $\mathbf{B}(k)$ asserted in Theorem 1, and so $b = B'$ be the unique symmetric equilibrium bid function for $\Gamma(\delta, M)$. When an *ordered population* $y_1, \dots, y_{n+1}$ (i.e., $n+1$ independent draws of buyers listed such that $y_1 \geq \dots \geq y_{n+1}$) plays $\Gamma(\delta, M)$ as predicted by the symmetric equilibrium, therefore, the auction price $p(\delta, M) = (b(y_1) + \dots + b(y_m))/m$. Taking expectation over all possible draws of populations, the *expected* auction price of $\Gamma(\delta, M)$ is

(15) $$Ep(\delta, M) = 1/m \int_0^1 (f_{1(n+1)}(w) + \dots + f_{m(n+1)}(w))\ B'(w)\ dw.$$

For the uniprice auction, on the other hand, it follows from Proposition 1 that $p(\delta, U) = y_{m+1}$, hence

$$(16) \quad Ep(\delta, U) = \int_0^1 (f_{(m+1)(n+1)}(w)V(w)\, dw,$$

a constant independent of δ, to which I will below refer as p^*.

Theorem 2 The expected auction price is p^* for all $\delta \in \Delta$ under both the multiprice and the uniprice rule.

Proof (observed by Bernard de Meyer) Multiply $B(k)$ on both sides by $(n+1)(1-u)/(m(1-k))$ to get

$$(17) \quad (n+1)/(m(1-k))(1-u)(F_{mn}B')' + ((n+1)k/m(1-k))f_{mn}B \\ = ((n+1)/m(1-u))f_{mn}V = f_{(m+1)(n+1)}V.$$

Upon integrating (17), the two terms on the left side by parts, and using $F_{mn}(0) = B(1) = 0$, one gets

$$(n+1)/(m(1-k)) \int_0^1 F_{mn}(w)B'(w)dw- \\ (n+1)k/m(1-k) \int_0^1 F_{mn}(w)B'(w)\, dw = \int_0^1 f_{(m+1)(n+1)}(w)V(w)\, dw,$$

and so

$$(18) \quad (n+1)/m \int_0^1 F_{mn}(w)B'(w)\, dw = \int_0^1 f_{(m+1)(n+1)}(w)V(w)\, dw.$$

The theorem now follows from (6), (15), and (16).

3.3 Shrinking range of bids

Having shown above that the auction price is the same in expectation for all $\delta \in \Delta = [0, 1)$ whether the auction is multiprice or uniprice, I next show in Theorem 3 below that the *range* of bids monotonically shrinks to 0, as the resale parameter δ increases from 0 to 1, in the multiprice auction. Recall from Proposition 1 that bidding is unaffected by δ in the uniprice auction.

Formally, consider the family Γ of auctions $\Gamma(\delta, R), \delta \in \Delta$. Let b_δ be the unique increasing symmetric equilibrium bid function for $\Gamma(\delta, M)$ and define $L(\delta, M) = b_\delta(0)$, $H(\delta, M) = b_\delta(1)$. Thus $L(\delta, M), H(\delta, M)$ are respectively the lowest-value bid and the highest-value bid, and naturally all bids fall in the range $[L(\delta, M), H(\delta, M)]$, for $\Gamma(\delta, M)$. Define $L(\delta, U), H(\delta, U)$ analogously.

Theorem 3 (i) $L(\delta, M)$ increases and $H(\delta, M)$ decreases as $\delta \in \Delta = [0,1)$ increases.

(ii) $L(0, M) = 0$, $H(0, M) < 1$ and $\lim_{\delta \to 1} L(\delta, M) = \lim_{\delta \to 1} H(\delta, M) = p^*$.

(iii) $L(\delta, U) = 0$, $H(\delta, U) = 1$ for all $\delta \in \Delta$.

Proof As mentioned, (iii) follows from Proposition 1. Let B_k be the unique solution of the differential equation $\mathbf{B}(k)$ as asserted in Lemma 1 and recall that $b_\delta = B_k'$ where $k = (\delta m - \delta) / (m - \delta) \in \Delta$.

The assertion in (ii) for the classical multiprice auction $\Gamma(0, M)$ that says $L(0, M) = 0, H(0, M) < 1$ is well known and easily checked. For the remaining assertion in (ii), check that $B_1(u) = c + du$ is the general solution of $B(1)$ for arbitrary constants c, d. By continuity, $B_k(u)$ approaches $B_1^*(u) = c^* + d^* u$ for some fixed constants c^*, d^* as $k \to 1$ and for all $u \in I$. Hence, $B_k'(u)$ approaches $B_1^{*\prime}(u) = d^*$ as $k \to 1$ for all $u \in I$. It follows from Theorem 2 that $d^* = p^*$. This concludes the proof of (ii) since $k \to 1$ as $\delta \to 1$.

To prove (i), we apply the perturbation method: Take any $k \in \Delta$ and $h = k + \varepsilon \in \Delta$ for ε sufficiently small. Then $B_h(u)$ is given by the perturbed solution $B_k'(u) + \varepsilon P(u) + O(\varepsilon^2)$, where $P(u)$ is a solution on I of the following differential equation (obtained by substituting for $B_h(u)$ in the perturbed equation (7) and equating the terms with coefficient ε),

(19) $$u\Phi_{mn} P'' + (1-u)^{m-1} P' + k(1-u)^{m-2} P = -(1-u)^{m-2}((1-u)V(u) + B_k(u)),$$

which additionally satisfies the condition

(20) $P(1) = 0$.

Proof of (i), hence the theorem, now follows from application of Lemma 3 stated below.

Lemma 3

(i) There exists a unique solution P of the differential equation (19) which is defined on I and which satisfies (20). Furthermore,

(ii) P' is decreasing on **I.**

(iii) $P'(0) > 0$, $P'(1) < 0$.

Proof The proof of (i) is identical to the proof of Lemma 1, and (ii) is proved in similar manner to Lemma 2. To prove (iii), simply observe that if $P(0) \leq 0$ then, in view of (ii), the expected auction price $Ep(h, M)$ would be less than p^*, contradicting Theorem 2. Similarly, if $P(1) \geq 0$ then, in view of (ii), $Ep(h, M)$ would be greater than p^*.

3.4 Illustration: Explicit solutions for two-unit auctions

This section is restricted to two-unit auctions $\Gamma(2, n, V, \delta)$ in which case the equilibrium equation $B(2, n, V, \delta)$ turns out to be the familiar Gauss *hypergeometric* equation. Taking V, further, to be the identity function $I(u) = u$, i.e., personal values to be uniformly distributed, the associated bid functions turn out to be expressible via the hypergeometric function, as I spell out below.

Let $b_{n\delta}$ be the symmetric equilibrium bid function for $\Gamma(2, n, I, \delta)$. Recalling the proof of Theorem 1 and differentiating (1), $b_{n\delta} = B'$ where B is the unique solution of the equation $B(2, n, I, k)$

$$(21)\quad (n - (n-1)u)uB'' + n(n-1)(1-u)B' + n(n-1)kB$$
$$= n(n-1)(1-k)u(1-u).$$

that satisfies the associated boundary conditions.

The general solution to (21) is the sum of two solutions

$B(u) = B^*(u) + A(u)$

where $B^*(u)$ is a particular solution to (21) and $A(u)$ is the general solution to the homogeneous equation

$$(22)\quad (n - (n-1)u)uB'' + n(n-1)(1-u)B' + n(n-1)kB = 0.$$

It is straightforward to find

$B^*(u) = (1/(n-1)(2n+2-nk)k)(-n(n-1)k - 2+$
$(n(n-1)k+2)ku + n(n-1)k(1-k)u^2)$.

To get the general solution to (22), let $z = ((n-1)/n)u$ and define $Q(z) = B(u)$. Upon this change of variable, (22) becomes

(23) $(1-z)zQ'' + (n-1-nz)Q' + nkQ = 0$,

which is the hypergeometric equation

$$(1-z)zQ'' + (n-1-(\alpha+\beta+1)z)Q' - \alpha\beta Q = 0$$

where

$\alpha = [(n-1) + ((n-1)^2 + 4nk)^{1/2}]/2,$

$\beta = [(n-1) - ((n-1)^2 + 4nk)^{1/2}]/2.$

Two linearly independent solutions of (23) are the hypergeometric series $F(\alpha, \beta, n-1, z)$ and $F(\alpha, \beta, n-1, z)\ln z$. The general solution of (22) is thus

$$B(u) = B^*(u) + (c + d\ln(((n-1)/n)u))F(\alpha, \beta, n-1, ((n-1)/n)u)$$

for arbitrary constants c, d. From the condition that our solution be defined at $u = 0$, it follows that $d = 0$, while from $B(1) = 0$, one has $c = -B^*(1)/F(\alpha, \beta, n-1, (n-1)/n)$. Thus

$$(24)\quad B^*(u) - B^*(1)F(\alpha, \beta, n-1, ((n-1)/n)u) / F(\alpha, \beta, n-1, (n-1)/n)$$

is the unique solution of $B(2, n, I, k)$ on the closed unit interval that satisfies $B(1) = 0$. Upon differentiating (24), one gets the bid function $b_{n\delta}$ in the explicit form

$$b_{n\delta}(u) = (1 + kn(n-1)/2 + (1-k)(n-1)u + (1-k)F(1+\alpha, 1+\beta, n, ((n-1)/n)u) / F(\alpha, \beta, n-1, (n-1)/n)) / ((n-1)(n+1-kn/2))$$

where $k = \delta/(2-\delta)$.

The graphs below of the equilibrium bid function $b_{n\delta}$ for several selected values of n, δ illustrate our results. (Legend: In Figures 1-3 $b_{n\delta}(0) \geq b_{n\delta'}(0)$ for $\delta \geq \delta'$ while in Figure 4 $b_{n\delta}(0) \geq b_{n'\delta}(0)$ for $n \geq n'$.)

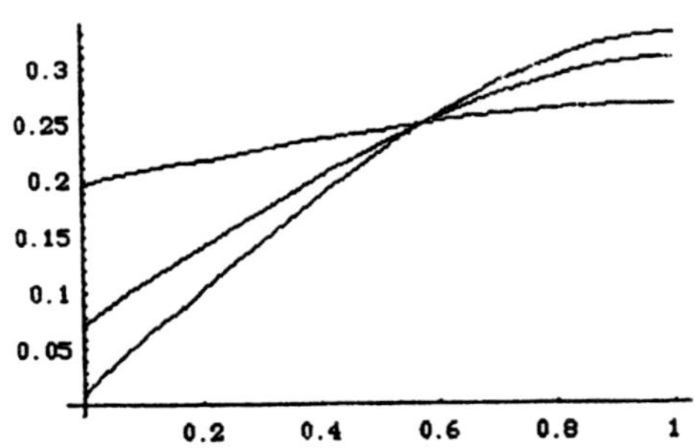

Figure 1: Bid functions when 3 buyers bid against 2 units with $\delta \in \{.1, .5, .9\}$

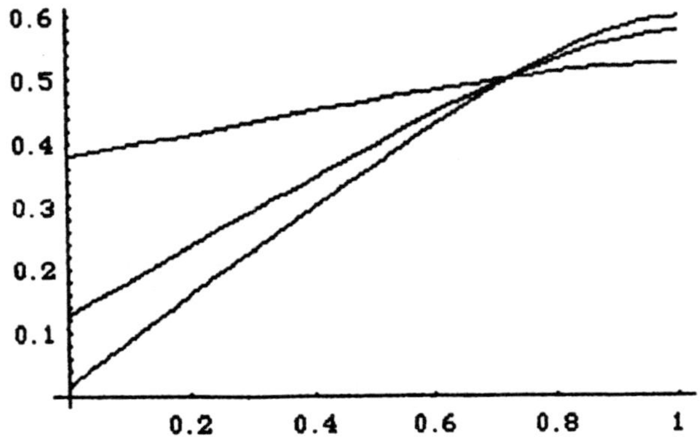

Figure 2: Bid functions when 4 buyers bid against 2 units with $\delta \in \{.1,.5,.9\}$

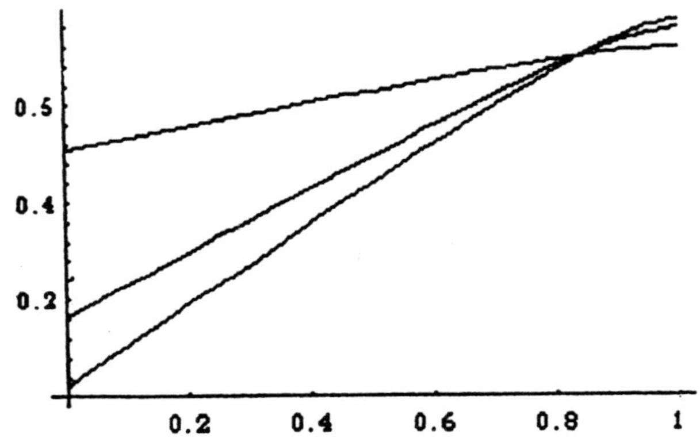

Figure 3: Bid functions when 8 buyers bid against 2 units with $\delta \in \{.1,.5,.9\}$

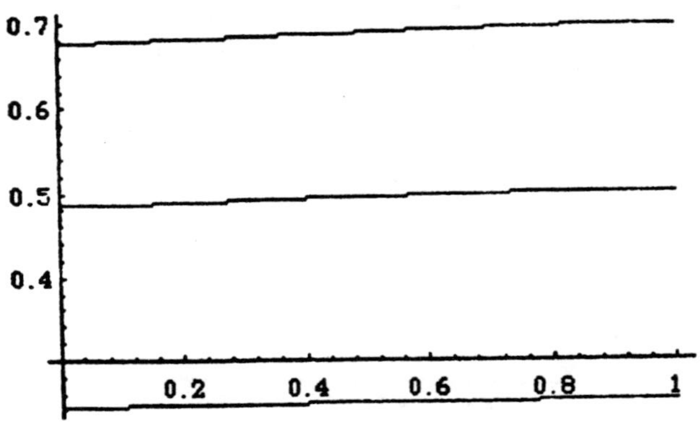

Figure 4: Bid functions when $n \in \{3,4,8\}$ buyers bid against 2 units with $\delta = .99$

4 Concluding Remarks

I have in this paper modelled an auction with resale as a one-period game that features an endogenous component in buyers' valuations as proxy for resale revenue and a parameter as measure of resale orientedness. I have shown that while there is a continuum of equilibria in the (purely speculative) game with *full* resale, a unique symmetric equilibrium exists for all other degrees of resale, under the multiprice rule. The findings then say that (i) the expected auction price is invariant for all degrees of resale under either payment rule and that (ii) the range of bids is narrower under the multiprice rule than under the uniprice rule, the more so the higher is the degree of resale. As policy implication for the seller, these results thus give support for the multiprice rule on grounds of price stability next to revenue equivalence.

To be sure, the two features in the model mentioned above are of an ad hoc nature. Regarding the first of these, let me conjecture that the model would bear nearly the same results if one were to postulate that resale occurs at the auction price plus a margin near zero. In the case of treasury auctions, Cammack (1991) has measured, for the period 1973 - 84, that the margin between the immediate post-auction market price and the auction price of the U.S. Treasury bill is 4 to 7 basis points (one basis point being one-hundredth of 1 percent.) Conjectured robustness together with empirical observation thus provides one level of the justification. One may, further, regard and tolerate the assumption that resale occurs at the auction price (plus a margin) as an equilibrium no-profit condition on participant profits.

One should also point out that expected revenue equivalence (Theorem 2) is likely to not hold if buyers' personal valuations were correlated to any degree and that expected revenue then is likely to emerge a degree higher under the uniprice rule. I conclude with a final conjecture that the shrinkage in the range of bids (Theorem 3) will then still continue to hold and offer support for the multiprice rule.

* This research was preceded and has been inspired by an empirical study of the Turkish Treasury Bill auctions (Alkan, (1989)). Previous versions of this paper have been presented in a seminar at Universite Libre de Brussels in February 1992, the Mathematical Economics Workshop at the Institute for Pure and Applied Mathematics, Rio de Janerio, in August 1993, the Bosphorous Economic Theory Workshop in September 1993, and the Economic Research Forum Financial Markets Conference, Beirut, in July 1994. I thank the participants, in particular Patrick Bolton, Hasan Ersel, Faruk Gül, and Matthew Jackson for discussions on the modelling aspect. My special thanks for his insightful efforts to Bernard de Meyer with whom I started the analysis of the model. Partial support by the Bogaziçi University Research Fund is gratefully acknowledged.

REFERENCES

Alkan, A., "Treasury Debt Auctions: An empirical study of the Turkish Case"(in Turkish), Bogazici University Research Paper, 1989

Bikhchandani, S. and C. Huang, "Auctions with resale markets: A model of treasury bill markets, *The Review of Financial Studies,* 1989, 2:3, 311-39

Bikhchandani, S. and C. Huang, "The economics of treasury bill markets, *The Journal of Economic Perspectives,* 1993, 7:3, 117-34

Chari, V. V. and R.J. Weber, "How the U.S. Treasury should action its debt", *Federal Reserve Bank of Minneapolis Quarterly Review,* 1992, 16:4, 1-12

Cammack, E., "Evidence on bidding strategies and the information in treasury bill auctions", *Journal of Political Economy,* 1991, 99:1, 100-30

Coddington E.A. and N.Levinson, *Theory of ordinary differential equations,* McGraw Hill, 1955

Friedman, M., *A program for monetary stability,* N.Y., Fordham University Press, 1960, 64-65

Milgrom, P and R. J. Weber, "A theory of auctions and competitive bidding," *Econometrica,* 1982, 50:5, 1089-122

U.S.Department of Treasury, Securities and Exchange Commision, and Board of Governors of the Federal System, Joint report on the government securities market, Washington D.C., 1992

Vickrey, W, "Counterspeculation, auctions, and competitive sealed tenders", *Journal of Finance,* 1961, 16, 8-37

Ahmet Alkan
Turkish Academy of Sciences
and
Bagazici University, Bebek 80815
Istanbul, Turkey

LARGE MONETARY TRADE, MARKET SPECIALIZATION AND STRATEGIC BEHAVIOUR

Meenakshi Rajeev[1]

Abstract: *This paper looks at the role of money as a medium of exchange in a competitive set-up. Together with this we have explored why, historically speaking, monetary trade and market specialization always go hand in hand. The set-up taken up for the purpose is derived from the well-known frame-work of Kiyotaki and Wright (1989). Our frame-work extends the above set-up to incorporate exchanges through trading posts for different pairs of goods. Here each agent is trying to choose his optimal strategy for trade given the best strategies of the others. The exercise reveals how a monetized trading post set-up can manifest itself through the agent's optimizing behaviour.*

1. Introduction

The theory of Walrasian equilibrium yields a set of prices at which the aggregate competitive demand for each commodity equals it aggregate competitive supply. Two important issues arise in this context. The first

is concerned with discovering the laws which guide the behaviour of the many economic variables, but especially prices, when the system is out of equilibrium. Walras (1890) tackled this problem by providing an algorithm for price adjustment which is well-known as the tatonnement scheme.

The other issue revolves around the function of an auctioneer as a clearing house for commodities. All agents are assumed to deposit their initial endowments with this auctioneer, who in turn reallocates them according to the pattern of excess demands. Thus, in the words of Starr (1972), "In a Walrasian pure exchange general equilibrium model, trade takes place between individual households and the *market*. Households do not trade directly with each other." Such an abstraction suppresses several important issues, in particular the problems of direct exchange between households due to a lack of mutual coincidence of wants even at market clearing prices. Transaction costs as well as a medium of exchange can become crucial in such cases. This paper is devoted to these issues.

When trade takes place between households in a decentralized fashion, it is likely that they would be restricted to those between pairs of agents. More importantly such pairwise meetings of a particular trader with different traders need to be separated in time. In the absence of a centralized agency, each agent going through such *sequential bilateral* trade will naturally insist on the value of his incomings to be at least as large as the value of his outgoings. In other words, trades should be bilaterally balanced in value terms after each meeting, or, equivalently, maintain a quid-pro-quo condition. However, in the absence of a perfect mutual coincidence of wants between the agents, this quid-pro-quo may have to be maintained by transferring a good to the creditor for which he has no Walrasian excess demand. The need for a medium of exchange in a competitive set-up can be best appreciated

[1] The author wishes to acknowledge many useful discussions with Dipankar Dasgupta.

against this back ground, for as soon as an agent accepts a good for which he does not have excess demand, it takes the form of a medium of exchange.

The earliest recognition of this problem came in Menger (1892). There have been many subsequent attempts to look into the role of money as a medium of exchange, but it was Hicks (1969) who posed it in the context of a Walrasian equilibrium. The last couple of decades saw further progress in this branch of the literature. In fact it has been a shared view (Starr (1972) Ostroy & Starr (1979), Kiyotaki and Wright (1989)), that money as a medium of exchange is an indispensable tool for attaining ones desired allocation. One then naturally asks which particular good can emerge as a medium of exchange through the agent's own optimizing behaviour aimed at transactions cost reduction. The *time* needed to attain ones desired allocation starting from the initial endowments is universally considered as a transaction cost which every agent wishes to minimize.

However, in order to deal with the problem, these authors have considered an institutionally vacuous economy insofar as trades are assumed to take place in the absence of specialized markets. All agents are assumed to gather in one place and meet in pairs to explore trading possibilities. In this context, together with developing the role of money as a medium of exchange, we would explore why, historically speaking, monetary trade and market specialization always go hand in hand. How far is it advantageous from the point of view of transactions cost to have monetary trade coupled with the social institution of markets ?

The set-up taken up for the purpose is derived from the well-known framework of Kiyotaki and Wright (1989) where each agent is trying to choose his optimal strategy for trade given the best strategies of the others. Here, the optimization is done on an agent's life time utility, with respect to transactions cost. Several Nash equilibria can be derived, revealing in the process the different goods which may play the role of a medium of exchange.

Our framework however extends the above set-up to incorporate exchanges through trading posts for the different pairs of goods. The exercise reveals how a monetized trading post set-up can manifest itself through the agents' optimizing behaviour.

More precisely, when the cost of establishing and running a market is not very high as against a marketless trading arrangement, the social institution of markets can reduce the transaction costs (through a reduction in time cost) to the extent of dominating all possible equilibria.

The next section reviews some of the earlier work in the literature. Section 3 describes the basic frame-work under consideration. The penultimate section looks into the equilibrium strategies and makes a comparison between a trading post and a market less set-up. The concluding section sums up the findings.

2. A brief review of the literature

Once the role of money as a medium of exchange had been emphasized, the literature began to ask which commodity could be a "good" choice as a medium of exchange, i.e., which commodity, if used as a medium of exchange, could lead the agents to their desired allocation within a reasonable *time span*. An attempt to characterize the commodity led Ostray and Starr to impose the following condition on the money good.

$$\sum_{c \neq M} p_c [z_{kc}]^+ \leq p_M w_{kM}$$

where c is the index for commodities, M the index for the money good, z_{kc} the excess demand of agent k for good c with $z_{kc} > 0 (< 0)$ if k is an excess demander of c (if k is an excess supplier of good c),

$$[z_{kc}]^+ = \max[0, z_{kc}] \text{ and}$$

w_{kM} the initial endowment of good M for agent k.

Thus, this condition implies that if there exists a commodity such that the initial endowment of it is large enough for each agent of the economy for them to back up all their desired purchases with this good, then its use as a medium of exchange allows agents to attain the equilibrium allocation for **any** competitive economy in at most *one round*; that is, after each agent visits each other agent once and only once.

Clearly, the above condition imposes a strong restriction on the initial endowment of the money good. However, it has been shown later by Starr (1976) that if one removes all the conditions on the money good and allows every good to be a means of payment, the time requirement can become unboundedly high.

In order to keep a balance between the two extremes of Ostroy and Starr (1974), Starr (1976), a need to look for a more plausible condition has been felt. In this respect we have shown that if a good is (finally) demanded in positive quantities (however small) by all the agents of the economy then using that good as a medium of exchange, equilibrium can be attained in *finite* time (see Dasgupta, D. and M. Rajeev (forthcoming)).

To emphasize the role of the social institution of markets in monetary trade, it has been further shown in the context of Ostroy-Starr (1974) framework (see M. Rajeev (1996)) that the time requirement to attain equilibrium allocation in any competitive economy is bounded above irrespective of the size (number of agents) of the economy, when trades take place through trading posts. Conversely, this time cost, in the absence of markets can become indefinitely high as population grows. This reveals the advantages of market specialization for large economics. The following discussion is going to address the same issues in a game theoretic framework.

3. The frame-work

As with Kiyotaki and Wright (1989) we consider a competitive economy in a state of equilibrium, consisting of three types of infinitely lived agents (type 1, 2 and 3) each specialized in consumption and production. Every type consists of an equal number of agents producing one unit of a specific good. There are three indivisible commodities, viz., goods 1, 2 and 3. Type k agents derive utility from the consumption of good k only and are able to produce k^* ($\neq k$). In our model we assume $1^* = 2$, $2^* = 3$, $3^* = 1$. As soon as a type k agent acquires good k he consumes it and produces one unit of k^*. Each good can be stored at a cost, but an agents capacity to store is restricted to one unit only. Let b_{kc} denote the cost (in terms of instataneous disutility) to the type k agents of storing good c. It is assumed that $0 < b_{k1} < b_{k2} < b_{k3}$. For a type k agent, let u_k denote the instantaneous utility from consumption of good k net of disutility of producing k^* and $\beta \in (0,1)$ the common discount factor. An economy with these features is denoted by E.

For the economy E we consider two types of trading arrangements viz., the marketless arrangement and the trading post set-up. In a marketless arrangement (Kiyotaki and Wright (1989), Aiyagari and Wallace(1991)), the agents meet each other randomly in pairs (irrespective of the goods they want to trade) and exchange of goods takes place when it is mutually agreeable.

On the other hand in a trading post set-up there exists three different markets to deal with good 1 against 2, good 1 against good 3 and good 2 against good 3. By the (c, c') trading post we refer to the market where good c is exchanged against good c'. Agents wishing to trade good c against good c' visit the (c, c') trading post where buyers and sellers (of c against c' can identify each other and meet and trade). It appears therefore, that trading post set up would be able to avoid meetings between agents who are unlikely to benefit from trade. However, though there is a saving of time cost in the economy, one needs to incur additional costs (above the storage costs) for the setting up and maintanance of a market system. More precisely, let $\gamma_{cc'}$ be the per period cost to be incurred by an agent trading in the (c, c') trading post to run the market (it includes eg. tax payable, electricity charges etc.). [2]

We would consider two types of alternative relations amongst the costs to be incurred in a trading post set-up.

$$CASEI \qquad b_{k2} + \gamma_{12} < b_{k1} + \gamma_{31} < b_{k3} + \gamma_{23}, \gamma_{12} \leq \gamma_{31} \leq \gamma_{23}.$$

$$CASEII \qquad b_{k2} + \gamma_{12} > b_{k1} + \gamma_{31} > b_{k3} + \gamma_{23}, \gamma_{12} \geq \gamma_{31} \geq \gamma_{23}.$$

All other possible relationships can be considered and delt with similarly. It is assumed that the net utility u_k is large enough compared to the costs (measured in terms of instantaneous disutility) so as not to induce any agent to drop out of the market economy. This may be ensured through the following sufficient condition.

$$u_k - \frac{b_{kk^*} + \gamma_{ck^*}}{1-\beta} \geq -\frac{b_{kc} + \gamma_{cc'}}{1-\beta}, \forall c \text{ and } c'.$$

In a set-up with trading posts a type k agent has two pure stategies: either to go for direct barter i.e., to exchange k against k^* directly or to go for indirect trade by exchanging k^* against some good c and then c against k. In the next section we would examine the possible equilibrium strategies for such a scenario.

4. Equilibrium Strategies:

4.1 Set-up of complete marketisation

Here we look for the steady state Nash Equilibrium strategies [3] for Cases I and II separately under the assumption that trades can be carried out *only* through the trading posts, i.e., marketless trading is not permitted. In both these situations fundamental strategies (see Kiyotaki and Wright (1989)) can be shown to be the equilibrium strategies. More precisely, it means that the type of agents for whom the storage cost of the goods they produce plus the running cost of the markets relevant for their direct barter trade is the highest, would opt for indirect trade by using a good with lesser transactions cost (a good which they neither produce nor use for final consumption) as the medium of exchange. Naturally, the trading post that would have been relevant for their direct barter will not function. Thus, we have:

Proposition 1: Under CaseI, the fundamental strategies in a trading post set-up forms a set of equilibrium strategies under the following sufficient condition:

$$\gamma_{13} - \gamma_{12} < \frac{\beta}{2} u_3$$

[2] Here we have made these costs market specific. Similar exarcise can be carried out if one makes these costs agent specific or alternatively dependent on the good one wants to *sell* in that market.

[3] A steady state Nash equilibrium is a set of trading strategies S_k one for each type k, together with a steady state distribution $\bar{p}$ which gives the proportion of type k agents with good c, that satisfies
(i) each individual k chooses S_k to maximize his expected utility given the best strategies of others and the distribution $\bar{p}$;
(ii) given S_k, $\bar{p}$ is the resulting steady state distribution.
The exercise holds even when we introduce an additional it once for all fixed establishment cost which satisfies the condition that an agent's expected life time utility covers the cost.

Proof : See Appendix.

The above condition ensures that the discounted net utility gain from direct barter trade for a type 3 agent exceeds the additional cost he has to incur to run the (1, 3) market.

In an exactly similar manner one can establish,

Proposition 2 : Under Case II, the fundamental strategies in a trading post set-up form a set of equilibrium strategies for all parameter values.

In the equilibrium under Case I of Proposition 1, the type 2 agents would go for indirect trade by using good 1 as a medium of exchange whereas in the equilibrium of Proposition 2, the type 1 agents would act as the intermediaries.

Let us now define the welfare derived by a type k agent as :

$$WF_k = (1-\beta)\sum_c p_{kc}V_{kc}$$

where, p_{kc} is the proportion of type k agents with good c in the steady state and V_{kc} is the utility derived by a type k agent by acquiring good c.
If we were to compare the steady state welfare levels (see Kiyotaki and Wright (1989)) of the equilibrium of Proposition 1 with that of corresponding fundamental equilibrium in a marketless economy we arrive at the following result :

Proposition 3 : For the economy E defined above the welfare of every agent is higher under the fundamental strategies in a trading post set-up as compared to that of the marketless trading arrangement, if the following conditions hold :

$$\frac{\beta u_1}{3} > \gamma_{12}, \frac{\beta u_3}{3} > \gamma_{13}, \frac{\beta u_2}{3} > \frac{1}{2}(\gamma_{12}+\gamma_{13})$$

Proof : See Appendix.

Thus, if u_k's are sufficiently large (and the discount rate is not very small) as compared to the cost of running a market, a trading post set-up would always dominate a marketless arrangement.

Kiyotaki and Wright have also shown that none of the equilibrium strategies in a marketless arrangement are Pareto-optimal. For, a nonimplementable strategy (to be called S) which directs every pair of agents to exchange the respective goods they possess (whenever they meet) can be (welfarewise) Pareto superior if u_k's are sufficiently high. But left to themselves, the traders would never opt for this strategy and hence it would not constitute an equilibrium. However, if $\gamma_{cc'}$'s are not very high as compared to the u_k's, fundamental strategies in a trading post set-up can even dominate S with respect to welfare.

4.2 Mix of trading post set-up and marketless arrangement

The above results (in particular, Propositions 1 and 2) are derived under the assumption that trades are to be carried out necessarily in the respective trading posts. This need not always be desirable especially if the cost of establishing and running a market is prohibitively high. Therefore, a natural question arises : If the option of trading in a marketless set-up is available together with a trading post set-up, will trading without some markets be preferred to trading through them, resulting thereby in the coexistence of marketless trading and exchange through a network of trading posts ? Under some restrictions on the parameter values, the answer to this question is in the affirmative.

Thus, consider the following set of strategies :

S_{k1} : type k agents go for direct barter through a trading post.
S_{k2} : type k agents go for indirect trade through a trading post.
S_{k3} : type k agents go for direct barter through marketless trade.
S_{k4} : type k agents go for indirect trade through marketless set-up.
S_{k5} : type k agents go for indirect trade first in a market and then in a marketless set-up.
S_{k6} : type k agents go for indirect trade by trading first in a marketless set-up and then through a trading post.

Thus we have the following result :

Proposition 4 : Under Case I, the strategy profile (S_{11}, S_{26}, S_{33}) constitutes a set of steady state Nash equilibrium strategies if the following conditions hold

$$p_1^{12}u_1 - \gamma_{12} > 0 \quad \text{and} \quad p_{31}\{\beta u_2 - (b_{21} + \gamma_{12})\} \geq b_{23} - b_{21}$$

where, p_1^{12} is the steady state probability of meeting an (type 2) agent with good 1 in the (1, 2) market by a type 1 agent and p_{31} is the probability of meeting a type 3 agent with good 1 in the marketless set-up.

Proof : See Appendix.

Under the strategy profile (S_{11}, S_{26}, S_{33}) the type 1 agents would go for direct barter in the (1,2) trading post and the type 3 agents would opt for direct trade in a marketless set-up. It is the type 2 agents who would act as the intermediaries by exchanging good 3 against good 1 in a marketless set-up and then buy good 2 for good 1 in the (1,2) market.

This equilibrium, however, will be Pareto non-comparable with the one of Proposition 1. This is because type 1 agents are going to be worse off in this new equilibrium as their complementary trading partners (i.e., the type 2 agents) are now going through a more time consuming trading process, whereas the type 3 agents would be better off if the running cost of the market relevant for them, i.e, γ_{31}, is sufficiently high. Thus we have

Proposition 5 : The equilibrium derived in Proposition 4 is Pareto non-comparable with that of Proposition 1 if the running costs of some markets (viz., (1,3) and (2,3)) are sufficiently high. However, if $\gamma_{cc'}$'s are sufficiently small, in particular $\gamma_{cc'}$ goes to 0 for all (c, c') and the welfare levels are positive, the equilibrium under complete marketization (of Proposition 1) is welfarewise Pareto superior to the equilibrium derived in Proposition 4.

Proposition 4 establishes our intution that the utility of trades through monetized markets cannot be dominated by monetized trade in the absence of markets. However, under reasonable assumptions, the former would in fact be superior.

5. Conclusion

This paper looks into the possibility of trade through a trading post set-up vis-a-vis a marketless trading arrangement. In this context, several interesting steady state Nash equilibria are derived and the steady state utility levels are compared. However, we are concerned here only with commodity money. The use of fiat money in the process of exchange is an important issue which needs detailed study too. The role of fiat currency in a search theoretic marketless framework has been discussed in Kiyotaki and Wright (1993, 1991, 1989).

In Kiyotaki and Wright (1989), *fiat money* is introduced as a commodity with least (in particular 0) storage cost. Due to the indivisibility of the real commodities and one unit storage space availability, it is not possible to hold fiat currency and a real commodity simultaneously. It is then shown how the lack of faith in fiat money makes it an unusable medium of exchange in Nash's sense. However, if on the contrary every one believes that others will accept fiat money then fundamentals (storage cost) and marketability both acting favourably together makes it in equilibrium a medium of exchange.

Let P units of fiat money be required to buy one unit of each of the real commodities and real balance is defined by $R = \frac{M}{P}$. Steady state utility level or welfare for a type i agent can be shown to be $\frac{\partial W_i}{\partial R}]_{R=0} > 0$, $\forall i$, as long as U_i's are not too large. Using fiat money reduces inefficient storage of real commodities. However, since introduction of fiat money in turn reduces real commodities, which can have an unfavourable effect on the frequency of consumption, welfare improvement cannot hold unconditionally.

In this context it would be interesting to examine how these conditions on U_i change with the introduction of the social institution of *markets* and its effect on the velocity of circulation of a fiat money.

References

Aiyagari, S.R. and N. Wallace (1991) : Existence of steady states with positive consumption in Kiyotaki-Wright model : *Review of Economic Studies*, **58**, 901-916.

Dasgupta, D. and M. Rajeev : A note on feasibility criteria in monetary trade, the *Japanese Economic Review*, forthcoming.

Hicks, J. (1967) : *Critical essays in monetary theory*, ELBS and Oxford University Press.

Kiyotaki, N. and R. Wright (1989) : On money as a medium of exchange, *Journal of Political Economy*, **97**, 927-954.

Kiyotaki, N. and R. Wright (1991) : A contribution to the pure theory of money, *Journal of Economic Theory*, **53**, 215-235.

Kiyotaki, N. and R. Wright (1993) : A search theoretic approach to monetary economics, *American Economic Review*, **83**, 63-77.

Menger, K. (1892) : On the origin of money, *Economic Journal* **2**, 239-255.

Ostroy, J.M. and R.M. Starr (1974) : Money and the decentralization of excahnge, *Econometrica*, **42**, 1093-1113.

Ostroy, J.M. and R.M. Starr (1990) : The transactions role of money, in *Handbook of Monetary Economics.*

Rajeev, M. (1996) : Money and Markets, Occasional Paper No. 157, CSSS, Calcutta.

Starr, R.M. (1972) : The structure of exchange in barter and monetary economics, *Quarterly Journal of Economics*, LXXXVI, 290-302.

Starr, R.M. (1976) : Decentralized non-monetary trade, *Econometrica* **44**, 1087-1089.

Walras, L. (1900) : *Elements of pure economics*, translated and edited by W. Jasse, Homewood Illinois, Irwin.

Appendix

Proof of Proposition 1

Let V_k^D and V_k^I denote respectively the (expected, discounted, lifetime) utility derived by a type k agent by going through direct and indirect trades respectively. We want to show $V_1^D > V_1^I$, $V_2^D < V_2^I$ and $V_3^D \geq V_3^I$

Let us first consider a type 1 agent. As soon as he decides to go through direct barter he has to pay the costs $b_{12} + \gamma_{12}$. Next period he would meet a type 2 agent with good 1 with probability p_{21} and attain net utility u_1 and the entire process starts again. With probability $(1 - p_{21})$ he has the option of choosing V_1^D or V_1^I whichever is larger.

$$V_1^D = -(b_{12} + \gamma_{12}) + \beta[p_{21}(u_1 + max(V_1^D, V_1^I)] + (1 - p_{21})max(V_1^D, V_1^I)]$$

When he decides to go through indirect trade, he would visit $(2,3)$ market and for entire life time would not meet any complementary trading partner given $V_2^D \leq V_2^I$ and $V_3^D \geq V_3^I$. Hence

$$V_1^I = -\frac{b_{12} + \gamma_{23}}{1 - \beta} \Rightarrow V_1^D > V_1^I$$

Similarly, one can show that V_2^I and V_3^D are also optimal strategies (under the condition on the parameters stated above). It can be easily checked that in the steady state, $p_{21} = \frac{1}{2}$.

Proof of proposition 3 :

For a marketless set-up (see Kiyotaki and Wright (1989))

$$WF_1 = \frac{\beta u_1}{6} - b_{12}, WF_2 = \frac{\beta u_2}{6} - \frac{1}{2}(b_{21} + b_{23}), WF_3 = \frac{\beta u_3}{6} - b_{31}$$

For a trading post set-up

$$\overline{WF_1} = (1 - \beta)V_1^D = -(b_{12} + \gamma_{12}) + \frac{\beta}{2}u_1 \text{ and } \overline{WF_3} = -(b_{31} + \gamma_{31}) + \frac{\beta}{2}u_3$$

For a type 2 agent let V_{21} and V_{23} denote respectively the indirect utilities of acquiring good 1 (by visiting the (1, 3) trading post) and of acquiring good 3 (by visiting (1, 2) trading post i.e. by acquiring good 2 and then producing good 3). We have $p_{21} = p_{23} = \frac{1}{2}$.

$$V_{23} = -(b_{23} + \gamma_{13}) + \beta V_{21},$$

$$V_{21} = -(b_{21} + \gamma_{12}) + \beta(u_2 + V_{23})$$

$$\overline{WF_2} = (1 - \beta)\left(\frac{1}{2}V_{21} + \frac{1}{2}V_{23}\right) = \frac{\beta u_2}{2} - \frac{(b_{23} + \gamma_{13}) + (b_{21} + \gamma_{12})}{2}$$

Comparing $\overline{WF_k}$ with WF_k we get the result.

Let WF_k^* denote the welfare derived by a type k agent under the strategy profile S. It can be shown that (see Kiyotaki and Wright (1989)) :

$$WF_k^* = \frac{\beta u_1}{3} - \frac{1}{3}(b_{13} - b_{12}) - b_{12}, \; WF_2^* = \frac{\beta u_2}{3} - \frac{b_{23} + b_{21}}{2} - \frac{1}{6}(b_{23} - b_{21})$$

$$WF_3^* = \frac{\beta u_3}{3} - \frac{1}{3}(b_{23} - b_{31}) - b_{31}$$

Thus, if

$$\frac{\beta u_1}{6} + \frac{1}{3}(b_{13} - b_{12}) - \gamma_{12}, \; \frac{\beta u_2}{6} - \frac{\gamma_{12} + \gamma_{13}}{2} + \frac{1}{6}(b_{23} - b_{21})$$

and $\frac{\beta u_3}{6} + \frac{1}{3}(b_{32} - b_{31}) - \gamma_{13}$ quantities are non-negative we get $WF_k^* \leq \overline{WF_k}, \forall k$. As can be seen if $Lt\ \gamma_{cc'} \to 0,\ \forall c, c'$ then $WF_k^* < \overline{WF_k}$ holds unconditionally.

Proof of Proposition 4 :

Let U_{ki} denote the (expected, discounted, lifetime) utility derived by a type k agent by adopting the strategy $S_{ki} (i = 1, 2, \ldots, 6)$.

$$\begin{aligned} U_{11} = -(b_{12} + \gamma_{12}) \;&+\; \beta[p_1^{12}(u_1 + max(U_{ki}, i = 1, 2, \ldots, 6)) \\ &+\; (1 - p_1^{12})max(U_{ki}, i = 1, 2, \ldots, 6)] \end{aligned}$$

where p_1^{12} is the probability of meeting a trader with good 1 in the (1, 2) trading post. $U_{12} = -\frac{b_{12} + \gamma_{23}}{1-\beta}$, $U_{13} = -\frac{b_{12}}{1-\beta}$

$$\begin{aligned} U_{14} = -b_{12} \;&+\; \beta[p'_{23}\{-b_{13} + p'_{31}(u_1 + max(U_{ki}, i = 1, 2, \ldots, 6)) + (1 - p'_{31})v_{13}\} \\ &+\; (1 - p'_{23})max(U_{ki}, i = 1, 2, \ldots, 6)] \end{aligned}$$

where p'_{23} is the steady state probability of meeting a type 2 agent with good 3 in a marketless set-up when a type 1 agent adopts S_{14}, p'_{31} is the steady state probability of meeting a type 3 agent with good 1 in a marketless set-up when a type 1 agent adopts S_{14}, v_{13} is the indirect utility of acquiring good 3 by a type 1 agent.

$$U_{15} = -\frac{b_{12} + \gamma_{23}}{1 - \beta}$$

$$U_{16} = -b_{12} + \beta[p''_{23}(-\frac{b_{13} + \gamma_{13}}{1 - \beta}) + (1 - p''_{23})max(U_{ki},\ i = 1, 2, \ldots, 6)]$$

p''_{23} is the probability of meeting a type 2 agent with good 3 when the type 1 agents opt for S_{16}.

Now given the optimal strategy of the type 2 agents, in the steady state $p'_{23} = 0$. Thus, U_{11} will be optimal if $-(b_{12} + \gamma_{12}) + \beta p_1^{12} u_1 > -b_{12}$

$\Rightarrow \beta p_1^{12} u_1 - \gamma_{12} > 0$.

Proceeding in a parallel fashion it can be shown that U_{26} would be optimal if $p_{31}\{\beta u_2 - (b_{21} + \gamma_{12})\} \geq b_{23} - b_{21}$. Similarly optimality of U_{33} can be shown.

Steady State Probability Distributions :

Let $\frac{N_1}{N}$ and $\frac{N_2}{N}$ be the steady state proportions of the type 2 agents in a marketless arrangement and in a trading post set-up respectively. Thus, in the steady state (for the equilibrium of Proposition 4) we get :

$$\begin{aligned} p_1^{12} = \tfrac{N_2}{N} \text{ and } p_{23} \;&=\; \text{probability of meeting a type 2 agent with good 3} \\ &\; \text{in the marketless arrangement (by a type 3 agent)} \\ &=\; \frac{N_1}{N + N_1} \end{aligned}$$

p_{31} : probability of meeting a type 3 agent with good 1 in the marketless arrangement (by a type 2 agent) $\Rightarrow p_{31} = \frac{N}{N+N_1}$

Also in the steady state $N_1.p_{31} = N_2$. Using these relations we get

$$p_{31} = \frac{\sqrt{5}-1}{2}, \quad p_1^{12} = \frac{2\sqrt{5}-4}{\sqrt{5}-1}, \quad p_{23} = \frac{3-\sqrt{5}}{2}$$

Proof of Proposition 5 :

We define welfare in an exactly similar manner as that of proposition 3. Let $\overline{\overline{W}}_k$ be the welfare derived by a type k agent under the equilibrium strategy of Propostion 4.

Then

$$\overline{\overline{W}}_1 \simeq -(b_{12}+\gamma_{12}) + (.38197)\beta u_1$$

$$< -(b_{12}+\gamma_{12}) + \frac{1}{2}\beta u_1 = \overline{W}_1, \; p_1^{12} = \frac{2\sqrt{5}-4}{\sqrt{5}-1} \simeq .38197$$

$$\begin{aligned} \Rightarrow \overline{\overline{W}}_3 &\simeq -(b_{31}) + \beta(.38197)u_3 \\ &= -b_{31} + \frac{1}{2}\beta u_3 - .11903\beta u_3 \\ \Rightarrow \overline{\overline{W}}_3 &< \overline{W}_3 \; if \; .11903\beta u_3 - \gamma_{31} > 0 \end{aligned}$$

and

$$\begin{aligned} \overline{\overline{W}}_2 &= \frac{N_1}{N}V_{21} + \frac{N_2}{N}V_{23} \\ V_{21} &= \frac{-[b_{23}+\beta p_{31}(b_{21}+\gamma_{12})] + \beta^2 p_{31}u_3}{1-\beta^2 p_{31}} \\ V_{23} &= -\frac{(\beta b_{23}+b_{21}+\gamma_{12}) + \beta u_2}{1-\beta^2 p_{31}} \end{aligned}$$

It can be checked in a straight forward manner that if

$$\gamma_{cc'} \to 0, \forall (c,c') \text{ and } \overline{\overline{W}}_k > 0, \forall k \text{ then } \overline{\overline{W}}_k < \overline{W}_k, \; \forall k.$$

Meenakshi Rajeev
Centre for Studies in Social Sciences
10, Lake Terrace, Calcutta - 29

THEORY AND DECISION LIBRARY

SERIES C: GAME THEORY, MATHEMATICAL PROGRAMMING AND OPERATIONS RESEARCH

Editor: S.H. Tijs, *University of Tilburg, The Netherlands*

1. B.R. Munier and M.F. Shakun (eds.): *Compromise, Negotiation and Group Decision.* 1988 ISBN 90-277-2625-6
2. R. Selten: *Models of Strategic Rationality.* 1988 ISBN 90-277-2663-9
3. T. Driessen: *Cooperative Games, Solutions and Applications.* 1988 ISBN 90-277-2729-5
4. P.P. Wakker: *Additive Representations of Preferences.* A New Foundation of Decision Analysis. 1989 ISBN 0-7923-0050-5
5. A. Rapoport: *Experimental Studies of Interactive Decisions.* 1990 ISBN 0-7923-0685-6
6. K.G. Ramamurthy: *Coherent Structures and Simple Games.* 1990 ISBN 0-7923-0869-7
7. T.E.S. Raghavan, T.S. Ferguson, T. Parthasarathy and O.J. Vrieze (eds.): *Stochastic Games and Related Topics.* In Honor of Professor L.S. Shapley. 1991 ISBN 0-7923-1016-0
8. J. Abdou and H. Keiding: *Effectivity Functions in Social Choice.* 1991 ISBN 0-7923-1147-7
9. H.J.M. Peters: *Axiomatic Bargaining Game Theory.* 1992 ISBN 0-7923-1873-0
10. D. Butnariu and E.P. Klement: *Triangular Norm-Based Measures and Games with Fuzzy Coalitions.* 1993 ISBN 0-7923-2369-6
11. R.P. Gilles and P.H.M. Ruys: *Imperfections and Behavior in Economic Organization.* 1994 ISBN 0-7923-9460-7
12. R.P. Gilles: *Economic Exchange and Social Organization.* The Edgeworthian Foundations of General Equilibrium Theory. 1996 ISBN 0-7923-4200-3
13. P.J.-J. Herings: *Static and Dynamic Aspects of General Disequilibrium Theory.* 1996 ISBN 0-7923-9813-0
14. F. van Dijk: *Social Ties and Economic Performance.* 1997 ISBN 0-7923-9836-X
15. W. Spanjers: *Hierarchically Structured Economies.* Models with Bilateral Exchange Institutions. 1997 ISBN 0-7923-4398-0
16. I. Curiel: *Cooperative Game Theory and Applications.* Cooperative Games Arising from Combinatorial Optimization Problems. 1997 ISBN 0-7923-4476-6

THEORY AND DECISION LIBRARY: SERIES C

17. O.I. Larichev and H.M. Moshkovich: *Verbal Decision Analysis for Unstructured Problems*. 1997 ISBN 0-7923-4578-9
18. T. Parthasarathy, B. Dutta, J.A.M. Potters, T.E.S. Raghavan, D. Ray and A. Sen (eds.): *Game Theoretical Applications to Economics and Operations Research*. 1997 ISBN 0-7923-4712-9

KLUWER ACADEMIC PUBLISHERS – DORDRECHT / BOSTON / LONDON

Lightning Source UK Ltd.
Milton Keynes UK
UKOW031414260911

179307UK00007B/49/A

9 780792 347125